Optical Fibers for Transmission

WILEY SERIES IN PURE AND APPLIED OPTICS

Advisory Editor

Stanley S. Ballard, University of Florida

Optical Fibers
for Transmission

JOHN E. MIDWINTER
Post Office Research Centre
Martlesham Heath
Ipswich
Suffolk
UK

John Wiley & Sons, New York / Chichester / Brisbane / Toronto

Library of Congress Cataloging in Publication Data

Midwinter, John E.
  Optical fibers for transmission.

  (Wiley series in pure and applied optics)
  Includes index.
  1.   Optical communications.   2.   Fiber optics.
I. Title.
TK5103.59.M52   621.38'0414   78-15797

ISBN 0-471-60240-X

Preface

This book is addressed to people who wish to learn about fiber communications systems and, in particular, about the properties of optical fibers. Some excellent theoretical treatises have already been published on this subject, but most require for assimilation a very considerable mathematical expertise on the part of the reader. This book assumes no more than the knowledge possessed by those having a bachelor's degree in physics, chemistry, or engineering, and further assumes that they will have forgotten much of that and will need some review. The early chapters serve this purpose.

The next group of chapters cover the major principles of propagation by theoretical analysis. The models presented have been selected for the physical "feel" that they give the reader, since propagation in multimode guides is so complex that simple models and physical understanding are generally of much greater assistance than a precise, exact analysis. Thus propagation in step-index and graded-index fibers and the effects that can occur from mode coupling are examined, using the LP mode and ray models. Considerable space is also devoted to discussion of the physical models of mode and ray propagation, and visualization of particular LP modes, and the equivalence of the different models.

A secondary function of the book is to give the reader some awareness of the practicalities of fibers: their manufacture, measurement, testing, and usage. To this end there are discussions of the various fiber forming processes, a range of measurement techniques, and cabling and system design. However, throughout all these sections I have attempted to deal in principles which will endure rather than absolutes which will become outdated.

I anticipate that the book will appeal to a number of different groups of readers. It is expected to be of value to the many people who are now facing the replacement of existing transmission media by optical fibers and who want a tutorial introduction to the subject. The book will also be of value to workers in the field, since the subject range now embraced by optical communications is so large that many of them find a need for a ready reference book on aspects in which they are not personally expert. Finally, it is expected that this book will provide the basis for master of science course on fiber communications, a need for which is felt in many educational centers around the world at this time.

J. E. MIDWINTER

Ipswich, England
May, 1978

Acknowledgments

I wish to thank all the authors cited throughout the text for their permission to use the figures taken from their work and also the publishers concerned for allowing reproduction. My thanks also go to the Director of the Post Office Research Centre for permission to publish this work. Although it does not necessarily represent the views of that organization, it does draw heavily on the experience I gained while working there. In particular I should like to mention Mr. F. F. Roberts, former Head of Optical Communications and sadly now deceased, for the encouragement and opportunity he gave me to learn about this exciting new field of endeavour. I thank my many friends within the Optical Communications Division for their help and stimulation, and the Production and Editorial Staff of John Wiley for their professional advice and assistance in preparing the copy for publication. Last but by no means least, I owe a great debt to my wife for her patience and encouragement over the long period spent in preparing this book and I thank her most sincerely.

J. E. MIDWINTER

Contents

Optical Fibers for Transmission

1

Introduction to Optical Communications

1.1. INTRODUCTION

The use of light for sending message information goes far back into time. One thinks of Indian smoke signals, or the use of mirrors reflecting the sun to attract attention. More recently signaling lamps served for secure communications between ships. What is probably not widely known is that the use of light for audio communication was seriously proposed and explored by Alexander Graham Bell.[1] He is credited with the invention of the telephone in 1876, and in 1880 he filed patents for the photophone. In retrospect it appears as a very crude device (see Figure 1.1), yet one cannot fail to be impressed by the fact that Bell made it work and with it communicated over a distance of 200 m. However, all these optical communication systems were bedeviled by a number of problems. Systems relying on the eye as detector were seriously limited in speed of communication, and thermal light sources, tungsten lamps, for example, could not have been modulated rapidly. Perhaps most serious of all, the atmosphere is not a reliable optical transmission medium, as anyone who has been stranded in fog or smog knows.

The systems that this book is concerned with owe their ancestry to a separate line of development whose start can be traced to the invention of the laser[2] in the early 1960s. The laser made available for the first time a coherent source of light. This in itself was a striking advance, although for many of the optical fiber systems that are now being developed it is, in retrospect, of rather slight importance. Perhaps the major reason that the invention of the laser should be viewed as the starting point for the development of the present systems is the fact that the laser encouraged researchers to view the optical frequency spectrum as an extension of the radio and microwave spectrum. The outcome of this new view was a dramatic increase in the study of all optical components; sources, modulators, detectors, lenses, mirrors, waveguides, and complete systems.

1

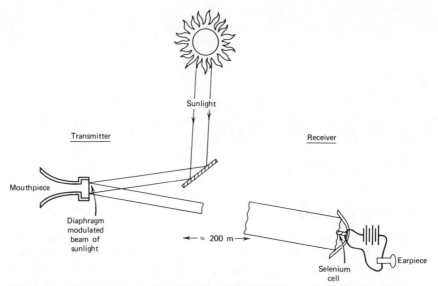

Figure 1.1. Alexander Graham Bell's Photophone, which used sunlight modulated by a diaphragm to transmit speech to a receiver over 200 m away.

The laser made available a coherent optical frequency carrier of the order of 10^{15} Hz frequency. Following the trend in radio and microwaves, it was natural to see in this development the basis of the ultimate communication system. A 0.1% bandwidth would correspond to a data bandwidth of 1000 GHz; the system should provide a communication capacity previously undreamed of, although some problems stood in the way of its implementation, not the least being that the use for such capacity was far from clear.

With this background it is interesting to jump to the present and to examine what is actually being built. Wideband optical communication systems have been developed for use in military, space, and civil applications areas. The biggest potential market probably lies in civil telecommunications such as the Bell Telephone System and the European PTT networks, for connections between switching centers where many individual telephone conversations become multiplexed together for transmission along a single route, leading to a wideband signal when multiplexed into a single channel.* The signaling rates for these systems lie in the range from 2 Mbit/sec to about 1 Gbit/sec, with most interest centered on the range from 10 to 300 Mbit/sec. Immediately we can see that what people

*See Appendix 6.

actually need and are willing to pay for is very different from what it might be possible to produce.

Before we leave this subject, it is worth noting that the contrast between the ultimate sophisticated system and the systems being developed is even greater than suggested above. The present day optical transmission system is more akin to the spark-gap transmitter and receiver used at the turn of the century. The reason is that the sources commonly used, the GaAs laser[3] and the light emitting diode (LED),[4] typically have linewidths of the order of 2 and 20 nm, respectively ($\sim 10^{12}$ and 10^{13} Hz), against which we see that the data bandwidth is trivial. We are almost using a noise source for our transmitter, impressing our data upon it merely by turning it on and off.

Why does such a crude system attract so much attention? Surely it must be possible to improve upon it. During the following discussion it will become clear that, despite its apparent crudity, much sophisticated study has gone into establishing the form of the systems emerging for use and that, despite their apparent simplicity, they represent very effective solutions to real system needs. Perhaps the "ultimate" systems will follow in the course of time, in one or two decades, but a dramatic increase in the need for ground communications will have to emerge before they can be utilized.

An indication of the components that might be used in such ultimate systems already exists. Very stable single-frequency laser oscillators have been known for some years. Wideband modulators operating off very low drive powers have been developed. Attempts are being made to integrate these technologies into compact optical circuits under the general topic heading of "integrated optics."[5] Here the objective has been to apply the technology of integrated circuits, in the form of planar masking, etching, and evaporation and film growth, to produce microoptical circuits and components on single planar substrates. The problems are considerable, particularly when a combination of materials is required with a single substrate to produce a common source and modulator, but the promise is clearly there. With such technologies we can imagine optical analogues of multicarrier frequency, frequency-division multiplex or time-division multiplex transmitters or receivers operating with coherent, phase sensitive receivers. But these remain largely dreams at present, and in the meanwhile simple systems are rapidly being developed for use in the field.

1.2. HISTORICAL BACKGROUND

We have already seen how the development of the laser in 1960–1961 stimulated widespread interest in the use of light for tasks previously considered the domain of radio and microwave frequencies. The high carrier frequency inevitably attracted the attention of communications engineers, to whom a carrier of 10^{15} Hz frequency spelled unlimited bandwidth (note that a microwave wavelength of 1 cm corresponds to a carrier frequency of approximately 3×10^{10} Hz). Accordingly much early work was carried out with wideband transmission in mind.

One transmission medium studied in depth was a gas filled pipe,[6] either with lenses placed at regular intervals (of the order of 100 m) to recollimate the beam or with flowing gas, in which the flow produced local gas heating at the walls, forming a simple, continuously distributed gas lens. Systems designed around such tubes offered phenomenal performance in terms of available bandwidth and repeater separations (many gigahertz and tens of kilometers) yet were exceedingly complex and expensive to build and did not meet a real need.

Optical transmission through the atmosphere was generally found to be too unreliable, subject to interference from fog, dust, rain, and so on. However, despite problems some modest data links have been built and are used for applications such as linking a television camera to its base vehicle in outside broadcast applications or for data links between buildings where distances of a few hundred meters are involved and a cable would be excessively difficult to install. Under these conditions the modulated, highly collimated optical beam can provide a simple way of transmitting the required information.

However, the present worldwide effort devoted to optical fiber communications stems very largely from two papers,[7,8] published almost simultaneously, proposing that optical fibers be used as dielectric waveguides in underground cables to carry telecommunications and other traffic. The fiber was seen as a replacement for coaxial, baseband, or carrier transmission systems.

In retrospect, it is interesting to see how close those original proposals were to the systems that are now emerging from the research laboratories and are going into field trial or service. The source was conceived of as a GaAs laser and generally remains so, although GaAs LEDs are used for systems with modest performance requirements. The wavelength of operation was therefore 840 to 900 nm and remains so, although there is a small but growing interest in operation at longer wavelengths (1.06 to 1.3 microns).

The fiber was conceived of as a single-mode transmission line, with a circular core of one glass material surrounded by a cladding of a lower refractive index glass to produce guidance. The index difference was to be about 1%, and the loss to be less than 20 dB/km to allow sufficiently long spacings between repeaters to be competitive with existing media. Today we find the single-mode fiber viewed as suitable for the most exotic systems, with most interest in multimode graded-index fibers having adequate bandwidth and more readily handled. The index difference remains around 1%, and the losses are expected to be less than 10 or even 5 dB/km.

The detector was expected to be a silicon photodiode[9] and will be so, although the development of the reach-through avalanche diode allows rather better performance than was originally envisaged. The data are most likely to be transmitted in binary digital format for civil telecommunications use, and the repeater spacing is now expected to be 5 to 10 km, rather than 1 to 3 as originally conceived.

The vision has proved essentially correct, and despite enormous efforts in the research laboratories few if any really major changes have occurred. However, the level of understanding has increased a thousandfold, perhaps a millionfold, and the technology now exists whereas then it did not. This is a tribute to the efforts of the many workers who devoted their skills to this undertaking in the intervening period. In the following chapters we will give only a broad picture of this activity, making no attempt to cover every aspect in detail.

1.3. ATTRIBUTES OF OPTICAL FIBER SYSTEMS

What are the attractions of the fiber system, when compared to existing transmission media, that justify the worldwide investment in its development? A number of features attract, with their relative importance varying among different users. Since the link between source and detector can be metal-free, containing only glass with polymer packaging, earth loop pickup problems can be eliminated in conditions in which electromagnetic interference is severe. This consideration is a major attraction to the military users of fiber links, which are envisaged as replacing metal cables in many applications in aircraft, ships, or vehicles where sensitive electronic systems have to be packaged alongside electrical power systems. From the civil user's viewpoint, protection against damage from lightning strike is probably the greatest benefit that accrues from this performance aspect.

We will see when we analyze its transmission performance that the fiber is potentially a very wideband transmission medium with a bandwidth potential of several gigahertz over 1 km and hundreds of megahertz over distances of up to 10 km without intervening electronics. Alternatively, the possibility is becoming apparent for transmission of more modest bandwidths over very long continuous lengths, a few megahertz over 20 km, for example. By comparison the losses in coaxial cable limit transmission at a few hundred megahertz to distances of the order of 1 km.

The fiber is small. It is slightly thicker than a human hair, so that very large numbers of fibers can be packaged into small spaces, or modest numbers of fibers can be packaged into cables that are very compact and whose size may even need to be artificially enlarged to make them more compatible with conventional handling and usage. Perhaps most important of all, in civil telecommunications usage, is the fact that these systems are expected to be cheaper to install, operate, and maintain than existing systems of comparable performance. This is so partly because the fiber itself is relatively cheap, but mainly because the very low losses and high bandwidths that it offers will make possible more efficient utilization of existing duct space and the use of fewer repeaters or electronics in systems of comparable length. One can also see the elimination of buried electronics in cities where the link lengths possible with optical fiber systems begin to compare with the longest distances between switching centers, so that the electronic components can be confined to the terminal buildings only.

1.4. CATEGORIES OF FIBER SYSTEMS

The preceding discussion has been based largely on the civil telecommunications system, as is much of the remainder of the book, and this is likely to be the biggest single market for fiber links. The civil telecommunications system is likely to take the format shown in Figure 1.2. The basic link consists of the transmit and receive modules, coupled by a single fiber. The data rate transmitted is likely to be in the range of 2 to 560 Mbit/sec, and the fiber cable length in the range of 5 to 10 km. As such, these systems generally pose more stringent requirements on the optical performance of the fiber in terms of bandwidth and loss than other systems currently envisaged.

Another civil system that could be of very widespread application and swamp all other applications in volume would be the distribution of wideband (frequency band) services to the individual customer. One might envisage a number of television services together with wideband video data, facsimile, and telephone, all interacting with the individual customer via fiber links, so that, where essentially every house now has a telephone wire pair cable entering, this might include a wideband fiber link. The sheer volume of fiber required for this application dwarfs all other possibilities, but it is clear that such developments must be regarded, from our present viewpoint, as being well into the future. Equally, it does not appear to have been shown that, even if such services were to be offered, carrying the wideband data by fiber would necessarily be the best way to do this, although the same arguments that apply to trunk transmission would tend to apply here also.

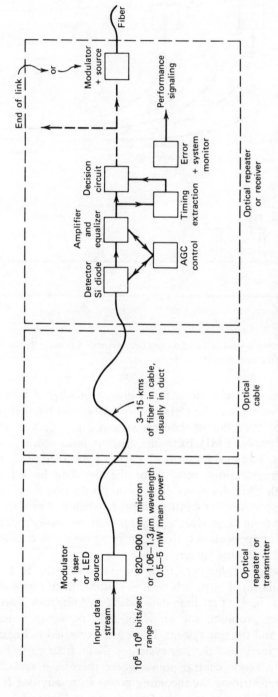

Figure 1.2. Schematic diagram of an optical fiber communications system for civil telecommunications use.

7

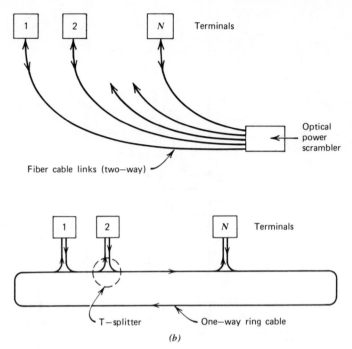

Figure 1.3. Schematic layout of two data bus fiber systems, (a) using the star configuration and (b) using the ring configuration.

Outside the civil telecommunications area, a number of uses are emerging. For example, there is the class of short data links between two points, where the major attraction of the fiber system is its freedom from electromagnetic interference (EMI). Examples are data links within power distribution networks, such as for critical parts of an Electricity Generating Authority communications network or the signaling lines from various sensors in tough EMI situations, such as on production lines, sending data to some microprocessor for control purposes when it is vital to protect the microprocessor from large electrical spikes that are easily picked up from motors. A similar application is that of carrying timing or data streams in a large telephone switching center.

Another class of applications is of the data-bus type. In these a large number (10 or more) of terminals wish to communicate with each other on a random-access basis at medium data rates (say kbit/sec). Two configurations have been proposed for servicing such networks,[10] using optical fibers: the star and the ring system. These are illustrated in Figure 1.3.

The primary feature of the star system is that a fiber cable runs to each terminal from a single central point, where there is a scrambler whose function is to redistribute the incoming power from any one fiber equally among all others. This can take the form shown in Figure 1.4, where fiber

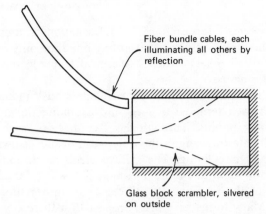

Fiber bundle cables, each illuminating all others by reflection

Glass block scrambler, silvered on outside

Figure 1.4. A star coupler for a fiber bundle data bus (Ref. 12).

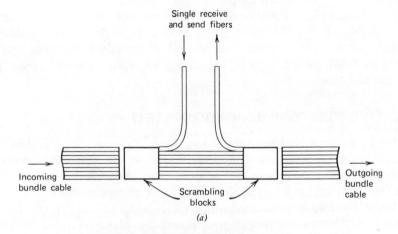

Single receive and send fibers

Incoming bundle cable

Scrambling blocks

Outgoing bundle cable

(a)

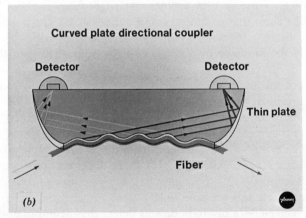

Curved plate directional coupler

Detector

Detector

Thin plate

Fiber

(b)

Figure 1.5. T-connectors for use in a ring data bus *(a)* for a bundle system and *(b)* for a single-fiber ring system. (Ref. 11).

bundles are assumed for each cable. Information is transmitted on a time-division multiplex basis, so that at any one time only one terminal is transmitting and all others "listen." A critical factor in the system is the scrambler performance.

The alternative multiaccess link is the ring data bus[12] indicated in Figure 1.3b, in which each terminal is T-ed into a unidirectional link of fiber, probably of fiber bundle. Through the T-connectors, each terminal can both access all data traveling in the ring and transmit its own additions to the data already circulating. The critical elements deciding the system performance now tend to be the T-connectors, which largely control the insertion loss of the ring and the efficiency of information capture and transmission. Many forms have been proposed for these elements; two are shown in Figure 1.5, one that is appropriate for a fiber bundle system[10] and one that is appropriate for a single-fiber system.[11]

Data links of both these types have been proposed for short-link-length use (of the order of 100 m to 1 km) in aircraft, ships, or other military vehicles or for ground based data links. The attractions are, once again, freedom from ground loop pickup and EMI, and a general saving in weight and space.

1.5. CIVIL TELECOMMUNICATIONS SYSTEMS

We have already indicated the general form of systems to be used in the transmission of telephone and other traffic in the civil environment. However, before describing the details of the fibers themselves, we will outline the typical parameters that apply in slightly more detail.

We will usually be concerned with a GaAs laser source, operating in the region of 840 nm wavelength and generating between 1 and 10 mW average power. [Note that the visible wavelength range is about 400 (violet) to 700 (red) nm, so that its output is not quite visible.] The laser output will be pulsed, usually to carry a binary digital signal with the two digits signified by the presence or absence of light, respectively, in the bit time interval.

The power will then be launched into a fiber, which will probably be a graded-index fiber having a core diameter of about 50×10^{-6} m and an overall diameter of cladding of about $100\text{--}120 \times 10^{-6}$ m. The fiber will be one of several in a cable (from perhaps 10 to as many as 100 or more) with each fiber carrying an independent communication channel. These fibers will run from 1 to 10 km or more to the receiver. The cables will be installed in lengths of 0.5 to as long as 2 km at a time and then jointed to form the complete transmission link.

At the receive end a photodiode will detect the incoming optical power, with each absorbed photon giving rise to a single photoelectron. If the detector is an avalanche detector, this photoelectron will be multiplied by a factor of 40 to 100 to generate a photocurrent which will provide the input to a low noise amplifier. The signal will be amplified to a level well above thermal noise and then processed. This processing may take the form of regeneration of the original bit-stream (with a few errors inevitably added because of the effects of noise in the system) for retransmission into another fiber link, in which case the regeneration stage would be termed a repeater, or the received data may be passed directly to the user, the switching center or whatever.

The data transmitted by the system will almost certainly not be the raw signal data but will be coded at the ends of the system to ensure faithful transmission, the code being chosen to match the electrical characteristics of the coded stream to those of the receivers, to allow for error detection, timing extraction, automatic gain control, and the like. In achieving this, the input data stream will have additional data added according to logic rules built into the processors.

In the context of telecommunications, input data will generally be delivered to the transmission system in well defined blocks, whose values will depend upon the geographical area—Europe, the United States, or Japan, each having preferred standard block sizes. These block sizes are listed in Appendix 6.

The discussion in this book is largely concerned with the use of sources in the band of 800 to 900 nm wavelength, the GaAs based sources. However, growing interest is apparent in operation at longer wavelengths, notably at about 1060 nm, near the silicon detector band edge, where more favorable fiber properties can be obtained, and some discussion of such systems is included (Section 15.3). Great interest also centers on the region around the 1.3 micron wavelength, where fiber loss and material dispersion both show minima, (Sections 5.9 and 8.3.1–8.3.4). Elsewhere in the book the reader is left to carry out his or her own interpretation in this matter.

1.6. PRESENTATION AND CONTENTS

In this book we attempt to discuss the transmission medium itself in detail, starting from a fundamental level. We examine the dielectric waveguide, discover the nature of the guidance mechanism and the form of the guided modes and the fields associated with them, and then move on to examine propagation in multimode guides, those carrying thousands of modes. This leads to a consideration of mode dispersion, graded-index guides, and mode coupling.

After covering the theoretical background, we examine some of the practical aspects of making fibers, controlling their losses, making graded-index fibers, and measuring and characterizing them, before discussing the problems associated with jointing and cable design. Finally, we turn to the system aspects of the problem, examine the general features of the optical receiver, and learn how it is integrated into a system, a study which naturally leads us to consider the optimization of the whole system.

No attempt has been made to present a complete review of the theoretical analysis of optical fibers or systems, for to do so would have led to a presentation many times longer and would have overlapped existing work. Chapters 2, 3, and 4 serve as introductory chapters, presenting material which could equally well have been included as appendices. Chapters 5, 6, and 7 present the background analysis of the fiber for the rest of the book. In each chapter methods of analysis have been selected which lead to analytical results linking closely to physical models, since in multimode fibers the problems are so complex that clear physical models are of inestimable value to the worker in understanding what is occurring in any particular fiber. Readers who are already familiar with the solutions of Maxwell's equations and with the properties of planar dielectric wave-guides may well choose to skip Chapters 2, 3, and 4 and to move directly to Chapter 5. However, the analysis of the planar guide serves a useful purpose in allowing one to build some useful physical models of propagation in dielectric guides for use when considering circular guides or fibers.

Chapters 8 and 9 discuss the materials and methods of preparation of low loss optical fibers. A discussion of the measurement techniques used in their study is presented in Chapters 10 and 11.

The design of cables is considered in Chapter 12, which presents a discussion of the general principles involved without attempting to set out firm rules, since these are not yet established, although the important factors to be considered in achieving a good design are rapidly becoming more clearly defined. A brief discussion of connectors and joints for cables follows in Chapter 13. In the last two chapters receiver design is examined, as are the effect of the transmission medium on the receiver error rate and the overall system optimization and trade-offs that inevitably follow in deciding on a particular design.

The sections on systems, cables, measurements, fiber materials and production, and fiber theory can largely be read independently without difficulty. They are supported by additional data which have been relegated to appendices covering material of interest that does not fit naturally into the main text.

1.7. NOTATION

A variety of different notations has emerged to describe optical fiber parameters. We have chosen to use a notation which attempts to emphasize the physical origin of the terms to be represented, since an objective of the presentation is to generate a physical awareness of the significance of particular effects. In particular, the use of subscripts and superscripts in this presentation should be noted. We have reserved subscripts to indicate the material used—1 for the guiding material, 2 and 3 for cladding material(s). A superscript preceding a symbol indicates the particular vector component under consideration; thus x, y, z, r, and ϕ all appear as superscripts. For example, the symbol ${}^x k_1^2$ indicates the square of the x component of the propagation vector in the guiding medium, medium 1.

Table 1.1 summarizes the main symbols used.

TABLE 1.1 NOTATION AND SYMBOLS USED (IN ORDER OF USAGE)

General analysis of planar and circular guides

x, y, z	Cartesian coordinates
r, ϕ, z	cylindrical polar coordinates; z aligned along fiber axis
u, v, w	Cartesian coordinates used in analysis of Goos-Haenchen shift (Section 3.5)
T	Superscript indicating transverse component, i.e., vector sum of x and y or r and ϕ components
k	wave vector (equal to $2\pi/\lambda$ in free space)
λ	free space wavelength of light
c	free space velocity of light
γ	complex wave vector (equal to ik)
l	azimuthal index of LP_{1m} mode, i.e., $2l$ spots on mode pattern around any circumference
m	radial index of LP_{1m} mode, i.e., m rings crossing any radius
$E, \mathbf{E}$	complex electric field
$H, \mathbf{H}$	complex magnetic field
$D, \mathbf{D}$	complex dielectric displacement
$B, \mathbf{B}$	complex magnetic displacement
ω	angular frequency (rad/sec)
f	frequency (Hz)
t	time (sec)
$\mathbf{i}, \mathbf{j}, \mathbf{k}$	unit vectors along x, y, z
$\mathscr{E}, \pmb{\mathscr{E}}$	real part of electric field
$\mu \ (\mu_0)$	magnetic susceptibility (of free space)
$\epsilon \ (\epsilon_0)$	dielectric susceptibility (of free space)
∇	Laplacian operator
θ	angle of ray from surface ($\theta = 0$ is grazing incidence)
Θ	angle of ray incidence ($\Theta = 0$ is normal incidence)
Φ	phase shift on reflection at dielectric interface

TABLE 1.1 (*Continued*)

General analysis of planar and circular guides

r_E, r_H reflection coefficients for TE and TM waves, respectively, at dielectric interface

n refractive index ($n^2 = \epsilon$)

P time averaged Poynting vector components, giving power flow

V_g, V_p group and phase velocities, respectively

N group refractive index

M single index mode number, equal to $2m + l$ (Section 7.2.1)

Δ fractional index difference, equal to $(n_1 - n_2)/n_2$

$0, 1, 2, 3$ subscripts used for free space, core, and claddings, respectively

L guide length

τ time delay in propagation

σ root mean square pulse width

V $(U^2 + W^2)^{1/2} = 2\pi a(n_1^2 - n_2^2)^{1/2}/\lambda$

U $^T k_1 a$

W $\gamma_2 a$ (Section 5.8)

a, a_{core} core radius

a_{clad} cladding radius

α index profile parameter or exponential attenuation coefficient

$\bar{E}$ pseudoenergy ⎫

$\bar{I}$ pseudoangular momentum ⎬

R radius at ray turning points ⎪ (Section 6.5)

α_0, γ_0 direction cosines ⎭

D power diffusion coefficient

A angular coupling coefficient

h $4.35\gamma_\infty$ loss due to microbending (dB/km) ⎬ special symbols for Section 7.4 and 7.5

σ, Θ used locally

s Laplace frequency

Strength Analysis: Chapter 12

G cumulative probability of fiber fracture

g probability of fiber fracture at specified values

f stress (kg/m^2)

f_a average stress over sample section

p, b, m power indices characterizing failure

l fiber test length

P probability of flaw of specified size

D density of flaws

x crack depth

ρ radius of crack tip

TABLE 1.1 (*Continued*)

Analysis of fiber perturbation by impressed forces—microbending:
Chapters 12 and 15

X	amplitude of deformation
V	amplitude of applied deformation
K	mechanical wavenumber of deformation
H	beam stiffness
E	Young's modulus
I	Moment of inertia of beam (fiber)
D	Young's modulus of surrounding material
F	relative propensity to microbending

Flow analysis of fiber coating: Chapter 12

V	velocity of liquid flow
V_0	fiber velocity
T	coating thickness
Q	volume of coating material applied per second
a	fiber outer radius
b	nozzle radius
F	filling factor of coating material (fraction of solid to solvent plus solid)

Optical receiver: Chapter 14

Gsn	Gaussian distribution
Psn	Poisson distribution
k	Boltzmann's constant
T	absolute temperature
H	amplifier noise factor
g	avalanche gain
$\bar{n}$	dark current—equivalent count
X	avalanche diode noise factor

System optimization: Chapter 15

$F(T)$	Total power available to system (dB) for time interval of operation of T
T	bit-interval (sec)
P_L	launched power (dBm)
ϕ_M	material dispersion parameter
$\Delta\lambda$	source linewidth
ψ	mode dispersion parameter
b	power law of mode dispersion with length (L^b) $1 \geqslant b \geqslant \frac{1}{2}$
F_{ISI}	power lost from the effects of intersymbol interference for constant error rate
R	Rayleigh scattering parameter

REFERENCES

1. See, for example, Robert V. Bruce, *Alexander Graham Bell and the Conquest of Solitude.* London: Gollancz, 1973.
2. T. H. Maiman, *Nature*, **187**, 493 (1960).
3. a) R. N. Hall, G. E. Fenner, J. D. Kingsley, T. J. Soltys, and R. O. Carlson, *Phys. Rev. Lett.*, **9**, 366 (1962).
   b) M. I. Nathan, W. P. Dumke, G. Burns, F. D. Hill, and G. J. Lasher, *Appl. Phys. Lett.*, **1**, 62 (1962).
   c) T. M. Quist, R. H. Rediker, R. J. Keyes, W. E. Krag, B. Lax, A. L. McWhorter, and H. J. Zeiger, *Appl. Phys. Lett.*, **1**, 91 (1962).
4. F. D. King, J. Straus, O. I. Szentisi, and A. J. Springthorpe, *Proc. IEE*, **123**, 619 (1976).
5. H. F. Taylor and A. Yariv, *Proc. IEEE*, **62**, 1044 (1974).
6. D. W. Berriman, *Bell Syst. Tech. J.*, **43**, 1469 (1964); **44**, 2117 (1965).
7. K. C. Kao and G. A. Hockham, *Proc. IEE*, **113**, 1151 (1966).
8. A. Werts, *Onde. Elec.*, **45**, 967 (1966).
9. a) A. D. Lucas, *Proc. IEE*, **123**, 623 (1976).
   b) P. P. Webb, R. J. McIntyre, and J. Conradi, *RCA Rev.*, **35**, 234 (1974).
10. F. L. Thiel, Paper WE2, *Optical Fiber Transmission.* Williamsburg, Va.: OSA/IEEE, January 1975.
11. W. J. Stewart, *2nd European Conference on Optical Communications, Paris, September 1976.* Paris: SEE.
12. See, for example, A. F. Milton, and L. W. Brown, Paper 3.8, Conference on Laser Engineering and Applications, Washington, D. C., May 1975.

2

Field Relations and
Maxwell's Equations

2.1. INTRODUCTION

In this chapter we bring together the background material that is used throughout the remainder of the theoretical section of this book. For readers who are well conversant with the vector relationships between the electromagnetic field components and their relationships with the wave equation this chapter will contain nothing new. It is included purely for completeness and as a ready source for reference.

We begin by deriving the field relationships for the planar guide geometry encountered in two-dimensional integrated optical guides, in which the guide film lies in the y-z plane. This is followed by an analysis of the fiber geometry, for which cylindrical polar coordinates are the most appropriate.

2.2. DERIVATION OF FIELD RELATIONS FROM MAXWELL'S EQUATIONS

The vector relationships between the electromagnetic fields in a medium with zero conductivity are well known and take the following form:

$$\nabla \times \mathbf{E} = \frac{-\partial \mathbf{B}}{\partial t} = -\mu\mu_0 \frac{\partial \mathbf{H}}{\partial t} \tag{2.2.1}$$

$$\nabla \times \mathbf{H} = \frac{\partial \mathbf{D}}{\partial t} = \epsilon\epsilon_0 \frac{\partial \mathbf{E}}{\partial t} \tag{2.2.2}$$

$$\mathbf{D} = \epsilon\epsilon_0 \mathbf{E} \tag{2.2.3}$$

$$\mathbf{B} = \mu\mu_0 \mathbf{H} \tag{2.2.4}$$

$$\nabla \cdot \mathbf{D} = 0 \quad \text{(no free charges)} \tag{2.2.5}$$

$$\nabla \cdot \mathbf{B} = 0 \quad \text{(no free poles)} \tag{2.2.6}$$

17

We will assume that our guiding structure guides along the z direction. Thus we look for solutions to problems in which the z dependence of the field is of the form

$$\mathbf{E} = \mathbf{E}_0(x,y)\exp i(\omega t - {}^z k \cdot z) \qquad (2.2.7)$$

$$\mathbf{H} = \mathbf{H}_0(x,y)\exp i(\omega t - {}^z k \cdot z) \qquad (2.2.8)$$

In this expression we note that ω is related to ν, the frequency, by the relation $\omega = 2\pi\nu$. Furthermore, in writing the time and space variation of the field in terms of the complex exponential, it is understood, but not written, throughout this book that when a field is to be evaluated (e.g., for measurement) the only physical meaningful part of this complex expression is the real part. Thus, if we denote a real measurable field by $\mathcal{E}$ and we wish to relate this to the theoretically derived field $\mathbf{E}$, which is given by an expression of the form of equation 2.2.7, then

$$\mathcal{E} = \mathrm{Re}[\mathbf{E}] \qquad (2.2.9)$$

The advantage of this approach is that expressions involving the exponential of a complex quantity are more readily manipulated than the equivalent expressions involving sine and cosine.

We will now use these relations to derive expressions for the field components in planar and cylindrical geometries before studying the solutions for guided waves in planar and cylindrical waveguides.

2.3. PLANAR GEOMETRY

The expression $\nabla \times$ in rectangular Cartesian coordinates is

$$\nabla \times \mathbf{A} = \begin{vmatrix} \mathbf{i} & \mathbf{j} & \mathbf{k} \\ \dfrac{\partial}{\partial x} & \dfrac{\partial}{\partial y} & \dfrac{\partial}{\partial z} \\ {}^x A & {}^y A & {}^z A \end{vmatrix} \qquad (2.3.1)$$

where $\mathbf{i}$, $\mathbf{j}$, $\mathbf{k}$ are unit vectors in the x,y,z directions. Thus from equation 2.2.1 we obtain for the x directed component of $\nabla \times \mathbf{E}$

$$\left(\frac{\partial^z E}{\partial y} - \frac{\partial^y E}{\partial z} \right) = -\mu\mu_0 \frac{\partial^x H_x}{\partial t} \qquad (2.3.2)$$

We can write similar equations for $\partial^y H/\partial t$, $\partial^z H/\partial t$, $\partial^x E/\partial t$, $\partial^y E/\partial t$, and $\partial^z E/\partial t$.

If we then substitute the field expressions of equations 2.2.7 and 2.2.8 into these expressions, we obtain the following results:

$$\frac{\partial^z E}{\partial y} + i^z k^y E = -i\mu\mu_0\omega^x H \qquad (2.3.3)$$

$$i^z k^x E + \frac{\partial^z E}{\partial x} = i\mu\mu_0\omega^y H \qquad (2.3.4)$$

$$\frac{\partial^y E}{\partial x} - \frac{\partial^x E}{\partial y} = -i\mu\mu_0\omega^z H \qquad (2.3.5)$$

$$\frac{\partial^z H}{\partial y} + i^z k^y H = i\omega\epsilon\epsilon_0^x E \qquad (2.3.6)$$

$$-i^z k^x H - \frac{\partial^z H}{\partial x} = i\omega\epsilon\epsilon_0^y E \qquad (2.3.7)$$

$$\frac{\partial^y H}{\partial x} - \frac{\partial^x H}{\partial y} = i\omega\epsilon\epsilon_0^z E \qquad (2.3.8)$$

Our analysis of planar structures will be restricted to infinite films that lie in the y-z plane. Thus, in addition to the assumption that the fields have the z dependence already postulated, we can further assume that the partial derivative with respect to y vanishes (hereafter abbreviated to $\partial/\partial y = 0$) for an infinite plane wave traveling in the z direction. With this assumption the above equations simplify and demonstrate a fundamental relationship about the fields in such a structure. The simplified relations are as follows:

$$i^z k^y E = -i\mu\mu_0\omega^x H \quad \text{(TE group)} \qquad (2.3.9)$$

$$i^z k^x E + \frac{\partial^z E}{\partial x} = i\mu\mu_0\omega^y H \quad \text{(TM group)} \qquad (2.3.10)$$

$$\frac{\partial^y E}{\partial x} = -i\mu\mu_0\omega^z H \quad \text{(TE group)} \qquad (2.3.11)$$

$$i^z k^y H = i\omega\epsilon\epsilon_0^x E \quad \text{(TM group)} \qquad (2.3.12)$$

$$-i^z k^x H - \frac{\partial^z H}{\partial x} = i\omega\epsilon\epsilon_0^y E \quad \text{(TE group)} \qquad (2.3.13)$$

$$\frac{\partial^y H}{\partial x} = i\omega\epsilon\epsilon_0^z E \quad \text{(TM group)} \qquad (2.3.14)$$

We see that the fields have now split into two separate groups, namely, $^y E$,

^xH, and ^zH are coupled and ^yH, ^xE, and ^zE are also coupled. The guided waves formed by the first group are described as TE modes (for transverse electric), and the latter are known as TM modes (for transverse magnetic). We can now use the above relations to derive simpler expressions for the transverse field components in terms of the ^zE and ^zH components only (e.g., eliminate ^yH between equations 2.3.10, 2.3.12, and 2.3.14 to obtain a relation for ^xE). This yields the following:

$$^xE = \left(\frac{-i^zk}{\omega^2 \epsilon \epsilon_0 \mu\mu_0 - {}^zk^2} \right) \frac{\partial {}^zE}{\partial x} \tag{2.3.15}$$

$$^yE = \left(\frac{i\omega\mu\mu_0}{\omega^2 \epsilon \epsilon_0 \mu\mu_0 - {}^zk^2} \right) \frac{\partial {}^zH}{\partial x} \tag{2.3.16}$$

$$^xH = \left(\frac{-i^zk}{\omega^2 \epsilon \epsilon_0 \mu\mu_0 - {}^zk^2} \right) \frac{\partial {}^zH}{\partial x} \tag{2.3.17}$$

$$^yH = \left(\frac{-i\omega\epsilon\epsilon_0}{\omega^2 \epsilon \epsilon_0 \mu\mu_0 - {}^zk^2} \right) \frac{\partial {}^zE}{\partial x} \tag{2.3.18}$$

Substituting these expressions into equations 2.3.9 to 2.3.14 yields two wave equations for propagation in the x direction:

$$\frac{\partial^2 {}^zE}{\partial x^2} - \left(\omega^2 \epsilon \epsilon_0 \mu\mu_0 - {}^zk^2 \right) {}^zE = 0 \tag{2.3.19}$$

$$\frac{\partial^2 {}^zH}{\partial x^2} - \left(\omega^2 \epsilon \epsilon_0 \mu\mu_0 - {}^zk^2 \right) {}^zH = 0 \tag{2.3.20}$$

These indicate that for the transverse dependence of the fields we should seek solutions of the form $\exp(i^xkx)$, where

$$^xk^2 = \omega^2 \epsilon\epsilon_0 \mu\mu_0 - {}^zk^2 = -({}^x\gamma)^2 \tag{2.3.21}$$

The significance of the variable $^x\gamma$ introduced here will become apparent later.

Notice that equation 2.3.21 could have been obtained much more straightforwardly by deriving the wave equation directly from equations 2.2.1 to 2.2.6, setting $\partial/\partial y = 0$, and substituting the field equations (equations 2.2.7 and 2.2.8). However, that route would not have yielded the detailed interrelationship between the vector components of the fields that we will need for finding the conditions for guided waves.

In summary, then, we have derived relationships between the vector components of the fields for a planar structure lying in the y-z plane, with a wave propagating in the z direction. These relations are summarized in equations 2.3.9 to 2.3.14. In addition we have shown that fields in such a structure take the general form

$$\mathbf{E} = \mathbf{E}(x) \exp i(\omega t - {}^zkz \pm {}^xkx) \tag{2.3.22}$$

$$\mathbf{H} = \mathbf{H}(x) \exp i(\omega t - {}^zkz \pm {}^xkx) \tag{2.3.23}$$

$$ {}^xk^2 = \omega^2 \epsilon \epsilon_0 \mu \mu_0 - {}^zk^2 \tag{2.3.24}$$

$$ n^2 k_0^2 = \omega^2 \epsilon \epsilon_0 \mu \mu_0 \tag{2.3.25}$$

2.4. CYLINDRICAL GEOMETRY

We now repeat the analysis of Section 2.3 but in cylindrical polar coordinates, since these are more appropriate to the analysis of optical fiber guides. The coordinates x, y, and z are now replaced by r, ϕ, and z. These are related to the Cartesian coordinates as follows:

$$x = r \cos \phi \tag{2.4.1}$$

$$y = r \sin \phi \tag{2.4.2}$$

$$z = z \tag{2.4.3}$$

Since we are still concerned with a structure that is expected to guide waves in the z direction, we postulate fields of the form

$$\mathbf{E} = \mathbf{E}(r, \phi) \exp i(\omega t - {}^zkz) \tag{2.4.4}$$

$$\mathbf{H} = \mathbf{H}(r, \phi) \exp i(\omega t - {}^zkz) \tag{2.4.5}$$

The expression for $\nabla \times \mathbf{A}$ in cylindrical polar coordinates is as follows:

$$\nabla \times \mathbf{A} = \begin{vmatrix} \dfrac{\mathbf{r}}{r} & \boldsymbol{\phi} & \dfrac{1}{r}\mathbf{k} \\[2mm] \dfrac{\partial}{\partial r} & \dfrac{\partial}{\partial \phi} & \dfrac{\partial}{\partial z} \\[2mm] {}^rA & r^\phi A & {}^zA \end{vmatrix} \tag{2.4.6}$$

where $\mathbf{k}$ is the unit vector in the z direction. We are now in a position to derive expressions for the field components by use of equations 2.2.1 to 2.2.6. We obtain the set of relations equivalent to equations 2.3.3 to 2.3.8

for the planar case:

$$\frac{1}{r}\left[\frac{\partial^z E}{\partial \phi} + i^z k(r^\phi E)\right] = -i\omega\mu\mu_0 {}^r H \tag{2.4.7}$$

$$i^z k^r E + \frac{\partial^z E}{\partial r} = i\omega\mu\mu_0 {}^\phi H \tag{2.4.8}$$

$$\frac{1}{r}\left(r\frac{\partial^\phi E}{\partial r} - \frac{\partial^r E}{\partial \phi}\right) = -i\omega\mu\mu_0 {}^z H \tag{2.4.9}$$

$$\frac{1}{r}\left[\frac{\partial^z H}{\partial \phi} + i^z k(r^\phi H)\right] = i\omega\epsilon\epsilon_0 {}^r E \tag{2.4.10}$$

$$i^z k^r H + \frac{\partial^z H}{\partial r} = -i\omega\epsilon\epsilon_0 {}^\phi E \tag{2.4.11}$$

$$\frac{1}{r}\left(r\frac{\partial^\phi H}{\partial r} - \frac{\partial^r H}{\partial \phi}\right) = i\omega\epsilon\epsilon_0 {}^z E \tag{2.4.12}$$

By elimination between these equations, we can now obtain expressions for the r and ϕ components only in terms of the z components. These are as follows:

$$^r E = \frac{-i}{{}^T k^2}\left({}^z k\frac{\partial^z E}{\partial r} + \omega\mu\mu_0\frac{1}{r}\frac{\partial^z H}{\partial \phi}\right) \tag{2.4.13}$$

$$^\phi E = \frac{i}{{}^T k^2}\left(\frac{{}^z k}{r}\frac{\partial^z E}{\partial \phi} - \omega\mu\mu_0\frac{\partial^z H}{\partial r}\right) \tag{2.4.14}$$

$$^r H = \frac{-i}{{}^T k^2}\left({}^z k\frac{\partial^z H}{\partial r} - \frac{\omega\epsilon\epsilon_0}{r}\frac{\partial^z E}{\partial \phi}\right) \tag{2.4.15}$$

$$^\phi H = \frac{-i}{{}^T k^2}\left(\frac{{}^z k}{r}\frac{\partial^z H}{\partial \phi} + \omega\epsilon\epsilon_0\frac{\partial^z E}{\partial r}\right) \tag{2.4.16}$$

$$^T k^2 = \omega^2\epsilon\epsilon_0\mu\mu_0 - {}^z k^2 = n^2 k_0^2 - {}^z k^2 \tag{2.4.17}$$

where $^T k$ is the total transverse component of the k vector in the guide.

2.5. THE ELECTROMAGNETIC WAVE EQUATION

For the sake of completeness, we note here the standard derivation of the wave equation, and we give the form of the Laplacian operator for rectangular Cartesian and cylindrical polar coordinates.

If we take the curl of equation 2.2.1, we obtain

$$\nabla \times (\nabla \times \mathbf{E}) = -\mu\mu_0 \left(\nabla \times \frac{\partial \mathbf{H}}{\partial t} \right) \quad (2.5.1)$$

Differentiating equation 2.2.2 with respect to time yields

$$\nabla \times \frac{\partial \mathbf{H}}{\partial t} = \epsilon\epsilon_0 \frac{\partial^2 \mathbf{E}}{\partial t^2} \quad (2.5.2)$$

We then make use of the vector identity

$$\nabla \times (\nabla \times \mathbf{E}) = -\nabla(\nabla \times \mathbf{E}) - \nabla^2 \mathbf{E}$$
$$= -\nabla^2 \mathbf{E} \quad (2.5.3)$$

since $\nabla \cdot \mathbf{E} = 0$ (equation 2.2.5).

Then it follows directly by substitution that

$$\nabla^2 \mathbf{E} = \mu\mu_0\epsilon\epsilon_0 \frac{\partial^2 \mathbf{E}}{\partial t^2} \quad (2.5.4)$$

and likewise

$$\nabla^2 \mathbf{H} = \mu\mu_0\epsilon\epsilon_0 \frac{\partial^2 \mathbf{H}}{\partial t^2} \quad (2.5.5)$$

These equations are both of the general form

$$\nabla^2 \mathbf{A} = \frac{1}{V^2} \frac{\partial^2 \mathbf{A}}{\partial t^2} \quad (2.5.6)$$

where V is the velocity of propagation (phase velocity) of the wave in the medium. It follows that

$$V_p = \frac{1}{\sqrt{\mu\mu_0\epsilon\epsilon_0}} \quad (2.5.7)$$

and that, for free space, we have the velocity of light, c, given by

$$c = \frac{1}{\sqrt{\mu_0\epsilon_0}} \quad (2.5.8)$$

The situations to be examined in this book all concern either planar waveguides, described by rectangular Cartesian coordinates, or circular fibers, described by cylindrical polar coordinates. For these two systems the Laplacian operator takes the forms

$$\nabla^2 A = \frac{\partial^2 A}{\partial x^2} + \frac{\partial^2 A}{\partial y^2} + \frac{\partial^2 A}{\partial z^2} \qquad (2.5.9)$$

$$\nabla^2 A = \frac{1}{r}\frac{\partial}{\partial r}\left(r\frac{\partial A}{\partial r}\right) + \frac{1}{r^2}\frac{\partial^2 A}{\partial \phi^2} + \frac{\partial^2 A}{\partial z^2} \qquad (2.5.10)$$

REFERENCE

For general background, see one of the many excellent books on electromagnetic theory, for example, the following:

1. B. I. Bleaney and B. Bleaney, *Electricity and Magnetism*. Oxford: Oxford University Press (1957).

3

Reflection and Refraction; Interactions of the Electromagnetic Field at a Plane Interface

3.1. INTRODUCTION

We will now use the planar field relations derived in Chapter 2 to examine a particularly simple situation, namely, the reflection and refraction of a plane wave at an infinite plane dielectric interface. This is a useful exercise in the use of the field relations and also will serve to give some physical insight into the mechanisms involved in the propagation of electromagnetic energy through a planar dielectric waveguide.

3.2. REFLECTION AT A PLANE INTERFACE; TOTAL INTERNAL REFLECTION

Basic to our understanding of propagation in dielectric waveguides is a thorough understanding of reflection at a simple interface between two different dielectric materials, and the particular case of total internal reflection.

The interface that we will consider is illustrated in Figure 3.1. The wave is taken to be incident at angle Θ_1 to the normal to the interface,* and the coordinates are defined so that the direction of incidence lies in the x-z plane with the surface forming the y-z plane. There is therefore no component of propagation in the y direction. Furthermore, we will assume that the structure and the waves are infinite in the y direction with propagation in the z direction only, so that we can set $\partial/\partial y = 0$ in the analysis (this was already done in Chapter 2).

Because we are primarily interested in guidance by a dielectric structure, we will set the index of refraction of the material in which the wave is

---

*Note. We use Θ measured from the normal and set $\theta = \pi/2 - \Theta$, measured from the waveguide propagation axis for later use.

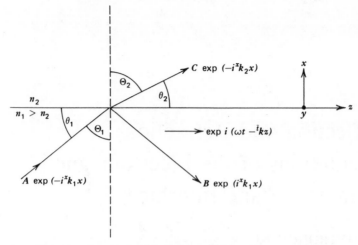

Figure 3.1. Planar interface analyzed in the text, defining the waves and angles involved.

incident, n_1, greater than that of the adjoining material, n_2. If we did not do this, total internal reflection could not occur and lossless guidance in the guide structure would not follow.

To find the solution for the amplitude and propagation characteristics of waves in such a structure, we proceed in the standard way by solving the wave equation for propagation in a homogeneous medium to find suitable waves in the two bounding media. We then apply the boundary conditions* appropriate to the interface to link them together and to determine their relative amplitudes and other properties.

The wave equation for a lossless medium in Cartesian coordinates is

$$\nabla^2 E = \mu\mu_0\epsilon\epsilon_0 \frac{\partial^2 E}{\partial t^2} = \frac{\partial^2 E}{\partial x^2} + \frac{\partial^2 E}{\partial y^2} + \frac{\partial^2 E}{\partial z^2} \qquad (3.2.1)$$

We postulate a time dependence of $\exp(i\omega t)$, and we assume that the waves are impinging in the x-z plane onto the interface, which lies in the y-z plane, so that we can assume that $\partial/\partial y = 0$. Since the phase fronts must

*Note that from equation 2.2.5 we may deduce that at a dielectric interface the normal component of D is continuous across that interface. Likewise, we may deduce that the tangential electric field E at the interface is constant as one crosses the interface. This result is obtained by considering a small rectangular path aligned with its long sides along the field direction (parallel to the interface) and its short sides normal to and cutting the interface. Consider the energy expended in traversing a point charge around the path as the short sides are allowed to shrink to zero. Evidently the energy also must approach zero, so that the fields on both sides of the interface must be equal.

match at all points along the interface in the z direction, we can ascribe to the waves the same z propagation constant, ^zk, so that they travel as $\exp(-i^zkz)$.

We postulate three waves in the structure, one of amplitude A incident upon the interface and two of amplitudes B and C, respectively, reflected and transmitted from the interface. The forms of the waves then follow directly:

$$A = A_0 \exp(-i^xk_1x)\exp i(\omega t - {}^zkz) \tag{3.2.2}$$

$$B = B_0 \exp(i^xk_1x)\exp i(\omega t - {}^zkz) \tag{3.2.3}$$

$$C = C_0 \exp(-i^xk_2x)\exp i(\omega t - {}^zkz) \tag{3.2.4}$$

where

$$^xk_1^2 = n_1^2k_0^2 - {}^zk^2 = -\gamma_1^2 \tag{3.2.5}$$

$$^xk_2^2 = n_2^2k_0^2 - {}^zk^2 = -\gamma_2^2 \tag{3.2.6}$$

$$n_1^2k_0^2 = \omega^2\epsilon_1\epsilon_0\mu_1\mu_0 \tag{3.2.7}$$

$$n_2^2k_0^2 = \omega^2\epsilon_2\epsilon_0\mu_2\mu_0 \tag{3.2.8}$$

$$\tan\theta_1 = \frac{{}^xk_1}{{}^zk} \tag{3.2.9}$$

$$\tan\theta_2 = \frac{{}^xk_2}{{}^zk} \tag{3.2.10}$$

For convenience we define the boundary as being at $x = 0$. We can now solve for the cases of the transverse electric (TE) and the transverse magnetic (TM) fields.

3.3. TRANSVERSE ELECTRIC MODES

We use the expressions in equations 3.2.2 to 3.2.4 to represent the y electric field components, ^yE. From these we can derive expressions for the tangential magnetic field component at the interface, ^zH, using equation 2.3.11. It follows then, by applying at the interface the boundary condition that the transverse E and H field components in medium 1 must equate to the transverse field components in medium 2, that (equating ^yE)

$$A_0 + B_0 = C_0 \tag{3.3.1}$$

and [equating ^zH, since $^zH = (i/\mu\mu_0\omega)(\partial^yE/\partial x)$] that

$$-^xk_1A_0 + ^xk_1B_0 = -^xk_2C_0 \tag{3.3.2}$$

By simple algebra we can now derive these results:

$$B_0 = A_0\left(\frac{^xk_1 - ^xk_2}{^xk_1 + ^xk_2}\right) = Ar_E \tag{3.3.3}$$

$$C_0 = A_0\left(\frac{2^xk_1}{^xk_1 + ^xk_2}\right) \tag{3.3.4}$$

We see that the reflected wave, B, is in phase with the incident beam but that its amplitude is less than A; it is partially reflected. This is always the case as long as $^xk_1$ and $^xk_2$ are both real. But as the angle Θ_1 increases, the component ^zk increases and both $^xk_1$ and $^xk_2$ decrease. A point is reached when $^xk_2$ goes to zero, so that

$$^xk_2^2 = n_2^2k_0^2 - ^zk^2 = 0 \tag{3.3.5}$$

$$^xk_1^2 = \left(n_1^2 - n_2^2\right)k_0^2 \tag{3.3.6}$$

With $^xk_1$ and $^xk_2$ both real, the angles θ_1 and θ_2 are given by the expressions

$$\tan\theta_1 = \frac{^xk_1}{^zk}; \qquad \sin\theta_1 = \frac{^xk_1}{n_1k_0} \tag{3.3.7}$$

$$\tan\theta_2 = \frac{^xk_2}{^zk}; \qquad \sin\theta_2 = \frac{^xk_2}{n_2k_0} \tag{3.3.8}$$

$$\frac{\sin\Theta_1}{\sin\Theta_2} = \frac{n_2}{n_1} = \frac{\cos\theta_1}{\cos\theta_2} \tag{3.3.9}$$

The expression in equation 3.3.9 is of course Snell's law of refraction. When $^xk_2$ is zero, θ_2 is zero. If θ_1 is further decreased, ^zk further increases and from equation 3.2.6 we see that $^xk_2$ becomes imaginary. We write it in the form $-i(\gamma_2)$. Throughout this, $^xk_1$ has remained real because of our assumption that $n_1 > n_2$, made at the outset. If we now insert in place of $^xk_2$ the term $-i(\gamma_2)$ in equation 3.3.3, we see that the modulus of the reflected wave, B, is now identical to the modulus of the incident wave, A. Thus the wave has been totally internally reflected (internally, because the model is usually applied to a glass-air interface and the reflection is thus inside or internal to the glass).

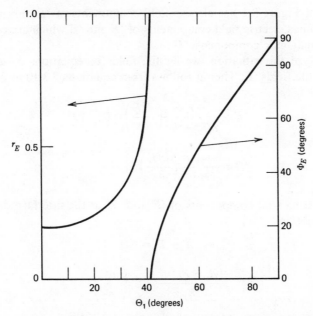

Figure 3.2. Reflection coefficient and phase shift on reflection for TE waves versus angle of incidence, $n_1 = 1.5$, $n_2 = 1.0$.

Thus we must also note that there is now a phase change in the reflected wave relative to the incident wave, that is,

$$B_0 = A_0 \left[\frac{{}^x k_1 + i(\gamma_2)}{{}^x k_1 - i(\gamma_2)} \right] = A_0 \exp(2i\Phi_E) \qquad (3.3.10)$$

where

$$\tan \Phi_E = \frac{\gamma_2}{{}^x k_1}$$

$$|B_0| = |A_0|$$

In Figure 3.2 we plot the amplitude reflection coefficient, $|r_E|$, and the phase, Φ_E, as a function of the angle of incidence, Θ_1, for an interface having indices of refraction $n_1 = 1.5$ and $n_2 = 1.0$.

3.4. TRANSVERSE MAGNETIC MODES

We showed in Chapter 2 that a wave reflected at a plane interface could be split into two groups of waves, TE and TM groups. The TE group in the notation of Figure 3.1 has electric field components of ^yE only, while the TM group has electric field components of ^xE and ^zE which are coupled to only one magnetic component, ^yH.

To analyze this situation, we let the fields of equations 3.3.2 to 3.3.4 represent the fields ^zE. Then if follows from equations 2.3.10 to 2.3.14 that

$$^xE = \frac{-i\,^zk}{n^2k_0^2 - \,^zk^2}\left(\frac{\partial\,^zE}{\partial x}\right) = \left(\frac{^zk}{^xk}\right)^zE \tag{3.4.1}$$

$$^yH = \frac{-i\omega n^2}{^xk^2}\left(\frac{\partial\,^zE}{\partial x}\right) \tag{3.4.2}$$

Equating tangential components of ^zE and ^yH at the interface defined by $x = 0$, we obtain

$$A_0 + B_0 = C_0 \tag{3.4.3}$$

$$\frac{(A_0 - B_0)n_1^2}{^xk_1} = C_0\left(\frac{n_2^2}{^xk_2}\right) \tag{3.4.4}$$

Hence we obtain as before the relations for reflection and transmission of the new polarization components:

$$B_0 = A_0\left(\frac{^xk_1n_2^2 - \,^xk_2n_1^2}{^xk_1n_2^2 + \,^xk_2n_1^2}\right) = A_0r_H \tag{3.4.5}$$

$$C_0 = A_0\left(\frac{2\,^xk_1n_2^2}{^xk_1n_2^2 + \,^xk_2n_1^2}\right) \tag{3.4.6}$$

The expression for r_H is very similar to that for r_E obtained from equation 3.3.3. However, with both $^xk_1$ and $^xk_2$ real, it is possible for r_H to be equal to zero, whereas under the same condition r_E could not.

The condition that $r_H = 0$ can be used to derive another well known relation, Brewster's condition. If $r_H = 0$, it follows that

$$^xk_2 \cdot n_1^2 = \,^xk_1 \cdot n_2^2 \tag{3.4.7}$$

Using this result, we can show that

$$\tan\Theta_1 = \frac{n_1}{n_2} \tag{3.4.8}$$

$$\tan\Theta_2 = \frac{n_2}{n_1} = \frac{1}{\tan\Theta_1} \tag{3.4.9}$$

so that $\Theta_1 + \Theta_2 = \pi/2$, and the reflected and refracted rays are at right angles to one another.

This last fact gives us a clear clue to the physical reason why the reflection coefficient for the wave at this angle is zero. It is simply that the E vector of the incident wave points directly along the beam direction for the reflected wave and therefore cannot couple in that direction. It follows also that under these conditions all the power must be transmitted. For a more thorough discussion of this subject, the reader is referred to the book by Marcuse,[2] who derives the full expressions for the power reflection and transmission coefficients at the plane dielectric interface.

Following our discussion of the TE reflection situation, we now consider once again the effect of increasing the angle of incidence, Θ_1. As before, an angle will be reached at which $^x k_2$ goes to zero and beyond which $^x k_2$ becomes imaginary. The modulus of r_H then becomes unity, and total internal reflection occurs as before but with a different phase change, that is,

$$B_0 = A_0 \exp(2i\Phi_H) \tag{3.4.10}$$

$$\tan \Phi_H = \left(\frac{n_1}{n_2}\right)^2 \left(\frac{\gamma_2}{^x k_1}\right) \tag{3.4.11}$$

In Figure 3.3 we plot the values of $|r_H|$ and Φ_H against the angle of incidence for an interface having $n_1 = 1.5$ and $n_2 = 1.0$. Notice that at the Brewster angle defined by equation 3.4.8 there is in addition to zero reflection, no change in phase of the reflected wave.

In summary, then, we have shown that a wave incident on a plane dielectric interface in general undergoes reflection and refraction. We have derived Snell's law of refraction and Brewster's Law, which shows that for one polarization component, incident at Brewster's angle, the reflection coefficient goes to zero. More important for our purposes, we have shown that as the angle of incidence increases, an angle is reached (the critical angle, Θ_c) at which the reflected amplitude equals the incident wave amplitude. A more detailed analysis readily shows that the power reflection coefficient is also unity. In our terminology the critical angle is characterized by the condition $^x k_2 = 0$. Using equation 3.3.9, we see that this corresponds to the condition

$$\sin \Theta_c = \sin \Theta_1 = \frac{n_2}{n_1} \tag{3.4.12}$$

For angles of incidence greater than Θ_c, the reflected wave undergoes a change of phase on reflection but its amplitude remains constant. The phase change depends upon both the polarization (TE or TM) and the

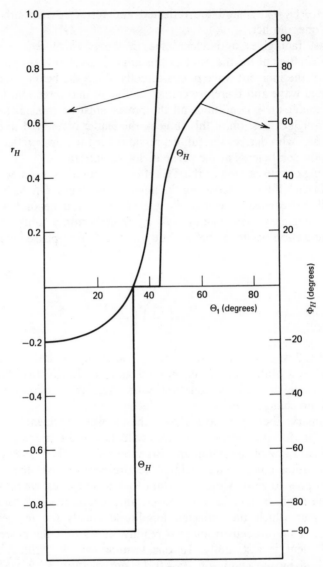

Figure 3.3. Reflection coefficient and phase shift on reflection for TM waves versus angle of incidence, $n_1 = 1.5$, $n_2 = 1.0$.

angle of incidence. In the next section we will discuss the significance of the phase change on reflection in much more detail.

So far, we have not commented on the form of the electric field in the medium of lower refractive index when the incident wave is totally reflected. Taking the form of the field of equation 3.2.4 and substituting $i(\gamma_2)$ in place of ${}^x k_2$, we obtain

$$C = C_0 \exp\left(-\gamma_2 x\right) \exp i(\omega t - {}^z kz) \qquad (3.4.13)$$

The field amplitude is seen to be exponentially decaying into the medium of lower index, away from the interface. The sign of γ_2 was chosen to yield this result. The opposite sign could have been chosen, but this would have led to an exponentially growing wave as the distance away from the interface increased. This is not a physically useful solution for the problems under consideration, since it would not represent a bound guided wave. However, in some other problems such a solution is important. An example is seen in the excitation of guided waves in a planar guide by the penetration of a wave from an adjoining medium.

The field in the lower index medium with exponentially decaying amplitude is termed an evanescent field. Such a field stores energy and carries energy in the direction of propagation (z), but it does not transport energy laterally (in the x direction in this case). To study this situation in more detail we should write down the Poynting vector for the energy flux and evaluate it for each particular case.

3.5. GOOS-HAENCHEN SHIFT

We have shown that under conditions of total internal reflection a wave incident upon the dielectric interface gives rise to a reflected wave of equal amplitude, but that the reflected wave undergoes a phase shift. The significance of this phase shift has not been discussed. Our theory also tells us that, although the wave is reflected at the dielectric interface, a field exists beyond the apparent plane of reflection, namely, in the medium of lower index. Experimentally this can be demonstrated to be so by bringing another high index medium into close proximity with the original one, but leaving a film of lower index material between them, as shown in Figure 3.4. Under these conditions a beam incident as shown gives rise to a transmitted beam, D, as well as to the expected beam, B. The relative power carried by beams B and D is a sensitive function of the film thickness, the angles of incidence, and the refractive indices. But clearly energy tunnels through the film layer.

To analyze the tunneling process is simple. Figure 3.4 shows the waves needed for the analysis. One proceeds as in the case of total reflection

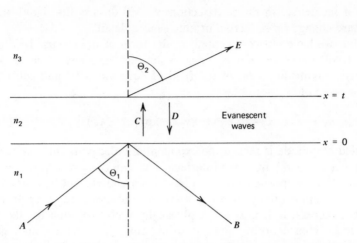

Figure 3.4. Definition of waves and angle used in the analysis of frustrated total reflection.

except that two additional equations are present, from matching E and H fields at the second interface, and these allow the elimination of the two additional variables, D and F. We will not carry through the algebra here.

We are left without a physical model to explain the phase shift on total reflection. However, a simple model exists which can be viewed in two different but equivalent ways. The model is illustrated in Figure 3.5, which is based upon the physical observation that a finite beam is shifted laterally on total reflection. This shift is known as the Goos-Haenchen shift, after its first observers.

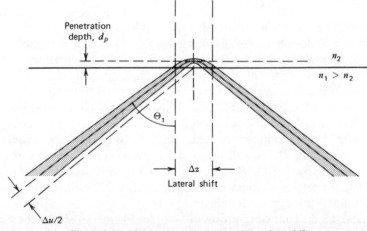

Figure 3.5. Physical model of Goos-Haenchen shift.

Our theory has shown that the phase shift on total reflection varies with the angle of incidence. This was illustrated in Figures 3.2 and 3.3. If we consider a Gaussian beam impinging at angle Θ to the interface normal, we can easily derive an expression for the shift.

We use the notation of Figure 3.6. The beam is incident along A and is expected to be reflected along B. We express the Gaussian beam in terms of the coordinates u and v, taking their origin at the point of incidence. The electric field takes the form

$$E(r) = E_0 \exp - \left(\frac{r^2}{R^2} \right) \qquad (3.5.1)$$

where

$$r^2 = u^2 + v^2$$

The same beam can be expressed as a sum of plane waves, by Fourier analysis, to yield

$$E(r) = \int_{-\infty}^{\infty} \int_{-\infty}^{\infty} E(^uk, {}^vk) \exp - i(\mathbf{k \cdot r}) \, d^uk \, d^vk \qquad (3.5.2)$$

where

$$|\mathbf{k}|^2 = {}^uk^2 + {}^vk^2 + {}^wk^2 = {}^Tk^2 + {}^wk^2 \qquad (3.5.3)$$

$$E(^uk, {}^vk) = \left(\frac{1}{4\pi} \right) E_0 R^2 \exp - \left(\frac{{}^Tk^2 R^2}{4} \right) \qquad (3.5.4)$$

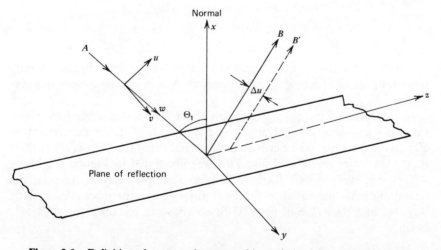

Figure 3.6. Definition of waves and terms used in analysis of Goos-Haenchen shift.

This beam travels at an angle Θ to the plane of reflection. Plane wave components traveling at angle $\Delta\Theta$ to this angle of incidence are described by values of ^uk given by

$$\Delta\Theta = \frac{^uk}{^wk} \qquad \text{(assuming that } ^uk \ll {^wk}) \qquad (3.5.5)$$

Now suppose that the phase of the reflected plane wave component shifts by an amount

$$\Phi_{\Delta\Theta} = \Phi_0 + \frac{\partial\Phi}{\partial\Theta}(\Delta\Theta) \qquad (3.5.6)$$

Each plane wave component in equation 3.5.2 now has to be multiplied after reflection by the new phase factor appropriate to the component. The result is that after reflection equation 3.5.2 takes the form

$$E(r) = \int_{-\infty}^{\infty}\int_{-\infty}^{\infty} E(^uk, {^vk})\exp -i\left[{^uk}\left(u - \frac{1}{^wk}\frac{\partial\Phi}{\partial\Theta}\right) + {^vkv} + {^wkw}\right]$$

$$\times \exp(i\Phi_0)\, d^uk\, d^vk \qquad (3.5.7)$$

and we see that the beam has apparently shifted laterally an amount where

$$\Delta u = \frac{-1}{^wk}\frac{\partial\Phi}{\partial\Theta} \qquad (3.5.8)$$

In practice, this shift is quite small and difficult to observe. But it is vitally important for our understanding of propagation in optical dielectric waveguides.

A simple model indicating the origin of the shift and also the total phase shift, ϕ, was advanced by Kapany and Burke.[1] It proposes that reflection effectively does not take place at the material interface but occurs slightly inside the lower index medium. This was illustrated in Figure 3.5. Using this model, we see that there should be a simple relation between the lateral shift Δz, the value of the total phase shift experienced by a plane wave, ϕ, and the value of $(\partial\Phi/\partial\Theta)$ used above to calculate the shift of a beam.

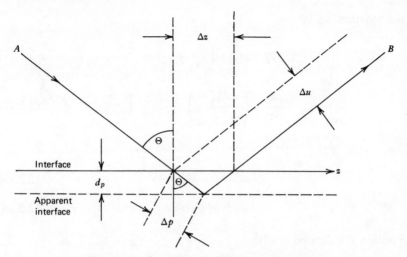

Figure 3.7. Lateral view of Goos-Haenchen shift.

Referring to Figure 3.7, we see that, by simple geometry, this is true, since the relations below follow directly:

$$\Delta p = d_p \sec \Theta \qquad\longrightarrow\qquad \Phi = 2^w k\, \Delta p$$
$$= 2^w k\, d_p \sec \Theta$$

$$\downarrow$$

$$\Delta z = 2 d_p \tan \Theta \qquad\qquad \frac{\partial \Phi}{\partial \Theta} = 2^w k\, d_p \sec \Theta \tan \Theta$$

$$\Delta u = \Delta z \sec \Theta \qquad\qquad = {}^w k \sec \Theta\, \Delta z$$

$$\therefore \Delta u = \Delta z \sec \Theta = \frac{1}{{}^w k} \frac{\partial \Phi}{\partial \Theta} \qquad (3.5.9)$$

Now, setting ${}^w k = n_1 k_0$, we have

$$d_p = \frac{\cos \Theta}{2 \tan \Theta} \frac{1}{n_1 k_0} \frac{\partial \Phi}{\partial \Theta} \qquad (3.5.10)$$

Finally, we will evaluate the expression for the shift. Very conveniently, $\partial \Phi / \partial \Theta$ can be easily calculated. Starting with the TE polarization, we have from equation 3.3.10

$$\Phi_E = \arctan\left(\frac{\gamma_2}{{}^x k_1}\right) \tag{3.5.11}$$

$$\frac{\partial\Phi}{\partial\Theta} = \left(\frac{{}^x k_1^2}{{}^x k_1^2 + \gamma_2^2}\right)\frac{\partial}{\partial\Theta}\left(\frac{\gamma_2}{{}^x k_1}\right) \tag{3.5.12}$$

Then, noting that

$${}^x k_1^2 = n_1^2 k_0^2(1 - \sin^2\Theta) \tag{3.5.13}$$

$$\gamma_2^2 = k_0^2(n_1^2\sin^2\Theta - n_2^2) \tag{3.5.14}$$

we obtain the desired result:

$$\Delta z_{TE} = \frac{2}{\gamma_2}\tan\Theta \tag{3.5.15}$$

$$d_{p(TE)} = \frac{1}{\gamma_2} \tag{3.5.16}$$

Similar expressions can be derived for the TM polarization. If we define a parameter q such that

$$q = \left(\frac{n_1\sin\Theta}{n_2}\right)^2 - \cos^2\Theta \tag{3.5.17}$$

we obtain the equivalent relations as follows:

$$\Delta z_{TM} = \frac{1}{q}\Delta z_{TE} \tag{3.5.18}$$

$$d_{p(TM)} = \frac{1}{q}d_{p(TE)} \tag{3.5.19}$$

3.6. SUMMARY

We have shown that, when a plane wave is "totally reflected" at a dielectric interface, the wave returns not from the interface but apparently from a layer spaced a distance d_p inside the lower index medium. For a plane wave this is manifest by the phase shift on reflection, but in the case of a restricted beam a lateral shift can be observed under carefully controlled conditions. These results are highly significant when we come to examine propagation in a planar dielectric guide, a thin film between two media of lower index, since it is perfectly possible for the penetration depth d_p to be comparable to or greater than the film thickness, $2f$, so that the wave spends more time outside the film than in it. Similar situations occur in circular dielectric guides. It follows that we cannot consider the guided wave to be traveling solely in the core medium and that therefore the properties of the cladding or substrate medium are equally important. It also follows that the thickness of the cladding medium is likely to be of importance, since in general it must be made thick enough so that for all practical purposes there is no field at the outer boundary of the layer.

REFERENCES

1. N. S. Kapany and J. J. Burke, *Optical Waveguides*. New York: Academic, 1972, p. 74.
2. D. Marcuse, *Theory of Dielectric Optical Waveguides*. New York: Academic, 1974.

4

Planar Dielectric Waveguide

4.1. INTRODUCTION

Having studied the reflection of a plane wave at a high index-low index interface, it is now a simple matter to see how a planar dielectric waveguide operates and to understand its most important features. In Figure 4.1 a planar guide is illustrated. The interface at $x = d$ corresponds to the interface previously studied. An incident plane wave gives rise to a reflected wave, which is in turn reflected again at the second interface at $x = d$. Total internal reflection is assumed to occur at both interfaces so that $n_1 > n_2$ and $n_1 > n_3$. Consequently the wave is guided along the structure in a zigzag path by successive reflections.

Another point is apparent in Figure 4.1. For the twice-reflected wave to be indistinguishable from the infinite plane wave, AA, the total phase shift in traversing the guide twice and being reflected twice must be equal to an integer multiple of 2π radians. We will see that this requirement imposes an additional condition upon the guidance process that was not present in reflection at a single interface. Reflection alone coupled the amplitudes and phases of all the waves involved at that interface, so that, given one, the others were totally specified. But in the guide we will find that solutions exist only for certain values of ^zk, which are called modes of propagation and which physically correspond to the condition that the round trip phase delay be a multiple of 2π radians.

To analyze the problems further, we can either write down the form of the waves intuitively, as shown in Figure 4.1 and in Table 4.1, by extrapolation from the analysis of total internal reflection, or start with the wave equation and deduce the form of the waves required.

In Cartesian coordinates the wave equation for electromagnetic waves in a nonconducting medium takes the form

$$\frac{\partial^2 A}{\partial x^2} + \frac{\partial^2 A}{\partial y^2} + \frac{\partial^2 A}{\partial z^2} = \mu\mu_0 \epsilon\epsilon_0 \frac{\partial^2 A}{\partial t^2} \qquad (4.1.1)$$

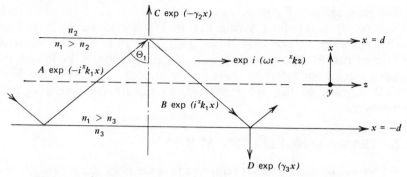

Figure 4.1. Diagram of a planar guide and the waves used in this analysis.

TABLE 4.1   WAVES USED IN PLANE WAVE ANALYSIS

Impinging on n_1/n_2 boundary at $x = d$

$A \exp -i(^x k_1 x + {}^z kz - \omega t)$

$C \exp (-\gamma_2 x) \exp i (\omega t - {}^z kz)$

Impinging on n_1/n_3 boundary at $x = -d$

$B \exp i(^x k_1 x - {}^z kz + \omega t)$

$D \exp (\gamma_3 x) \exp -i({}^z kz - \omega t)$

If we now postulate that we are interested only in solutions propagating in the z direction at frequency ω and that the structure is infinite in the y direction, so that we can set $d/dy = 0$, it follows that the wave must have a z dependence of the form $\exp \pm(i^z kz)$, and we obtain the result

$$\frac{\partial^2 A}{\partial x^2} + \left(\mu\mu_0 \epsilon\epsilon_0 \omega^2 - {}^z k^2\right) A = 0 \qquad (4.1.2)$$

The general solution of this equation is simply

$$A = A_0 \exp \pm \left(i^T kx\right)$$
$$^T k = \sqrt{\mu\mu_0 \epsilon\epsilon_0 \omega^2 - {}^z k^2} = \sqrt{n^2 k_0^2 - {}^z k^2} \qquad (4.1.3)$$

and we see that the waves that we have already postulated by an intuitive argument have been derived from first principles.

Referring again to Figure 4.1, we consider the wave A incident at angle Θ on the n_1-n_2 interface, giving rise to a reflected wave, B, and an evanescent wave, C. The wave B is in turn reflected at the n_1-n_3 interface and gives rise to an evanescent wave, D, in medium 3. Following the analysis carried out before to relate the amplitudes A, B, and C and B, C, and D at each of the two interfaces, we equate the tangential field components.

4.2. TRANSVERSE ELECTRIC MODES

We let the field expressions represent first the field component ^yE. We know from equations 2.3.9 and 2.3.11 that ^xH and ^zH follow directly, and we equate components of ^yE and ^zH at the boundary $x = d$ to relate A, B, C and at the boundary $x = -d$ to relate B, C, D. This yields the following expressions:

$$A_0 \exp\left(-i^xk_1d\right) + B_0 \exp\left(i^xk_1d\right) = C_0 \exp\left(-\gamma_2d\right) \quad (4.2.1)$$

$$-{}^xk_1 A_0 \exp\left(-i^xk_1d\right) + {}^xk_1 B_0 \exp\left(i^xk_1d\right) = i\gamma_2 C_0 \exp\left(-\gamma_2d\right) \quad (4.2.2)$$

$$A_0 \exp\left(i^xk_1d\right) + B_0 \exp\left(-i^xk_1d\right) = D_0 \exp\left(\gamma_3d\right) \quad (4.2.3)$$

$$-{}^xk_1 A_0 \exp\left(i^xk_1d\right) + {}^xk_1 B_0 \exp\left(-i^xk_1d\right) = -i\gamma_3 D_0 \exp\left(\gamma_3d\right) \quad (4.2.4)$$

Four equations containing four unknowns have a solution only if the matrix of their coefficients is zero. In this manner we find an equation which is known as the eigenvalue equation, because it leads to specific values or eigenvalues for ^zk (or ^xk or γ), and which must be satisfied if a solution or guided wave is to be obtained:

$$\begin{vmatrix} \exp(-i^xk_1d) & \exp(i^xk_1d) & \exp(-\gamma_2d) & 0 \\ -{}^xk_1\exp(-i^xk_1d) & {}^xk_1\exp(i^xk_1d) & i\gamma_2\exp(-\gamma_2d) & 0 \\ \exp(i^xk_1d) & \exp(-i^xk_1d) & 0 & \exp(\gamma_3d) \\ -{}^xk_1\exp(i^xk_1d) & {}^xk_1\exp(-i^xk_1d) & 0 & -i\gamma_3\exp(\gamma_3d) \end{vmatrix} = 0$$

$$(4.2.5)$$

This equation reduces by straightforward algebra to the surprisingly simple form

$$\tan(2^xk_1d) = \frac{(\gamma_2/^xk_1) + (\gamma_3/^xk_1)}{1 - (\gamma_2/^xk_1)(\gamma_3/^xk_1)} \quad (4.2.6)$$

Now, by using the expansion for $\tan(a+b)$, we see that this can be written as

$$(2^x k_1 d) \pm n\pi = \tan^{-1}\left(\frac{\gamma_2}{^x k_1}\right) + \tan^{-1}\left(\frac{\gamma_3}{^x k_1}\right) \qquad (4.2.7)$$

Reference to equation 3.5.12 shows immediately that this is the phase condition already deduced on physical grounds, the terms on the right-hand side being the phase changes on total internal reflection and the left-hand side representing the phase change on traversing the guide.

In the case of the symmetrical guide in which $n_2 = n_3$, we see that further simplification follows:

$$^x k_1 d \pm \frac{n\pi}{2} = \tan^{-1}\left(\frac{\gamma_2}{^x k_1}\right) \qquad (4.2.8)$$

We already know that the amplitudes of the incident waves and reflected waves (A and B in Figure 4.1) are equal from our analysis of total internal reflection, and the same result follows directly from equations 4.2.1 and 4.2.2. Consequently, using equation 4.2.8, we see that the resultant field in the guide layer is given by

$$^y E = A_0\left[\exp i\left(^x k_1 x - \frac{n\pi}{2}\right) + \exp - i\left(^x k_1 x \frac{n\pi}{2}\right)\right] \qquad (4.2.9)$$

For values of the integer, n, equal to $0, 2, 4$, etc., this reduces to $\cos(^x k_1 x)$, and for odd values of n it has the form $\sin(^x k_1 x)$. The guided modes fall into two categories: even and odd, those which have a null in the center of the guide and those which do not, but both derive directly from the plane wave analysis. Alternatively we could have assumed the sine or cosine form for the field directly and used that to obtain our solution.

4.3. TRANSVERSE MAGNETIC MODES

The preceding analysis was carried through for the TE polarization. We can repeat it for the orthogonal TM polarization and obtain a similar result. Using the expressions of Figure 4.1 to represent $^y H$ in place of $^y E$ and deducing expressions for $^z E$, we equate the fields at the interfaces and solve the determinant of coefficients to obtain, using equation 2.3.10, 2.3.12, and 2.3.14,

$$\tan(2^x k_1 d) = \left(\frac{\gamma_2 n_1^2}{^x k_1 n_2^2} + \frac{\gamma_3 n_1^2}{^x k_1 n_3^2}\right)\left[1 - \left(\frac{\gamma_2 n_1^2}{^x k_1 n_2^2}\right)\left(\frac{\gamma_3 n_1^2}{^x k_1 n_3^2}\right)\right]^{-1} \qquad (4.3.1)$$

or

$$2^x k_1 d - n\pi = \tan^{-1}\left(\frac{\gamma_2 n_1^2}{^x k_1 n_2^2}\right) + \tan^{-1}\left(\frac{\gamma_3 n_1^2}{^x k_1 n_3^2}\right) \qquad (4.3.2)$$

These solutions again lead to an expression for the resultant field in the guide of the form $\cos(^x k_1 x)$ or $\sin(^x k_1 x)$.

4.4. EIGENVALUE EQUATION AND CUTOFF

The solution of the eigenvalue equation, 4.2.7 or 4.3.2, for any particular guide (i.e., values of d, n_1, n_2, n_3) generally requires a numerical analysis on a computer if a great deal of tedious arithmetic is to be avoided. However, in Figure 4.2 we indicate the graphical solution of equation 4.2.7 for a guide supporting two modes. The modes are characterized by the integer n in equation 4.4.1 and would therefore be described as the TE_0 and TE_1

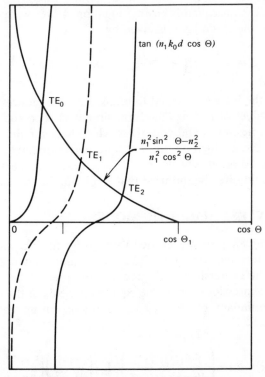

Figure 4.2. Graphical solution of the eigenvalue equation, equation 4.2.7.

modes. For the purposes of plotting, the variables $^x k_1$ and γ_2 have been replaced by the single variable Θ, defined as in Figure 3.1, that is,

$$\tan\left(^x k_1 d - \frac{n\pi}{2}\right) = \frac{\gamma_2}{^x k_1} \qquad (4.4.1)$$

becomes

$$\tan\left(n_1 k_0 d \cos\Theta - \frac{n\pi}{2}\right) = \sqrt{\frac{n_1^2 \sin^2\Theta - n_2^2}{n_1^2 \cos^2\Theta}} \qquad (4.4.2)$$

This plot shows quite clearly that for a finite value of d there are only a finite number of solutions, in this case two. As the guide thickness increases, more modes are allowed to propagate. However, it is interesting to note that in this case at least one mode (the TE_0) can always propagate since there will always be one intersection.

The reason that higher order modes cannot propagate is simply that there is not sufficient thickness of guide, within the range of propagation angles allowed for total internal reflection, for the phase shift in traversing the guide in one direction to vary by a greater number of integer multiples of π radians. One mode is always allowed to propagate because, as the critical angle is approached, the phase shift on internal reflection falls to zero, so that for a vanishingly small guide the traversal phase shift can also approach zero and still allow a solution to the eigenequation.

Notice that at cutoff in the planar dielectric guide the wave is impinging on one or both of the film-cladding boundaries at the critical angle, and either γ_2, γ_3 or both are equal to zero. This contrasts with the metal clad waveguide, in which cutoff is associated with a wave which should travel forward but merely reflects back and forth across the guide at normal incidence to the walls. Also, we note here that cutoff in a dielectric guide of nonplanar symmetry is not necessarily associated with the critical angle. In a circular guide, as we shall see later, the angle of incidence for cutoff may depart slightly from the critical angle.

4.5. VISUALIZATION OF PLANAR GUIDE MODES

From the formulas developed for the fields of the guide modes, we can now easily form a picture of the appearance of the field for the planar guide. The two oppositely traveling waves in the guide interfere to form a standing wave pattern of sine or cosine form. The integer n is equal to the number of field maxima across the guide. The field does not go to zero at the guide-cladding interface, but it changes from oscillating spatial form to

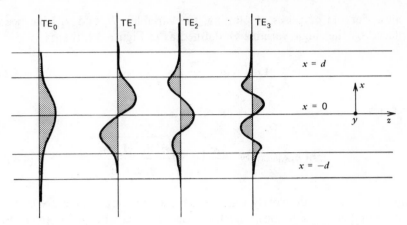

Figure 4.3. Field patterns for the lower order modes of a planar guide.

exponential decay, decreasing to smaller values as the distance from the guide layer increases. The rate of decrease of this field depends upon the parameters γ_2 and γ_3. As cutoff for a mode is reached, one or both of these parameters approaches zero and the field penetrates further into the cladding region. At cutoff, one of them is zero and that field no longer decays. In Figure 4.3 we have drawn schematically the field patterns for some of the low order modes of a symmetrical planar guide to illustrate the above points.

4.6. ENERGY DISTRIBUTION IN PLANAR GUIDE

Evidently the guided wave traveling along the planar guide carries energy. Some of that energy will certainly be carried within the guide layer; but since the field penetrates into the cladding, it seems inevitable that some will also be carried in the cladding. The respective amounts are readily calculated using the Poynting vector to estimate the energy flow. We will illustrate this in terms of the field expressions for a symmetrical guide carrying a symmetrical TE mode. Thus the fields for the guide layer are given by

$$^yE = A \cos(^xk_1 x) \tag{4.6.1}$$

and the field for the cladding is given by

$$^yE = A \exp -|\gamma_2 x| \left[\frac{\cos \left(^xk_1 d \right)}{\exp -|\gamma_2 d|} \right] \tag{4.6.2}$$

with the associated magnetic field given by

$$^xH = \sqrt{\frac{\epsilon\epsilon_0}{\mu\mu_0}} \; ^yE \tag{4.6.3}$$

The guide carries energy in the z direction, so that we must evaluate the z component of the time averaged Poynting vector. This is given by

$$^zP = \frac{1}{2} \int (^xE^yH - {}^yE^xH)\,dx \tag{4.6.4}$$

for unit guide width in the y direction. Evaluating this expression, we find that for a symmetrical guide ($n_2 = n_3$)

$$P_{core} = \frac{A^2}{2} \sqrt{\frac{\epsilon\epsilon_0}{\mu\mu_0}} \left[d + \frac{\sin(2^xk_1 d)}{^xk_1} \right] \tag{4.6.5}$$

$$P_{clad} = A^2 \sqrt{\frac{\epsilon\epsilon_0}{\mu\mu_0}} \; \frac{\cos^2(^xk_1 d)}{2\gamma_2} \tag{4.6.6}$$

(adding both cladding components together).

At this point, one has to solve the eigenvalue equation for γ_2 in terms of $^xk_1$ to obtain the result

$$\gamma_2 = {}^xk_1 \tan(^xk_1 d) \tag{4.6.7}$$

for this particular case, and then to insert numbers to find the particular situation for a given guide. However, casual inspection shows that, as the parameter γ_2 goes to zero (cutoff), the proportion of the power in the cladding will approach unity, and, conversely, for a well guided mode in a large guide (large d and also γ_2) the proportion of power in the cladding will approach zero. We will see later that similar considerations apply to the guided modes of circular fibers, although with some interesting changes due to the detailed properties of the structure.

4.7. ENERGY FLOW IN PLANAR DIELECTRIC GUIDE AND GROUP VELOCITY

The analysis we have presented so far gives a clear physical picture of propagation in a planar guide, in terms of plane waves being totally reflected at the two boundary walls and hence zigzagging down the guide.

We have shown that the two waves (zig and zag!) maintain a constant phase relationship to each other, so that the field distribution within the guide will appear to an observer as a standing wave pattern across the section of form sine or cosine of the displacement from the center. [The optical center can be displaced from the geometric center in an asymmetric guide $(n_2 \neq n_3)$.]

We also saw in an earlier section that energy tunnels into the cladding of the guide and an apparent shift occurs on reflection. Taking these facts together, we can picture the propagation in the guide in the form illustrated in Figure 4.4. Figure 4.4a illustrates our model of a ray traveling along by successive reflection. In Figure 4.4b we show the spatial distribution of the field deduced in equation 4.2.9 with a sine or cosine variation across the guide layer and an exponential decay into the cladding. But we have not discussed the velocity at which energy propagates along this structure, a vital parameter in any communications system.[1]

If we follow a particular phase point of a pure single-frequency electromagnetic wave traveling through the structure, that phase point, be it node or antinode, will travel at the phase velocity, which in our notation is given by

$$V_p = \frac{\omega}{{}^z k}$$

A pulse of energy is necessarily composed of the sum of plane wave

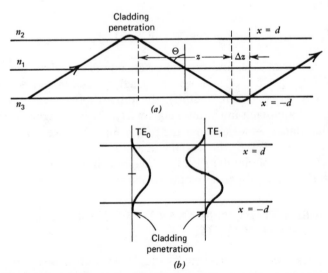

Figure 4.4. Physical model of propagation in the planar dielectric guide showing (a) ray paths and (b) wave field pattern.

components of different frequencies, and in general each travels at a different phase velocity with the result that the wave packet or pulse is observed to travel at the group velocity

$$V_g = \frac{\partial \omega}{\partial {}^z k} \qquad (4.7.1)$$

This is the velocity that is of greatest importance in the study of the transmission characteristics of a fiber or planar guide.

If we consider propagation in an infinite medium of refractive index n, we have the following relations:

$$k = \frac{n\omega}{c} \qquad (4.7.2)$$

$$V_p = \frac{c}{n} \qquad (4.7.3)$$

$$V_g = \frac{c}{m} = \frac{c}{n + \omega(\partial n / \partial \omega)} \qquad (4.7.4)$$

The parameter m is known as the group index.

We can define an analogous notation for our waveguide, in terms of the parameters N and M, the effective refractive and group indices of the guide. Then we have

$$N = \frac{c}{V_p} \qquad (4.7.5)$$

$$M = \frac{c}{V_g} \qquad (4.7.6)$$

We can see directly from Figure 4.4 that $N = n_1 \sin \Theta$.

Regarding propagation in the guide in terms of the zigzag ray model, it is tempting to write the group velocity for propagation as

$$V_g = \left(\frac{c}{m_1} \right) \sin \Theta = \frac{cN}{n_1 m_1} \qquad (4.7.7)$$

This leads to the conclusion that, as the angle Θ becomes smaller, the velocity of propagation necessarily becomes slower, yet we know that for a thin guide, as this happens, more of the guided energy travels in the cladding and proportionately less in the guiding film. Thus we might suppose the exact opposite, since the cladding will have a lower index. The latter can in fact be the case, and the reason for the discrepancy lies in our neglect of the processes associated with total reflection. We will now derive a more precise relationship for the group velocity which predicts the true behavior.

Referring to Figure 4.4, we note that the following relations apply:

$$z_1 = d \tan \Theta \tag{4.7.8}$$

$$\Delta z = \frac{\partial \Phi}{\partial^z k} \tag{4.7.9}$$

$$\Delta \tau = \frac{-\partial \Phi}{\partial \omega} \tag{4.7.10}$$

where $\Delta \tau$ is the time taken to traverse the distance Δz.

The group velocity will be given by the total time taken to traverse the guide divided into the total distance traveled in the propagation direction, that is,

$$V_g = \frac{z_1 + \Delta z}{\tau_1 + \Delta \tau} \tag{4.7.11}$$

While the energy is traveling in medium 1, the film, we expect normal behavior, so that

$$\tau_1 = \frac{m_1 d}{c \cos \Theta} \tag{4.7.12}$$

The expressions for $\Delta \tau$ can be derived using the waveguide eigenvalue equation, 4.7.16, and equation 4.7.10. This yields these results:

$$\Delta \tau(\text{TE}) = \frac{k_0 \left({}^x k_1^2 n_2 m_2 + \gamma_2^2 n_1 m_1 \right)}{c \, {}^x k_1 \gamma_2 \left({}^x k_1^2 + \gamma_2^2 \right)} \tag{4.7.13}$$

$$\Delta \tau(\text{TM}) = \frac{k_0 \left[{}^x k_1^2 n_2 m_2 + \gamma_2^2 n_1 m_1 - 2(m_1/n_1 - m_2/n_2) {}^x k_1^2 \gamma_2^2 / k_0^2 \right]}{c q \gamma_2 {}^x k_1 \left({}^x k_1^2 + \gamma_2^2 \right)} \tag{4.7.14}$$

Using the result already derived for z_1 (see equation 4.7.8), we can now write the total expression for the group velocity by substituting in equation 4.7.11.

We can easily derive two limiting cases of the group velocity. In the first we consider the effect of the guide becoming large and of propagation in the TE_0 mode. In this case $\Theta \rightarrow 90$, $\Delta z \rightarrow 0$, and $V_g = V_g(\text{film})$. The energy propagates with a velocity equal to or tending toward that of a free wave in the film medium.

The second case concerns the opposite limit, namely, $d \to 0$. Now we see that both z_1 and $\tau_1 \to 0$, and V_g is given by $\Delta z / \Delta \tau$. The expressions for $\Delta \tau(\text{TE})$ and Δz can be simplified in this case by using equation 4.6.7 for $^x k_1$ and γ_2 and by noting that $\Theta \to \Theta_c$. We then find that, as $d \to 0$,

$$V_g \to \frac{c}{m_2} \tag{4.7.15}$$

In Figure 4.5 we show an evaluation of the group velocity for a guide comprised of a film of a material such as ZnS with a cladding of fused silica. The group and refractive indices are plotted for the TE_0 mode as a function of guide thickness. The film index (our n_1) is indicated by n_f, and

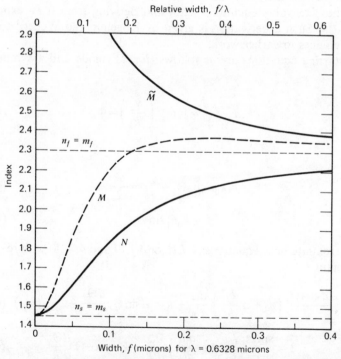

Figure 4.5. Calculation of the parameters for TE mode propagation in a ZnS guide on a fused silica substrate showing the phase index $N = c/V_p$ and the group index $M = c/V_g$ versus guide thickness. The upper curve labeled $\tilde{M}$ shows the group index that would be deduced by a ray model that did not include the Goos-Haenchen shift. Reproduced with permission from H. Kogelnic and H. P. Weber, *J. Opt. Soc. Am.*, **64**, 174 (1974).

the cladding, n_2, by n_s. Also plotted is the group index, $\tilde{M}$, predicted by the simplified model, equation 4.7.6, clearly demonstrating the importance of taking into account the Goos-Haenchen shift. Similar results apply for the TM modes.

It follows from the above analysis and results that the group velocity in a dielectric waveguide is a function of the properties of all of the materials used in its construction, the dimensions of the guide, the frequency of the propagating wave, and the mode of the guide that is carrying the energy. Since, in a practical case, more than one mode can often carry energy at a time, several distinct group velocities may be involved in the propagation of a pulse of energy along the structure, leading not only to dispersion of the pulse energy through the dispersion associated with each mode, an unavoidable effect, but also to the generally much larger dispersion between the different modes.

The analysis presented so far has been based upon a largely intuitive argument. The same result can be derived in a more rigorous manner by taking the eigenvalue equation (4.2.6) and deriving from it an expression for $\partial\omega/\partial^z k$. Such a derivation is given by Kogelnic and Weber,[1] together with references to earlier work.

The starting equations are as follows (for TE mode and a symmetrical guide):

$$^x k_1 d - \frac{n\pi}{2} = \tan^{-1}\left(\frac{\gamma_2}{^x k_1}\right) = \Phi_E \tag{4.7.16}$$

or

$$n_1 k_0 d \cos\Theta_1 - \Phi_E = \frac{n\pi}{2} \tag{4.7.17}$$

$$^z k = k_0 n_1 \sin\Theta_1 \tag{4.7.18}$$

We differentiate both equations (4.7.16 or 4.7.17 and 4.7.18) with respect to $^z k$ to obtain

$$\frac{d\cos\Theta_1}{c}\left(n_1 + \omega\frac{dn_1}{d\omega}\right)\frac{d\omega}{d^z k} - k_0 n_1 d \sin\Theta_1 \frac{d\Theta_1}{d^z k} - \frac{d}{d^z k}(\Phi_E) \tag{4.7.19}$$

$$1 = \frac{\sin\Theta_1}{c}\left(n_1 + \omega\frac{dn_1}{d\omega}\right)\frac{d\omega}{d^z k} + k_0 n_1 \cos\Theta\frac{d\Theta}{d^z k} \tag{4.7.20}$$

If we now recall that the group index, m_1 is given by $m_1 = n_1 + \omega(dn_1/d\omega)$ and we eliminate $d\Theta/d^z k$ between equations 4.7.19 and 4.7.20, using the result also that $d\Phi/d^zk = \partial\Phi/\partial^zk + [(\partial\Phi/\partial\omega)(\partial\omega/\partial^zk)]$, we obtain the result

$$V_g = \frac{d\omega}{d^zk} = \left(d\tan\Theta + \frac{\partial\Phi_E}{\partial^zk}\right)\left(\frac{m_1 d}{c\cos\Theta_1} - \frac{\partial\Phi_E}{\partial\omega}\right)^{-1} \quad (4.7.21)$$

This is exactly the result, equation 4.7.11, that we wrote down earlier on the basis of an intuitive argument.

4.8. SUMMARY

We have examined the path taken by energy propagating along a planar dielectric guide. From the relations for the reflection of energy at a single interface we have shown how to derive the conditions for a guided mode. The energy tunneling into the cladding during total internal reflection manifests itself as a shift on reflection. This fact makes possible propagation of the TE_0 mode in any planar guide regardless of thickness. It also allows us to understand the energy distribution between core and cladding, as well as the physical reasons for the group velocity expression that is derived most directly from the eigenvalue equation, but is rather lacking in any obvious physical interpretation.

The relations given in this chapter apply to planar guides of infinite extent in the y-z plane. However, most of the physical principles carry over to more complex structures, the only major difference being greater mathematical complexity. In the circular guides that we will analyze next, however, some new principles emerge.

REFERENCE

1. H. Kogelnic and H. P. Weber, *J. Opt. Soc. Am.*, **64**, 174 (1974).

5

Circular Dielectric Waveguide— The Optical Fiber

5.1. INTRODUCTION

In this chapter we build upon the basic understanding that we have acquired in the preceding chapters to analyze the circular dielectric guide. While optical fibers for use in transmission are not necessarily of circular cross section or symmetry, the majority have this configuration, and it seems likely that those to be used in practice will also, as shown in Figure 5.1.

The theory presented here is simplified by the assumption that the difference in index between the core and cladding materials is very small. This results in considerable mathematical simplification and very great simplification in the description of the mode fields, since it leads us to the linearly polarized (LP) field notation developed by Gloge,[1] rather than to the more precise notation for the generalized circular guide, which includes HE, EH, and TE modes. In practice, no serious sacrifice in accuracy is caused by this simplification; for the guides that are normally used the index difference is usually less than 3%, and under these conditions the HE, EH, and LP modes can be separated only by experiments of great sophistication that have little or no relevance to the systems person.

All the theory presented so far has been concerned with the planar dielectric waveguide, despite the fact that the subject of this book, the optical fiber, is a dielectric waveguide of circular symmetry. Studying the planar guide is useful since it provides a great deal of insight into the guidance process in the circular guide. The latter is rather more complex to analyze mathematically, although the analysis follows well trodden paths. However, physical interpretation of the analysis is more difficult for reasons that will become clear as we proceed.

We begin with the wave equation for a homogeneous medium, which is now to be solved for a structure of cylindrical symmetry. Thus, in cylindri-

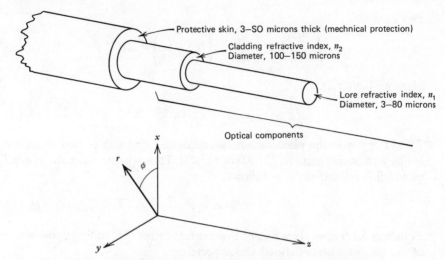

Figure 5.1. Schematic diagram of the circular dielectric waveguide fiber.

cal polar coordinates, the required equation is

$$\nabla^2 A = \frac{1}{r}\frac{\partial}{\partial r}\left(r\frac{\partial A}{\partial r}\right) + \frac{1}{r^2}\frac{\partial^2 A}{\partial \phi^2} + \frac{\partial^2 A}{\partial^2 z} = \mu\mu_0\epsilon\epsilon_0\frac{\partial^2 A}{\partial t^2} \tag{5.1.1}$$

We now postulate a solution of the following general form:

$$A = F_1(r)F_2(\phi)F_3(z)F_4(t) \tag{5.1.2}$$

As in the case of a planar guide, we try solutions for the z and t dependence of the form

$$F_3(z)F_4(t) = \exp i(\omega t - {}^z kz) \tag{5.1.3}$$

Substituting these functions leads to the result

$$\frac{1}{r}\frac{\partial}{\partial r}\left[r\frac{\partial(F_1 F_2)}{\partial r}\right] + \frac{1}{r^2}\frac{\partial^2(F_1 F_2)}{\partial \phi^2} - {}^z k^2(F_1 F_2) = \mu\mu_0\epsilon\epsilon_0\omega^2(F_1 F_2)$$

$$\tag{5.1.4}$$

We now try the solution for $F_2(\phi)$:

$$F_2(\phi) = \exp \pm (il\phi) \tag{5.1.5}$$

and obtain

$$\frac{\partial^2 F_1}{\partial r^2} + \frac{1}{r}\frac{\partial F_1}{\partial r} + \left({}^T k^2 - \frac{l^2}{r^2}\right) F_1 = 0 \tag{5.1.6}$$

Here we have set

$$^T k^2 = \mu\mu_0\epsilon\epsilon_0\omega^2 - {}^z k^2 = n^2 k_0^2 - {}^z k^2 \tag{5.1.7}$$

By analogy with the planar guide, we recognize that one of two situations is likely to occur, namely, ${}^T k^2 \geqslant 0$ or ${}^T k^2 < 0$. Thus we introduce the symbol γ, which is related to ${}^T k$ as follows:

$$^T k^2 = -\gamma^2 \tag{5.1.8}$$

Equation 5.1.6 now takes one of two general forms, depending upon which of the two situations defined above pertains:

$$\frac{\partial^2 F_1}{\partial r^2} + \frac{1}{r}\frac{\partial F_1}{\partial r} + \left({}^T k^2 - \frac{l^2}{r^2}\right) F_1 = 0 \tag{5.1.9}$$

or

$$\frac{\partial^2 F_1}{\partial r^2} + \frac{1}{r}\frac{\partial F_1}{\partial r} - \left(\gamma^2 + \frac{l^2}{r^2}\right) F_1 = 0 \tag{5.1.10}$$

Both equations, 5.1.9 and 5.1.10, can be recognized as Bessel equations. The general solution to the first takes the form

$$F_1 = A_1 J_l({}^T k \cdot r) + A_2 Y_l({}^T k \cdot r) \tag{5.1.11}$$

and the solution to the second, the form

$$F_1 = A_3 I_l(\gamma \cdot r) + A_4 K_l(\gamma \cdot r) \tag{5.1.12}$$

where A_1, A_2, A_3, and A_4 are constants; $J_l({}^T k \cdot r)$ is a Bessel function of the first kind, and $Y_l({}^T k \cdot r)$ of the second kind; and $I_l(\gamma \cdot r)$ and $K_l(\gamma \cdot r)$ are modified Bessel functions of the first and second kinds. The constants will be determined by the boundary conditions to the problem, which we have not yet specified.

Of the four constants, A_1 to A_4, two can be set to zero without difficulty. In fact, A_2 must be zero, since the function $Y_l({}^T k \cdot r)$ goes to infinity as r goes to zero and this cannot be allowed physically to describe the field in the core. The constant A_3 must also be set to zero because the function $I_l(\gamma \cdot r)$ goes to infinity as r goes to infinity, and we are interested only in solutions which represent bound waves. Thus we choose the Bessel function $J_l(^Tk\cdot r)$ for the core solution, and the modified Bessel function $K_l(\gamma\cdot r)$ to represent the field in the cladding region, since it behaves as $\exp(-\gamma r)$ for large r.

We are now in a position to seek the full solution of the equations by applying the boundary conditions at the core-cladding interface and evaluating the arbitrary amplitude terms, as well as solving the eigenvalue equation that results to obtain the allowed values of propagation constants that correspond to guided modes. This analysis has been discussed in detail by many authors. However, the results which follow are based upon a recent analysis by Gloge[1] which leads to simplified expressions for the fields and has the great attribute that these are readily visualized. For optical waveguides having a small index difference, the results are sufficiently accurate.

5.2. LINEARLY POLARIZED (LP) MODES

The analysis starts by looking for LP mode solutions, by analogy with the planar guide already discussed. The latter has such solutions because of the infinite y extension, which allows the condition $d/dy = 0$ to be used. The circular dielectric guide does not have any such condition that can be used, but in the limit of a very small index difference the waves in the structure propagate at a small angle to the axis and the polarizations become largely separated.

We postulate, following the results of equations 5.1.9 to 5.1.12, expressions for the ^yE field components in the core and cladding of the following form:

$$^yE_{core} = E_l^{co}\left[\frac{J_l(^Tkr)}{J_l(^Tka)}\right]\cos(l\phi) \simeq \left(\frac{z_0}{n}\right)^x H^{co} \qquad (5.2.1)$$

$$^yE_{clad} = E_l^{cl}\left[\frac{K_l(\gamma r)}{K_l(\gamma a)}\right]\cos(l\phi) \simeq \left(\frac{z_0}{n}\right)^x H^{cl} \qquad (5.2.2)$$

where $z_0 = \sqrt{\mu\mu_0/\epsilon\epsilon_0}$. We note from equation 5.1.5 that the choice of $\cos(l\phi)$ is arbitrary and that $\sin(l\phi)$ could equally well have been chosen, so that two degenerate sets of modes would exist, rotated $\pi/2$ from one another.

From these expressions we can immediately derive expressions for the z and ϕ components of the fields. We assume that the ^xE component is very much smaller than the ^yE component and show that this assumption leads to no inconsistency provided that $\Delta \ll 1$, where

$$\Delta = \frac{n_1 - n_2}{n_2} \simeq \frac{n_1 - n_2}{n_1} \tag{5.2.3}$$

These results can be obtained by seeking the full solutions for the guided modes and combining them in appropriate pairs to cancel unwanted field components. However, this route leads to substantially greater analysis and ultimately reaches the same conclusion.

Making use of our assumption that $^xE \ll {^yE}$ and hence that $^yH \ll {^xH}$, we obtain from equations 2.3.5 and 2.3.8 the following:

$$^zE \approx \frac{i}{\omega \epsilon \epsilon_0} \frac{\partial \,^xH}{\partial y} \tag{5.2.4}$$

$$^zH \approx \frac{i}{\omega \mu \mu_0} \frac{\partial \,^yE}{\partial x} \tag{5.2.5}$$

Then, with some algebra and the following standard results:

$$\frac{\partial}{\partial x} = (\cos\phi)\frac{\partial}{\partial r} - \left(\frac{\sin\phi}{r}\right)\frac{\partial}{\partial \phi} \tag{5.2.6}$$

$$\frac{\partial}{\partial y} = (\sin\phi)\frac{\partial}{\partial r} + \left(\frac{\cos\phi}{r}\right)\frac{\partial}{\partial \phi} \tag{5.2.7}$$

$$\frac{\partial}{\partial z}(J_l(z)) = \tfrac{1}{2}\left[J_{l-1}(z) - J_{l+1}(z)\right] \tag{5.2.8}$$

$$\left(\frac{2l}{z}\right)J_l(z) = J_{l-1}(z) + J_{l+1}(z) \tag{5.2.9}$$

we obtain expressions for the z components of the electric and magnetic fields in both the core and the cladding:

$$^zE^{co} = \frac{-iE_l^{co}\,{}^Tk_1}{2k_0 n_1}\left[\frac{J_{l+1}({}^Tkr)}{J_l({}^Tka)}\sin(l+1)\phi + \frac{J_{l-1}({}^Tkr)}{J_l({}^Tka)}\sin(l-1)\phi\right] \tag{5.2.10}$$

$$^zE^{cl} = \frac{-iE_l^{cl}\gamma_2}{2k_0 n_2}\left[\frac{K_{l+1}(\gamma_2 r)}{K_l(\gamma_2 a)}\sin(l+1)\phi + \frac{K_{l-1}(\gamma_2 r)}{K_l(\gamma_2 a)}\sin(l-1)\phi\right] \tag{5.2.11}$$

$$^zH^{co} = \frac{-iE_l^{co\,T}k_1}{2\mu_1\mu_0\omega}\left[\frac{J_{l+1}(^Tk_1r)}{J_l(^Tk_1a)}\cos(l+1)\phi - \frac{J_{l-1}(^Tk_1r)}{J_l(^Tk_1a)}\cos(l-1)\phi\right]$$

$$(5.2.12)$$

$$^zH^{cl} = \frac{-iE_l^{co}\gamma_2}{2\mu_2\mu_0\omega}\left[\frac{K_{l+1}(\gamma_2r)}{K_l(\gamma_2a)}\cos(l+1)\phi + \frac{K_{l-1}(\gamma_2r)}{K_l(\gamma_2a)}\cos(l-1)\phi\right]$$

$$(5.2.13)$$

We note two features of these expressions. The first is that the components ^zE are very much smaller than the components ^yE because of the factors $(^Tk/n_1k_0)$ and (γ_2/n_2k_0). The second feature is that they all include terms in both $l+1$ and $l-1$ which can be traced back to the fact that these LP modes are formed by combining pairs of individual modes, although this has not been explicitly stated in the analysis. We can also derive from the above results expressions for ^xE and ^yH and show that they too are small, as originally postulated.

To obtain solutions for the guided modes of the structures, we must equate the tangential components of the fields at the interface. Thus we calculate expressions for $E(\phi)$ and $H(\phi)$, using the results above and equations 2.4.14 to 2.4.16:

$$^\phi E^{co} = E_l^{co}\left[\frac{J_l(^Tk_1r)}{J_l(^Tk_1a)}\right][\cos(l+1)\phi + \cos(l-1)\phi] \qquad (5.2.14)$$

$$^\phi E^{cl} = E_l^{cl}\left[\frac{K_l(\gamma_2r)}{K_l(\gamma_2a)}\right][\cos(l+1)\phi + \cos(l-1)\phi] \qquad (5.2.15)$$

$$^\phi H^{co} = \frac{-E_l^{co}n_1k_0}{2\mu_1\mu_0\omega}\left[\frac{J_l(^Tkr)}{J_l(^Tka)}\right][\sin(l+1)\phi - \sin(l-1)\phi] \qquad (5.2.16)$$

$$^\phi H^{cl} = \frac{-E_l^{co}n_2k_0}{2\mu_2\mu_0\omega}\left[\frac{K_l(\gamma_2r)}{K_l(\gamma_2a)}\right][\sin(l+1)\phi - \sin(l-1)\phi] \qquad (5.2.17)$$

Now, if we solve equations 5.2.14 to 5.2.17 without the imposed assumption that $n_1 \simeq n_2$, the eigenvalue equation is obtained as follows:

$$\left(\frac{^Tk_1a}{n_1}\right)\left[\frac{J_{l\pm1}(^Tk_1a)}{J_l(^Tk_1a)}\right] = \pm\left(\frac{\gamma_2a}{n_2}\right)\left[\frac{K_{l\pm1}(\gamma_2a)}{K_l(\gamma_2a)}\right] \qquad (5.2.18)$$

This equation now leads to the solutions for the EH, TE, TM, and HE modes, from which the LP modes are formed. The $(l+1)\phi$ term leads to the $HE_{l+1,m}$ modes, while the $(l-1)\phi$ term leads to the $EH_{l-1,m}$, TE_m, and TM_m modes.

In the limit that $\Delta \ll 1$ and hence $n_1 \simeq n_2$, equation 5.2.18 reduces further to the form

$$ {}^Tk_1 \left[\frac{J_{l-1}({}^Tka)}{J_l({}^Tka)} \right] = -\gamma_2 \left[\frac{K_{l-1}(\gamma_2 a)}{K_l(\gamma_2 a)} \right] \qquad (5.2.19) $$

Just as in the planar 'guide, mode cutoff corresponds to the condition $\gamma_2 = 0$, which in turn leads to the condition

$$ J_{l-1}({}^Tka) = 0 \qquad (5.2.20) $$

It follows that the lowest order mode, characterized by $l=0$, has a cutoff given by the lowest root of the equation

$$ J_{-1}({}^Tka) = -J_1({}^Tka) = 0 \qquad (5.2.21) $$

Hence ${}^Tk_1 a = 0$; hence $a = 0$.

Thus the lowest order mode cuts off only when the guide ceases to exist ($a = 0$). This mode is labeled the LP_{01}, having a field pattern described by $l=0$ and a cutoff characteristic of the first zero of the Bessel function. The next mode of the type $l=0$ cuts off when J_1 next equals zero, that is, when ${}^Tk_1 a = 3.83$. This mode is labeled the LP_{02}. Similarly, the modes with a field characterized by $J_1({}^Tk_1 r)$ have cutoffs when $J_0({}^Tk_1 a) = 0$. Thus the LP_{11} mode cuts off when ${}^Tk_1 a = 2.405$. In Figure 5.2 we show the regions in which a given mode is the highest one allowed of a given l value group, labeled in LP mode notation. Also shown on the figure are the associated HE, EH, TE, and TM mode notations which are necessary when the approximations related to the assumption $\Delta n \ll 1$ that were made in our analysis cannot be made. The more rigorous analysis and derivation of the precise cutoff conditions for each of these modes has been discussed at length by Marcuse.[6]

The attractive feature of the LP mode theory given above is that it shows that there is a complete set of modes of the guides of interest in optical communications in which only one electric and one magnetic field component are significant. The E vector can be chosen to lie along any arbitrary radius with the H vector along a perpendicular radius; having made this choice, there will always be a second independent mode with E and H orthogonal to the first pair. But with only a single E vector, each mode is

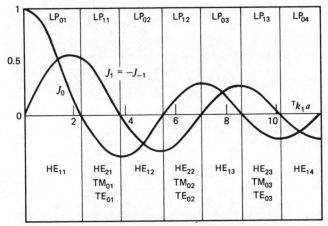

Figure 5.2. Allowed regions for LP modes plotted versus V value for a step-index guide ($V = {}^{T}k_1 a$) indicating also the equivalent HE, EH, TE, and TM modes. Reproduced with permission from D. Gloge, *Appl. Opt.*, **108** 2252 (1971).

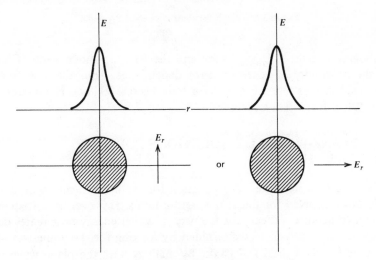

Figure 5.3. Field distribution of the LP_{01} mode.

readily visualized and drawn. To illustrate this, we show in Figures 5.3 and 5.4 the allowed forms of just two LP modes, the LP_{01} and the LP_{11}. Note that, since each of the two possible polarization directions can be coupled with either a $\cos l\phi$ or a $\sin l\phi$ azimuthal dependence, four experimentally discrete mode patterns are obtained for a single LP_{lm} label. For the sake of completeness, we note that the LP_{lm} mode is formed from a linear

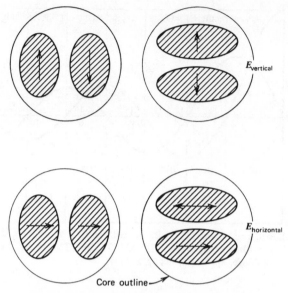

Figure 5.4. Field distribution of the LP$_{11}$ mode.

combination of the HE$_{l+1,m}$ mode and the EH$_{l-1,m}$ mode, each of which has the possibility of $\cos l\phi$ or $\sin l\phi$ dependence. Thus the new labeling system substitutes four LP modes for four discrete EH and HE modes, and each system forms a complete set.

5.3. PLANE-WAVE-LIKE SOLUTIONS FOR THE CIRCULAR GUIDE

We have seen how the wave equation for the cylindrically symmetrical guide can be solved for the z, t, ϕ, and r terms. However, it is frequently difficult to acquire a "feel" for the way in which energy propagates down the guide. Such a feel is greatly aided by looking for the plane-wave-like solutions for the cylindrical guide, by analogy with the plane wave solutions of the planar guide, and then seeing at what angle to the major guide axes these waves travel.

We begin with the wave equation, which with the usual substitutions reduces to the r-dependent equation as follows:

$$\nabla^2 A = \mu\mu_0\epsilon\epsilon_0 \frac{\partial^2 A}{\partial t^2} \qquad (5.3.1)$$

We put

$$A(r,\phi,z,t) = A(r)\exp \pm (il\phi)\exp i(\omega t - {}^zkz)$$

Then

$$\frac{\partial^2 A(r)}{\partial r^2} + \frac{1}{r}\frac{\partial A(r)}{\partial r} + \left(n^2 k_0^2 - {}^z k^2 - \frac{l^2}{r^2}\right) A(r) = 0 \qquad (5.3.2)$$

We see that the z-dependent solution is already in plane wave form, and the ϕ solution can be written in plane wave form as follows:

$$F(\phi) = \exp \pm (il\phi) = \exp \pm \left[i^\phi k(\phi r) \right] \qquad (5.3.4)$$

$$^\phi k = \frac{l}{r} \qquad (5.3.5)$$

We now seek a similar plane-wave-like solution to the r-dependent equation. If we try the following:

$$F(r) = (r)^{-1/2} \exp \pm i({}^r k_r r) \qquad (5.3.6)$$

we find, substituting in equation 5.3.2 and collecting terms, that

$$^r k^2 + \frac{1}{4r^2} + n^2 k_0^2 - {}^z k^2 - \frac{l^2}{r^2} = 0 \qquad (5.3.7)$$

or

$$^r k = \left(n^2 k_0^2 - {}^z k^2 - \frac{l^2 + \frac{1}{4}}{r^2} \right)^{1/2} \qquad (5.3.8)$$

This is an approximate solution because in the substitution we assumed that $^r k$ is not a function of r in making the differentiation. Thus we can expect the solution to be reasonably valid provided that

$$^r k \gg r \frac{\partial\, ^r k}{\partial r} \qquad (5.3.9)$$

Since $r(\partial\, ^r k / \partial r)$ is given by

$$r \frac{\partial\, ^r k}{\partial r} = \left(\frac{l^2 - \frac{1}{4}}{r^2} \right) \left(n^2 k_0^2 - {}^z k^2 - \frac{l^2 - \frac{1}{4}}{r^2} \right)^{-1/2} \qquad (5.3.10)$$

we see that the approximation is going to be generally valid unless $^r k \rightarrow 0$ or $r \rightarrow 0$. Thus we cannot expect the model to predict the fields in the center of the guide, and great care must also be taken as cutoff is approached. However, in the regions in which the skew or cylindrical nature of the structure is not dominant, the model predicts waves which are easily visualized and match those of the exact solution well.

To examine the form of waves which are present in the extremes or as $r \to 0$ and in which the skew nature dominates we switch to another solution. Assuming that $(l^2 - \frac{1}{4})/r^2 \gg (n^2 k_0^2 - {}^z k^2)$, we find that a valid solution to the wave equation (5.3.2) takes the form

$$E = E_0 \left(\frac{r}{a} \right)^{\pm l} \tag{5.3.11}$$

In the core center, as $r \to 0$, the appropriate solution is $E_0 (r/a)^l$, and in the cladding region and near cutoff, where l^2/r^2 dominates, the appropriate solution is $E_0 (r/a)^{-l}$.

For the core wave, as r increases, the solution changes from the r^l form to the $\cos({}^r kr)$ form as ${}^r k$ goes through zero. In the cladding, and close to cutoff, the solution changes from being dominated by the r^{-l} form to $(1/\sqrt{r}) \exp(-\alpha r)$, where $\alpha = \sqrt{{}^z k^2 - n_2^2 k_0^2}$.

Thus, for large radius, the fields behave in much the same manner as those in a planar dielectric guide. However, in the center of the guide and in the region of the core-cladding interface at cutoff, a new behavior is observed. The r^l form in the center of the guide for the skew modes fits in well with the fields predicted by the Bessel function solution, since for $l \neq 0$ the $J_l(x)$ functions all start at 0 for $x = 0$ and rise according to a power law until they reach the oscillating part of their characteristic, corresponding in this model to the region ${}^r k = 0$ or positive and real.

5.4. WEAKLY LEAKY MODES

In the region of cutoff the core-cladding interface fields show an interesting effect. Once the value of ${}^z k$ is such that $|{}^z k^2 - n_2^2 k_0^2| \ll (l^2 - \frac{1}{4})/a^2$, the field in the cladding close to the interface become essentially independent of the propagation constant ${}^z k$ and switches the the r^{-l} form. It holds this form right through cutoff (${}^z k = n_2 k_0$) and changes only when the inequality above is broken. However, for larger r there is an abrupt change in the field form as cutoff is passed, since the field at $r \gg a$ switches from the exponentially decaying form to the oscillating $[\cos({}^r kr)]$ form and represents a leaky wave. Because the field in the region of the core-cladding interface remains unchanged and has the r^{-l} form, the power leakage through this free wave may be very small. Thus, for modes with large l values (highly skew modes), cutoff is not necessarily associated with any dramatic change in the form of the cladding field near the core, nor is it necessarily associated with a large power loss to the guided wave.

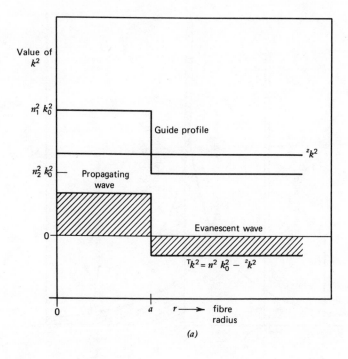

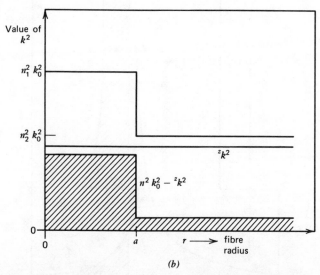

Figure 5.5. Plots of the components of k^2 for (a) a guided meridional mode $(l=0)$ and (b) a meridonal mode $(l=0)$ beyond cutoff.

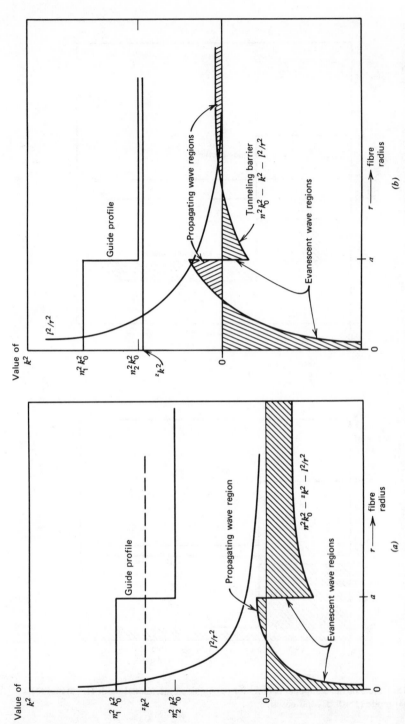

Figure 5.6 Plots of the components of k^2 for a skew mode ($l \neq 0$) in the situation clear of cutoff, that is, (a) guided and (b) in the tunneling leaky mode situation. In the refracting situation, $^zk^2$ would be smaller still, so that the cladding propagating field region would extend to the core boundary.

66

We should note also that for the skew modes just beyond cutoff there is a characteristic radius given by

$$R_c = \frac{\sqrt{l^2 - \frac{1}{4}}}{\sqrt{k_0^2 n_0^2 - {}^z k^2}} \tag{5.4.1}$$

at which the value of ${}^\prime k = 0$.

At cutoff, $R_c = \infty$. The mode then remains weakly leaky as ${}^z k$ decreases until R_c finally is equal to a, the core radius. The field in the cladding is now a fully propagating one from the core all the way to infinity without the benefit of a region having r^{-l} dependence to insulate it, so that the loss from the mode is now immense and it is no longer guided in any sense. Between these two extremes the mode is in a region of intermediate loss, being subject neither to perfectly lossless guidance nor to such high loss as to be unguided. The effects on the guide's performance of these modes have yet to be fully established.

Figures 5.5 and 5.6 illustrate the effects discussed above for a $l = 0$ mode and for a $l \gg 0$ mode.

5.5.   MODE CHARACTERIZATION

A particularly elegant method for observing these leaky modes and modes close to cutoff has been described by Stewart[2] and is illustrated in Figure 5.7. The full aperture of the fiber is excited by an incoherent source to energize all the guided and weakly leaky modes. Energy then leaks from all the modes in the region of cutoff when the fiber is immersed in a liquid

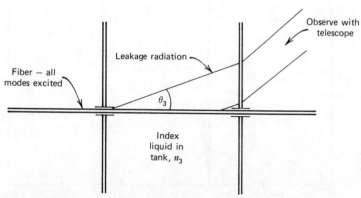

Figure 5.7.   Apparatus for observing the ${}^z k$ values of modes close to cutoff.

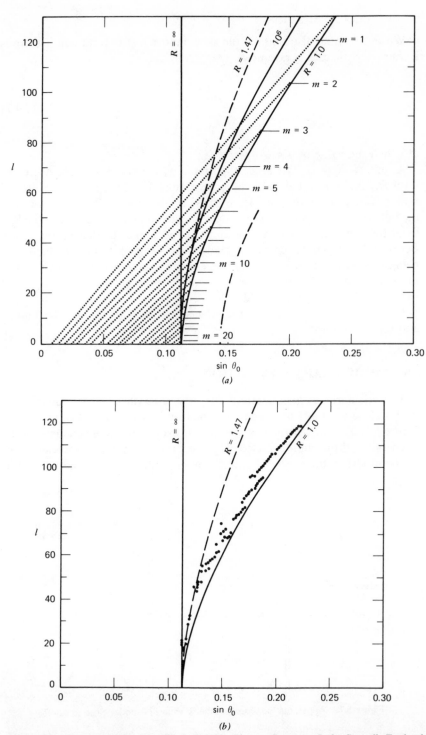

Figure 5.8. Results obtained by W. J. Stewart, Plessey Company Ltd., Caswell, England, using the type of apparatus in Figure 5.7. (a) the values of θ or $^z k$ for the modes of a step-index guide by calculation compared to the measured values for a step guide.

tank. In particular, the energy leaks at an angle to the fiber axis given by

$$\theta_3 = \text{arc } \cos\left(\frac{{}^z k}{n_3 k_0}\right) \tag{5.5.1}$$

where n_3 is the index of the liquid. By observing the far-field pattern of the radiation leaking from a long length of fiber through a telescope, very high angular resolution can be obtained in the θ_3 plane, thus allowing very small differences in k_z to be detected.

Figures 5.8a and 5.8b illustrate some results obtained by Stewart.[3] Figure 5.8b shows the measured values of θ for each of the modes detected at a particular wavelength (633 nm) in an experiment using a step-index Corning fiber with a core diameter of the order of 60 microns. The measured points are plotted against the l value for the mode observed, which could be assessed by counting the number of azimuthal nodes. Figure 5.8a shows the calculated points for the same fiber. Clearly evident are two points. The only modes that leak are those close to the curve labeled $R = 1$, which correspond to the condition given above of $R_c = a$. The vertical line labeled $R = \infty$ corresponds to $R_c = \infty$, the theoretical cutoff point for the modes. Thus the experiment clearly illustrates the fact that the modes which have a high skew component, $l \gg 1$, leak rapidly only well beyond their mathematical cutoff point. Equally, it shows that the low order modes remain tightly bound, as one would expect.

5.6. PLANE-WAVE-LIKE PROPAGATION ANGLES AND MODE VISUALIZATION

We have shown that the plane-wave-like solution can be characterized by k vectors in the z, r, and ϕ directions. It follows that we can immediately write down angles of skew and elevation for the plane wave normal directions at a local point in the guide. By reference to Figure 5.9 we have the following relations: (see Section 5.3)

$$\psi_s = \text{arc } \tan\left(\frac{{}^\phi k}{{}^z k}\right) \tag{5.6.1}$$

$$\psi_e = \text{arc } \tan\left(\frac{{}^r k}{{}^z k}\right) \tag{5.6.2}$$

By analogy with the planar guide, $\theta = \pi/2 - \Theta \simeq \sqrt{\psi_e^2 + \psi_s^2}$. For a mode having $l = 0$, the angle $\psi_s = 0$, and we can picture a wave traversing the guide, being totally reflected at the core-cladding interface, and passing through the center of the guide, there giving rise to a field maximum. However, for a skew mode, $l \neq 0$, such a plane wave travels along the guide

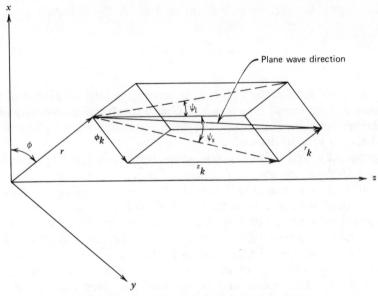

Figure 5.9. Notation used in plane-wave analysis.

in a helical path, reflecting off the core-cladding interface, not in the plane containing the guide axis but in a path cutting through that plane. Thus the wave corkscrews along the structure and never passes through the guide center, so that no field is observed there. This is in complete agreement with the wave amplitude being of the form r^l in the guide center and thus having zero amplitude there.

If we now turn our attention to the cladding region, for a guided mode $n_2 k_0 < {}^z k$, so that ${}^T k$ is always imaginary. It follows that the cladding field is of exponentially decaying evanescent form for all r, just as in the case of the planar guide.

5.7. POWER FLOW IN THE CIRCULAR STEP-INDEX GUIDE

We can derive expressions for the power flow along the guide in the cladding and core materials for a given mode in terms of the guide electric field parameters. This is done using the expression for the z component of the Poynting vector and integrating over the section of interest. Thus the power in the core and the cladding, respectively, is given by

$$ {}^z P^{co} = \frac{1}{2} \int_0^a \int_0^{2\pi} r({}^x E^y H^* - {}^y E^x H^*)\, d\phi\, dr \tag{5.7.1} $$

$$ {}^z P^{cl} = \frac{1}{2} \int_a^\infty \int_0^{2\pi} r({}^x E^y H^* - {}^y E^x H^*)\, d\phi\, dr \tag{5.7.2} $$

Using the expressions for ^yE and $^xH^*$ from equations 5.2.1 and 5.2.2, which are

$$^yE^{co} = \left(\frac{Z_0}{n_1}\right) {}^xH^{co} = E^{co}\left[\frac{J_l(^Tk_1r)}{J_l(^Tk_1a)}\right]\cos(l\phi) \tag{5.7.3}$$

$$^yE^{cl} = \left(\frac{Z_0}{n_2}\right) {}^xH^{cl} = E^{cl}\left[\frac{K_l(\gamma_2r)}{K_l(\gamma_2a)}\right]\cos(l\phi) \tag{5.7.4}$$

substituting for the fields, and discarding the components $^xE^yH^*$ as negligible, we obtain

$$^zP^{co} = \frac{\pi}{2}\frac{n_1(E^{co})^2}{Z_0J_l^2(^Tk_1a)}\int_0^a rJ_l^2(^Tk_1r)\,dr \tag{5.7.5}$$

$$^zP^{cl} = \frac{\pi}{2}\frac{n_2(E^{cl})^2}{Z_0K_l^2(\gamma_2a)}\int_a^\infty rK_l(\gamma_2r)\,dr \tag{5.7.6}$$

Of course $E^{cl} = E^{co} = E$. Both integrals are tabulated, so that we can express the above results as

$$^zP^{co} = \frac{\pi}{4}\frac{n1E^2a^2}{Z_0J_l^2(^Tka)}\left[J_l^2(^Tk_1a) - J_{l-1}(^Tk_1a)J_{l+1}(^Tk_1a)\right] \tag{5.7.7}$$

$$^zP^{cl} = \frac{\pi}{4}\frac{n_2E^2a^2}{Z_0K_l^2(\gamma_2a)}\left[K_{l+1}(\gamma_2a)K_{l-1}(\gamma_2a) - K_l^2(\gamma_2a)\right] \tag{5.7.8}$$

Setting

$$\mathcal{K} = \frac{K_l^2(\gamma_2a)}{K_{l-1}(\gamma_2a)K_{l+1}(\gamma_2a)} \tag{5.7.9}$$

and using the eigenvalue equation, we can now simplify the two power equations to the following form:

$$^zP^{co} = A\left[1 + \left(\frac{\gamma_2}{^Tk_1}\right)^2\frac{1}{\mathcal{K}}\right] = P_{core} \tag{5.7.10}$$

$$^zP^{cl} = A\left(\frac{1}{\mathcal{K}} - 1\right) = P_{clad} \tag{5.7.11}$$

$$A = \frac{\pi a^2nE^2}{4Z_0} \tag{5.7.12}$$

$$P_{\text{total}} = {}^z P^{\text{co}} + {}^z P^{\text{cl}} = \frac{A}{\mathcal{K}} \left[1 + \left(\frac{\gamma_2}{{}^T k_1} \right)^2 \right] \qquad (5.7.13)$$

$$= \frac{A}{\mathcal{K}} \left(\frac{V^2}{{}^T k_1^2 a^2} \right) \qquad (5.7.14)$$

Now, using the standard notation

$$W = \gamma_2 a, \qquad U = {}^T k_1 a, \qquad V^2 = U^2 + W^2$$

we immediately have the relations

$$\frac{P_{\text{core}}}{P_{\text{total}}} = 1 - \frac{U^2}{V^2}(1 - \mathcal{K}) \qquad (5.7.15)$$

$$\frac{P_{\text{clad}}}{P_{\text{total}}} = \frac{U^2}{V^2}(1 - \mathcal{K}) \qquad (5.7.16)$$

An approximate expression for $\mathcal{K}$ which is valid for modes not too close to cutoff is

$$\mathcal{K} = 1 - \frac{1}{\sqrt{W^2 + l^2 + 1}} \qquad (5.7.17)$$

A very well guided mode has $U < V_1$ so that $P_{\text{core}} \gg P_{\text{clad}}$, as one would expect. A mode approaching cutoff has $W < V$ and $U \approx V$, so that

$$\frac{P_{\text{core}}}{P_{\text{total}}} \to 1 - \frac{1}{\sqrt{l^2 + 1}} \qquad (5.7.18)$$

$$\frac{P_{\text{clad}}}{P_{\text{total}}} \to \frac{1}{\sqrt{l^2 + 1}} \qquad (5.7.19)$$

This result shows that for a nonskew mode ($l \approx 0$) the mode's power moves almost entirely into the cladding at cutoff. However, for skew modes ($l \gg 0$) the power remains concentrated in the core even at or just beyond cutoff. The physical reason for this is discussed in Section 5.3.

5.8. MODE DISPERSION IN A STEP-INDEX MULTIMODE GUIDE

In designing a system, it is vitally important to know something about the dispersion of the transmission medium[1] in addition to the model properties already discussed. The velocity with which the energy in a pulse travels down a waveguide—or through any other medium, for that matter —is called the group velocity and is characterized by the expression

$$V_g = c\left(\frac{\partial^z k}{\partial k_0}\right)^{-1} \qquad (5.8.1)$$

so that the time delay for the mode is given by

$$\tau_g = \frac{l}{c}\frac{\partial^z k}{\partial k_0} \qquad (5.8.2)$$

By definition, we have

$$U = a\sqrt{n_1^2 k_0^2 - {}^z k^2} = {}^T k_1 a \qquad (5.8.3)$$

$$W = a\sqrt{{}^z k^2 - n_2^2 k_0^2} = \gamma_2 a \qquad (5.8.4)$$

$$V^2 = U^2 + W^2 \qquad (5.8.5)$$

$$\Delta = \frac{n_1 - n_2}{n_2} \qquad (5.8.6)$$

It follows directly that we can write $^z k$ in the form

$$^z k = n_2 k_0 (b\Delta + 1) \qquad (5.8.7)$$

$$b = 1 - \frac{U^2}{V^2} = \frac{{}^z k^2 / k_0^2 - n_2^2}{n_1^2 - n_2^2} \qquad (5.8.8)$$

Then, by using the fact that $V = an_2 k_0 \sqrt{2\Delta}$, we see that

$$\frac{\partial^z k}{\partial k_0} = \frac{\partial}{\partial k_0}(n_2 k_0) + \frac{\partial}{\partial k_0}(n_2 k_0 b\Delta)$$

$$= \frac{\partial}{\partial k_0}(n_2 k_0) + n_2 \Delta \frac{\partial(Vb)}{\partial V}$$

$$= N_2 + n_2 \Delta \frac{\partial(Vb)}{\partial V} \qquad (5.8.9)$$

$$N_2 = n_2 + k_0 \frac{\partial n_2}{\partial k_0} = \text{group index} \qquad (5.8.10)$$

We have assumed that $\partial n_1/\partial k_0 = \partial n_2/\partial k_0$, so that $\partial\Delta/\partial k_0 = 0$.

It is now apparent from equation 5.8.9 that the group velocity has two components, one of which is related to the material dispersion and is characterized by the group index N, and a second which is related to the mode parameters and is strongly mode dependent in general.

To evaluate the mode dispersion term, we take the characteristic equation and differentiate both sides with respect to V:

$$\frac{\partial}{\partial V}\left[U\frac{J_{l-1}(U)}{J_l(U)}\right] = \frac{\partial}{\partial V}\left[\frac{-WK_{l-1}(W)}{K_l(W)}\right] \tag{5.8.11}$$

to obtain the characteristic equation in the simple form

$$\frac{dU}{dV} = \frac{U}{V}(1-\mathcal{K}) \tag{5.8.12}$$

where

$$\mathcal{K} = \frac{K_l^2(W)}{K_{l-1}(W)K_{l+1}(W)} \tag{5.8.13}$$

as before. Then we have

$$\frac{\partial(Vb)}{\partial V} = \frac{\partial}{\partial V}\left(V - \frac{U^2}{V}\right) \tag{5.8.14}$$

$$= 1 - \frac{U^2}{V^2}(1-2\mathcal{K}) \tag{5.8.15}$$

Thus, by using equations 5.8.2 and 5.8.9, we obtain

$$\tau_g = \frac{L}{c}\left\{N_2 + n_2\Delta\left[1 + \frac{U^2}{V^2}(1-2\mathcal{K})\right]\right\} \tag{5.8.16}$$

In this form the group velocity for a given mode can be quickly evaluated. However, for multimode guides the material dispersion component is the same for all modes, while the mode-dependent component usually is large and varies with mode number, so that the pulse dispersion of the guide is controlled by the variation of group velocity over the modes launched. The material dispersion term controls only the time delay of the whole mode group relative to the launch point and thus can usually be neglected except in nearly optimum graded index guides (Section 6.3.1).

For a single-mode guide, evaluating equation 5.8.16 would lead to a single time delay for the mode. This appears to suggest that there is no pulse broadening in such a guide. However, it is important not to overlook the fact that the pulse is transmitted on a carrier. The frequency spread of the transmitted radiation is at least the reciprocal pulse width, and in general is much greater, so that the single-mode pulse transmission will be controlled by the value

$$\tau_s = \left(\frac{\partial \tau_{gr}}{\partial k} \right) \delta k \tag{5.8.17}$$

where δk is the width of the transmitted optical beam in wavenumbers. In such guides the dispersion is very low and can be made exceptionally low since, by careful choice of guide parameters, it is possible to arrange that the material and guide dispersion terms are of opposite sign and cancel. The reader is referred to the literature for further discussion of this point.[4]

We note that the analysis above has been based entirely upon step-index multimode guides. We will obtain another expression in the following section on graded guides for the pulse dispersion or broadening of a multimode guide in the form of the rms pulse spreading, σ, for a uniformly excited guide. By setting the grade profile parameter, α, to infinity, we obtain an expression for the step guide which we repeat here for completeness, namely,

$$\sigma = \frac{LN_1\Delta}{c} \left(\frac{1}{2\sqrt{3}} \right) = \text{pulse spreading (rms)} \tag{5.8.18}$$

while the relative delay is given by

$$\tau_m = \frac{N_1 L}{c} \left[1 + \Delta \left(\frac{m'}{M'} \right) + \frac{3}{2} \Delta^2 \left(\frac{m'}{M'} \right)^2 \right] \tag{5.8.19}$$

Since we can associate (m'/M') with $(U/V)^2$ (see Sections 7.2.1 and 7.2.2 on mode numbers), equation 5.8.16 reduces to

$$\tau_g = \frac{NL}{c} \left[1 + \Delta \left(\frac{m'}{M'} \right) + \Delta \left(1 - \frac{2m'\mathcal{K}}{M'} \right) \right] \tag{5.8.20}$$

showing an obvious similarity. Good agreement is not expected since each result includes different approximations.

Equation 5.8.18 tells us the amount by which a delta function pulse is broadened as a result of transmission through the step-index guide. Inserting numbers in the formula for a typical optical fiber, we find the following result:

$$\sigma = \frac{(10^3)(1.5)(0.01)}{(3 \times 10^8)2\sqrt{3}} \tag{5.8.21}$$
$$= 15 \text{ ns/km}$$

If we transmit a pulse of width σ_0, the received pulse will be of the order of $\sqrt{\sigma_0^2 + \sigma^2}$, so that a 10 ns pulse will be broadened to about 18 ns after 1 km. We should note that these results assume that the energy is transmitted uniformly among the available guided modes and that there is no mode mixing. In practice, the higher order modes tend to lose energy more rapidly and some mode mixing occurs, both effects leading to some reduction in the pulse dispersion. Nevertheless, for a telecommunications system which may well be required to transmit 100 Mbit/sec over a continuous 5 km fiber, such dispersion is unacceptable.

We will discuss in Chapter 6 how this dispersion in a multimode guide can be greatly reduced. However, before doing so, it is worth noting the physical reason for the dispersion. We discussed in Section 5.3 the varying routes by which energy travels down the guide. With that in mind, if we simply concentrate on the LP_{0m} mode group, as illustrated in Figure 5.10, we see that, as the mode number m increases, so also does the angle ψ_e. The total path traversed in the core medium increases from L for the lowest order mode traveling directly down the center of the guide to $L/\cos\psi_e$, with a maximum value when the critical angle for total internal reflection is reached. We have already noted that, in addition to these modes, there are the skew modes of the form LP_{lm}, which follow helical paths through the guide. Obviously, these also travel on longer physical paths than do the LP_{01} modes and hence arrive later at the end of the guide.

In Figure 5.11 we show the values for the parameter $d(Vb)/dV$ plotted versus V, that is to say, the relative mode delay. The curves illustrate two features. For increasing V values the mode delay approaches the lower asymptote, representing the fact that, as the guide becomes larger, the ray

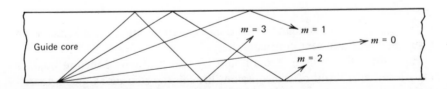

Figure 5.10.. Ray paths for LP_{0m} modes.

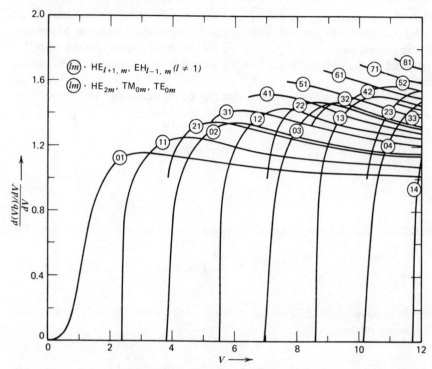

Figure 5.11. Relative mode delay, $d(Vb)/dV$, plotted versus V for a step-index guide showing the effects of longer ray path at large V values and the Goos-Haenchen shift as cutoff is approached. Reproduced with permission from D. Gloge, *Appl. Opt.* **10**, 2252 (1971).

angle for a given mode become shallower and more nearly travels directly down the center of the guide. As the guide shrinks, the ray angle increases, the path increases, and the delay increases until a new effect starts to dominate. This is the Goos-Haenchen shift, already discussed under planar guides. As the mode approaches cutoff, the Goos-Haenchen shift becomes a significant part of the total distance traversed per reflection. Alternatively, we may say that the guided mode energy starts to shift from the core progressively into the cladding material, where it travels faster. In either case we conclude—and observe—that the group delay rapidly decreases again as cutoff is approached.

5.9. CONDITIONS FOR ZERO MATERIAL DISPERSION

We have already derived for the pulse dispersion in a fiber an expression which splits into two parts, one which is traceable to the dispersion arising from the linewidth of the source and the dispersion of the material from which the fiber is fabricated, and a second which arises from the dispersion of the guided modes (see equation 5.8.16). Here we are concerned with the first part of the expression and with a simple way in which it can be set to zero.

We recall from equation 5.8.9 that the pulse delay due to the material component of dispersion in distance L is given by

$$\tau_{mat} = \frac{L}{c} \frac{\partial}{\partial k_0}(nk_0) \tag{5.9.1}$$

$$k_0 = \frac{2\pi}{\lambda} \tag{5.9.2}$$

It follows that the spreading of the pulse from this mechanism is given by

$$\delta\tau_{mat} = \frac{L}{c} \frac{\partial^2}{\partial k_0^2}(nk_0)\delta k_0 \tag{5.9.3}$$

Rewriting the above differential in terms of the wavelength, λ, leads to the following relationships:

$$\frac{\partial^2}{\partial k_0^2}(nk_0) = \frac{-\lambda^2}{2\pi} \frac{\partial}{\partial\lambda}\left[\frac{-\lambda^2}{2\pi} \frac{\partial}{\partial\lambda}\left(\frac{2\pi n}{\lambda}\right)\right]$$

$$= \frac{\lambda^3}{2\pi}\left(\frac{\partial^2 n}{\partial\lambda^2}\right) \tag{5.9.4}$$

Thus, setting $\delta k_0 = (-2\pi/\lambda^2)\delta\lambda$, we obtain the result for the pulse spreading in a length L of fiber:

$$\delta\tau_{mat} = -\left(\frac{L\lambda}{c}\right)\frac{\partial^2 n}{\partial\lambda^2}\delta\lambda \tag{5.9.5}$$

Evidently, if we can find a value of λ for which $\partial^2 n/\partial\lambda^2 = 0$, this source of pulse spreading will disappear. In Figure 5.12 we show the results of one such examination.[4] The solid curve is based upon the refractive index data for silica, and the measured points are for a phosphosilicate fiber. The plot is of $(\lambda/c)(\partial^2 n/\partial\lambda^2)$ versus λ and clearly shows the existence of a zero value for the material pulse spreading in the region of $\lambda = 1.2$ to 1.3 microns. Accurate measurements of this effect do not appear to have been given, and casual inspection of the measured points of Figure 5.12 suggests that the true zero material dispersion for the fiber in question may well be at a longer wavelength than the solid curve would suggest. However, it is

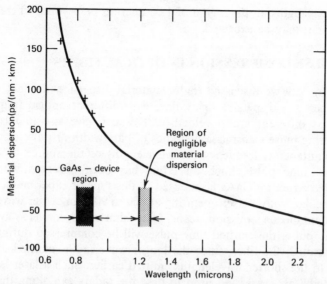

Figure 5.12. Material dispersion versus wavelength for silica with experimental points obtained from phospho-silicate fiber measurements. Note the zero of material dispersion near a wavelength of 1300 nm. Reproduced with permission from D. Payne and W. A. Gambling, *Electron. Lett.*, **11**, 176 (1975)

certain that for all optical materials such a zero point will exist at the point where the dispersion ceases to be dominated by the electronic band edge, ultraviolet absorptions and starts to be dominated by the infrared lattice absorptions.

The implications of these effects can be judged from Figure 5.12. The units on the vertical axis are picoseconds per nanometer per kilometer. In the region of operation of the GaAs devices, the material dispersion of a silica based fiber is in the region of 100 ps/(nm·km), so that for a typical LED source having a 20 nm linewidth and operating over a link of 5 km length the material dispersion is about 10 ns. For a low data rate system this could be neglected, but it effectively prevents the use of LED sources at the 140 Mbit/sec rate.

On the other hand, in a laser based system poorly frequency controlled GaAs lasers may oscillate over a 1 nm linewidth, so that on a 1 km fiber length a pulse dispersion of 0.1 ns may arise from material dispersion. Since the best graded-index fibers show mode dispersion in this region, we see that this effect will be a limiting factor in high bit-rate systems, even with laser sources, until true single-frequency sources are available. Even then, since a 1 nm linewidth corresponds, in electrical terms, roughly to 4 GHz, a single-frequency source carrying 1 GHz data would be comparable. Thus we find that there is a growing interest in the use of sources at longer wavelengths, in the region of 1.06 microns or beyond, where a large improvement may be expected.

5.10. PULSE COMPRESSION IN OPTICAL FIBERS

In Section 5.9 we discussed under material dispersion the fact that an optical fiber is a dispersive delay line, with different optical frequencies traveling at different group velocities. This can give rise to misleading effects during pulse transmission studies of bandwidth or pulse dispersion, if the instantaneous frequency (or wavelength) of the source used shifts during the time pulse. Such an effect has been observed with some single-heterostructure GaAs lasers used in this type of experiment.

In particular, with a laser operating at 800 to 900 nm center wavelength, if its instantaneous emission scans from shorter to longer wavelength during the pulse, the emitted time pulse will be compressed during transmission through the fiber, since the longer wavelengths emitted later will catch up to the shorter wavelengths emitted earlier. Such a laser is said to be "chirped," an expression used to describe radars exploiting this effect and derived from the nature of a bird's chirp.

Experimentally, the effect can be very serious in studying fibers having very low mode dispersion, since then the total dispersion is dominated by the material dispersion. Under these conditions measurement of the time integrated linewidth of the laser and use of this figure to calculate the material dispersion over a 1 or 2 km length can be grossly misleading, by as much as a factor of 2. Such effects have been observed and quantified by Wright and Nelson.[5]

Apart from the complexity that such a chirp effect produces in interpreting the results of pulse dispersion measurements, it may possibly offer a way of overcoming material dispersion in fibers, since a deliberately chirped signal could be used to remove the effect over a line of given length. However, the technology for controllably chirping a source is not yet developed, and it seems more likely that the source wavelength will be shifted to the 1.2 to 1.3 micron region, where the effect becomes negligible.

REFERENCES

1. D. Gloge, *Appl. Opt.*, **10**, 2252 (1971).
2. W. J. Stewart, *Optical Fiber Communication*, IEE Conf. Publ. 132, London, 1975, p. 21.
3. W. J. Stewart, Private communication.
4. D. Payne and W. A. Gambline, *Electron Lett.*, **11**, 176 (1975).
5. J. V. Wright and B. P. Nelson, *Electron Lett.*, **13** (1977).
6. D. Marcuse, *Theory of Dielectric Optical Waveguides*. New York: Academic, 1974.

6

Graded-Index Fibers

6.1. INTRODUCTION

Graded-index fibers have assumed considerable importance in current system design studies because of the attractive possibility that they offer of multimode propagation in a relatively large core fiber coupled with low or very low mode dispersion. This combination makes it possible to exploit the attractive features of the multimode fiber, such as ease of launching and jointing, while still transmitting high data rates over long link lengths. Typical parameters might be 140 Mbit/sec over a link of 8 or 10 km in conjunction with a GaAs laser source, compared to an upper limit of perhaps 10 to 20 Mbit/sec using a step-index fiber.

The reason that the graded-index fiber can offer this performance is readily perceived. Referring to Figure 6.1, we see that three propagation paths or "rays" have to be considered in examining propagation in such a medium. Since the index distribution is graded, with a high index in the center surrounded by a steadily decreasing index, the medium behaves similarly to a series of lenses. A ray traversing the center of the medium follows the axis, traveling in high index (slow) material all the way but traversing the shortest possible physical path from end to end.

A ray leaving the center of the fiber at an angle to the axis is bent by the index profile (lenslike structure) to curve around after some characteristic distance and cross the axis once again. It continues in a sinusoidal path. It thus travels a greater distance than the center ray, but much of its path is in low index (fast) material.

The third ray path is a helical path, formed when light is launched at a skew angle along the surface of a constant radius cylinder at some intermediate radius and index. Once again, the physical path length is longer than that taken by the central ray, but it is all traversed in lower index and hence faster material.

Evidently, the increased path length in the second and third cases is to some extent compensated for by the fact that a ray travels faster when in

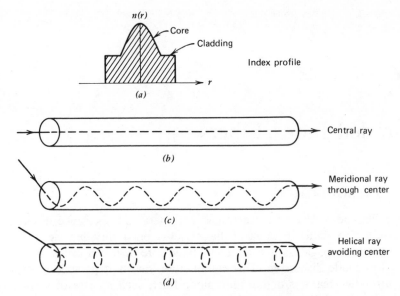

Figure 6.1. Typical ray paths in graded-index fibers.

lower index material. The remainder of this chapter is concerned with analyzing the extent to which these different paths can be made equal in group delay by careful control of the refractive index profile. It is the analysis, measurement, and production and control of such profiles that dominates present day studies of this family of optical fibers.

6.2. MODE THEORY FOR THE GRADED-INDEX GUIDE

A number of ray analyses have been published which set out to deduce the optimum index distribution as a function of radius for either skew rays following a helical path or for meridional rays following a path which cuts the central axis of the fiber and proceeding in sinusoidal fashion. These analyses proceed from the precepts of classical optics and suffer the limitations of ray analyses but are extremely useful in many cases.

A second approach, which has found favor and attracted wide interest, is a wave analysis of the graded-index medium. In the interests of mathematical tractability, it is again an approximate analysis, making use of the Wentzel-Kramers-Brillouin (WKB) approximation widely used in quantum mechanics. This route approaches the problems from the other side, as it were, in that it uses a wave approach but makes approximations also that are essentially the same as those for the "ray" analysis. However, it has the attraction that it allows analysis in terms of a profile that is easily expressed mathematically and that predicts low dispersion for both skew and central "rays."

The WKB analysis is most easily presented as a straight extension of the plane wave analysis discussed in Section 5.3. We postulate a multimode fiber in which the refractive index[1,2] is given by

$$n(r) = n_0 \left[1 - 2\Delta \left(\frac{r}{a} \right)^\alpha \right]^{1/2} \qquad r < a \qquad (6.2.1)$$

$$n(r) = n_0 (1 - 2\Delta)^{1/2} \qquad r \geqslant a \qquad (6.2.2)$$

This profile has two attractions. The first is simply that it can be solved analytically by making only entirely reasonable assumptions. The second is that, by setting $\alpha = \infty$, we obtain the step-index profile, so that a whole range of profiles of potential interest, from step to parabolic, is included in the analysis without further effort.

We will make several assumptions in the following anaysis:

1. The core diameter is large, say of the order of 100 wavelengths, so that many modes will be expected to propagate.
2. The index difference is small, so that modes can be considered transverse electromagnetic (equivalent to the LP mode assumptions made in Chapter 5).
3. The index variation is very small over distances of a wavelength, so that the local plane wave approximation can be used.

At any local point in the guide, we can associate with the electromagnetic field a local wave vector

$$\mathbf{k}(r) = \left[\frac{2\pi n(r)}{\lambda} \right] \mathbf{w} \qquad (6.2.3)$$

aligned along the direction of local "plane wave normal" $\mathbf{w}$. From the plane wave analysis already given, we know that the components of the vector $\mathbf{k}(r)$ are given (quite generally) by

$$\mathbf{k}(r) = \begin{vmatrix} {}^r k \\ {}^\phi k \\ {}^z k \end{vmatrix} = \begin{vmatrix} \left[n(r)^2 k_0^2 - {}^z k^2 - \dfrac{l^2 + \frac{1}{4}}{r^2} \right]^{1/2} \\[2ex] \left(\dfrac{l^2 + \frac{1}{4}}{r^2} \right)^{1/2} \\[2ex] {}^z k \end{vmatrix} \qquad (6.2.4)$$

We have already noted that transversely propagating waves exist only over the range of r in which $'k$ is real. Thus, for a given value of ^zk, two limiting values of r are given by the solution of

$$n^2(R)k_0^2 - {}^zk^2 = \frac{l^2 + \frac{1}{4}}{R^2} \qquad (6.2.5)$$

and labeled R_1 and R_2.

Between these limiting values of r, the field varies smoothly radially with sinusoidal form. From this observation we can say that

$$m\pi \simeq \int_{R_1}^{R_2} {}^rk\, dr \qquad (6.2.6)$$

The more precise WKB analysis predicts that

$$\left(m + \tfrac{1}{4}\right)\pi = \int_{R_1}^{R_2} {}^rk\, dr \qquad (6.2.7)$$

Obviously, if the values of l and m are large, the $\frac{1}{4}$ appearing with both in equation 6.2.7 can be neglected, giving a simplified result:

$$m\pi \simeq \int_{R_1}^{R_2} \left[n^2(r)k_0 - {}^zk^2 - \frac{l^2}{r^2} \right]^{1/2} dr \qquad (6.2.8)$$

Waves are fully guided provided that R_1 and R_2 are less than a, and ^zk is greater than $2\pi n_0(1 - 2\Delta)^{1/2}/\lambda(= k_c)$.

We see also that the integer m labels the number of radial maxima in the electric field of the mode, while $2l$ is the number of azimuthal maxima. Thus these integers closely correspond to the integers l and m used in labeling an LP_{lm} mode, so that an LP_{lm} is equivalent to an lm mode in the graded guide. It also follows that each pair of integers l and m characterizes a discrete mode pattern; but, just as with the LP modes of the step-index guide, each pattern can have two distinct orientations and, with each, two distinct polarizations. Thus the number of possible modes of the guide is given by

$$I = 4\Sigma\left[m(l)\Sigma(l) \right]_{\text{guided}} \qquad (6.2.9)$$

where the subscript "guided" means that the only values allowed are those that fall within the conditions for guidance.

This summation has been performed by assuming that l and m are sufficiently large that replacing the integer summation by an integral is a reasonable approximation:

$$I = \frac{4}{\pi} \int_0^{l_{\max}(r)} \int_{R_1(l)}^{R_2(l)} \left[n^2(r)k_0^2 - k_c^2 - \frac{l^2}{r^2} \right]^{1/2} dr\, dl \qquad (6.2.10)$$

Here we have replaced $^z k$ of equation 6.2.8 by k_c since the largest value of l for a given m always occurs for the mode closest to cutoff, or with $^z k \approx k_c$. Now $l_{\max}(r) = r[n^2(r)k_0^2 - k_c^2]^{1/2}$, so that by changing the order of integration we obtain

$$I = \frac{4}{\pi} \int_0^a \int_0^{l_{\max}(r)} \left[n^2(r)k_0^2 - k_c^2 - \frac{l^2}{r^2} \right]^{1/2} dl\, dr$$

$$= \int_0^a \left[n^2(r)k_0^2 - k_c^2 \right] r\, dr \qquad (6.2.11)$$

$$= \left(\frac{2\pi}{\lambda} \right)^2 \int_0^a \left[n^2(r) - n_c^2 \right] r\, dr$$

We can also write equation 6.2.11 in a different form by restricting the summation to values of $^z k_z$ up to $k_{m'}$ which then fall within a radius R', where

$$n^2(R')k_0^2 - k_{m'}^2 - \frac{l^2}{(R')^2} = 0 \qquad (6.2.12)$$

to obtain the number of modes having values of $^z k_{m'}$ from $^z k = n_0 k_0$ to $^z k = k_{m'}$; thus

$$m'(k_{m'}) = \int_0^{R'} \left[n^2(r)k_0^2 - k_{m'}^2 \right]^{1/2} r\, dr \qquad (6.2.13)$$

Now for the special case of the power series profile described by equations 6.2.1 and 6.2.2, the above expressions can be evaluated to yield the following results:

$$M' = \left(\frac{\alpha}{\alpha+2} \right) a^2 n_0^2 k_0^2 \Delta \qquad (6.2.14)$$

$$m'(k) = a^2 \Delta n_0^2 k_0^2 \left(\frac{\alpha}{\alpha+2} \right) \left(\frac{n_0^2 k_0^2 - k_{m'}^2}{2\Delta k_0^2 n_0^2} \right)^{(\alpha+2)/\alpha} \qquad (6.2.15)$$

The relation between m', M' and l, m for LP modes is discussed in Section 7.2.2.

By using equations 6.2.14 and 6.2.15, we can arrive at a simple expression for $k_{m'}$ for the m'th mode:

$$k_{m'} = k_0 n_0 \left[1 - \left(\frac{m'}{M'} \right)^{\alpha/(\alpha+2)} (2\Delta) \right]^{1/2}$$ (6.2.16)

The m'th mode therefore propagates in the z or guide direction with a propagation constant $k_{m'}$ or in the form $\exp(ik_{m'}z)$. We note also that, although the analysis has been presented in terms of the graded-index guide, setting $\alpha = \infty$ yields the step-index profile, so that the results include that case.

6.3. DISPERSION IN THE MULTIMODE GRADED FIBER

In the analysis of the multimode graded core fiber having an index distribution across its section described by

$$n(r) = n_0 \left[1 - 2\Delta \left(\frac{r}{a} \right)^{\alpha} \right]^{1/2} \qquad r < a$$ (6.3.1)

$$n(r) = n_0 (1 - 2\Delta)^{1/2} \qquad r \geq a$$ (6.3.2)

we found that the value of the propagation constant for the m'th mode was given by

$$^z k_{m'} = k_0 n_0 \left[1 - \left(\frac{m'}{M'} \right)^{\alpha/(\alpha+2)} (2\Delta) \right]^{1/2}$$ (6.3.3)

$$M' = \left(\frac{\alpha}{\alpha+2} \right) a^2 n_0^2 k_0^2 \Delta$$ (6.3.4)

$$= \text{total number of guided modes}$$

The group velocity for a given mode, m', is defined as

$$V_g = \left[\frac{1}{c} \left(\frac{\partial k_{m'}}{\partial k_0} \right) \right]^{-1}$$ (6.3.5)

$$\tau_{m'} = \frac{L}{c} \left(\frac{\partial k_{m'}}{\partial k_0} \right)$$ (6.3.6)

where $\tau_{m'}$ is the time taken for energy propagating in the m'th mode to travel a distance L.

We can derive a simple expression for $\tau_{m'}$ by differentiating equation 6.3.2 with respect to k_0 and inserting in equation 6.3.6. In performing the differentiation, we must remember that Δ, n_0, and M' are all functions of k_0. We obtain the result[4]

$$\tau_{m'} = \frac{N_0 L}{c} \left\{ 1 - \left[\frac{\alpha-2-y}{\alpha+2} \Delta \left(\frac{m'}{M'} \right)^{\alpha/(\alpha+2)} \right] \right.$$

$$\left. + \left[\frac{3\alpha-2-2y}{2(\alpha+2)} \Delta^2 \left(\frac{m'}{M'} \right)^{2\alpha/(\alpha+2)} \right] + O(\Delta^3) \right\} \qquad (6.3.7)$$

In deriving equation 6.3.7, we used the approximation $(1-x)^{-1/2} = (1+x/2)$ and set

$$y = \frac{-2n_0}{N_0} \frac{\lambda}{\Delta} \left(\frac{\partial \Delta}{\partial \lambda} \right) \qquad (6.3.8)$$

$$N_0 = n_0 + \lambda \frac{\partial n_0}{\partial \lambda} = \text{group index} \qquad (6.3.9)$$

6.3.1 Root Mean Square Pulse Width and Optimum Profile

Using the relationship connecting mode delay to pulse broadening derived by Personick,[3] we find that the rms pulse broadening is given by[4]

$$\sigma = \left[\langle \tau_{m'}^2 \rangle - \langle \tau_{m'} \rangle^2 \right]^{2} \qquad (6.3.10)$$

$$= \frac{L N_0 \Delta}{2c} \left(\frac{\alpha}{\alpha+1} \right) \left(\frac{\alpha+2}{3\alpha+2} \right)^{1/2}$$

$$\times \left[c_1^2 + \frac{4 c_1 c_2 \Delta (\alpha+1)}{(2\alpha+1)} + \frac{4 c_2^2 \Delta^2 (2\alpha+2)^2}{(5\alpha+2)(3\alpha+2)} \right]^{1/2}$$

where

$$c_1 = \frac{\alpha-2-y}{\alpha+2} \qquad (6.3.11)$$

$$c_2 = \frac{3\alpha-2-2y}{2(\alpha+2)} \qquad (6.3.12)$$

The expression for σ leads to a particularly interesting result when it is evaluated as a function of α for a particular guide. The result of such an evaluation is shown in Figure 6.2, which uses values of the parameters

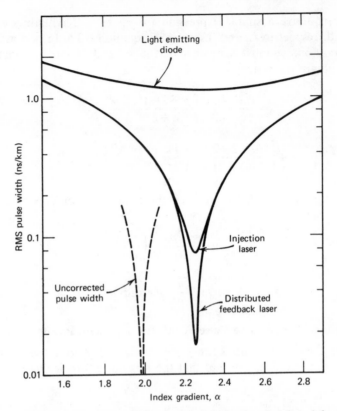

Figure 6.2. Calculated rms pulse spreading in a graded fiber as a function of the parameter α. The uncorrected curve assumes that $y=0$ and includes only mode dispersion. The other curves assume material dispersion for sources of 15, 1, and 0.2 nm linewidth in turn, as being representative of an LED, a laser and an improved laser, respectively, all operating at 900 nm. Reproduced with permission from R. Olshansky and D. B. Keck, *Appl. Opt.*, **15**, 483 (1976).

characteristic of the Corning titania doped silica guide, of numerical aperture 0.16 and with $\lambda = 900$ nm.

We see that there is a very sharp minimum in the pulse broadening very close to the point $\alpha = 2$, which would correspond to a parabolic distribution. However, the minimum is shifted significantly from that point, the exact amount being dependent upon the value of the parameter y in equation 6.3.12. For $y = 0$ the minimum falls just short of $\alpha = 2$. When $y \neq 0$,

$$\alpha_{\min} = 2 + y - \Delta \left[\frac{(4+y)(3+y)}{5+2y} \right] \qquad (6.3.13)$$

The significance of the curve of Figure 6.2 is clear when we examine the numerical values plotted. It shows that with an LED source, exciting all the guided modes of a large graded-index guide, the pulse dispersion can only be in the region of 1 ns/km, set by the material dispersion of the medium from which the guide is constructed. With a laser source having a linewidth roughly one-twentieth that of the LED, the minimum dispersion under the condition of all modes excited can fall as low as 0.1 to 0.2 ns/km. Such low dispersions allow the transmission of high data rates over very long fiber lengths without repeaters. For example, the transmission of 100 Mbit/sec has been reported by Corning over a single fiber of length 10 km, without a repeater and with a total insertion loss of 50 dB. The ability to achieve such excellent performance, while retaining much of the ease of jointing of a step-index multimode fiber, represents a very attractive combination to the systems designer.

Before leaving the subject, we note also that the dispersion of the step-index guide can be obtained immediately from equation 6.3.10 by setting $\alpha = \infty$ to yield the result ($y = 0$)

$$\sigma_{\text{step}} = \frac{LN_1\Delta}{2c} \frac{1}{\sqrt{3}} (1 + 3\Delta + 3\Delta^2) \qquad (6.3.14)$$

$$\approx \frac{LN_1\Delta}{2\sqrt{3}\,c}$$

The corresponding value of σ for a guide having a graded profile with $\alpha = 2$ and $y = 0$ is given by

$$\sigma_{(\alpha=2)} = \frac{LN_0\Delta}{c} \left(\frac{\sqrt{3}\,\Delta}{24} \right) \qquad (6.3.15)$$

and hence

$$\frac{\sigma_{\text{step}}}{\sigma_{\alpha=2}} = \frac{4}{\Delta} \qquad (6.3.16)$$

A more precise evaluation[10] leads to a result of $\sim 10/\Delta$ for equation 6.3.16.

For a value of $\Delta = 0.01$, we see that σ_{step} is typically 15 ns/km, while the value for the corresponding graded guide could be as low as 0.015 ns/km. Such a low dispersion is unlikely to be achieved in practice, however, because the source linewidths of the GaAs laser and LED typically limit the dispersion to the region of 0.08 and 1 ns/km, respectively. Nevertheless, dispersions have been measured with laser sources of 0.25 ns/km or a little less in well graded guides. It is evident from Figure 6.2 that, as the gain factor increases and the mode dispersion is further reduced, the tolerance for the allowed values of α becomes progressively smaller. It has yet to be established what tolerance can be achieved in mass production and at what additional cost.

We should also take note, before passing on to other matters, that all the above results were derived under the assumption that all modes of the guide carry equal amounts of energy. In practice such an assumption will be invalid, since the modes closest to cutoff always carry less power than the modes that are well guided. This fact means that the spread of τ values to be included in the summation or averaging calculation for the rms pulse broadening will be changed from that used here, so that for a given α value the results given here may be pessimistic or optimistic.

6.3.2   Profile Dispersion

During the preceding analysis we clearly separated two effects: material dispersion, which was discussed in Section 5.9, and mode dispersion, which has been discussed both in this chapter and in the preceding one. We have seen in this chapter how, by careful control of the refractive index profile, mode dispersion can be minimized, and this has led us to the power law profile characterized by the parameter α and to the expression for the optimum value of α given by equation 6.3.13. However, if we refer to that expression and note the form of the parameter y used, it is immediately clear that the optimum value of α depends not just upon simple materials parameters but also upon the wavelength at which we wish to use the fiber. There is not a single ideal refractive index profile, but for the highest performance the fiber must be optimized to operate at a single wavelength. Thus a fiber designed to operate at 840 nm wavelength with a GaAs laser source will in general not be optimum for operation at 1060 nm wavelength. The extent of this effect is illustrated in Figure 6.3, which shows the calculated values for the optimum profile values of α for a number of fibers made from different materials systems.[9] It is evident that the differences are significant, and the wavelength variation is large in some cases. The effect is the same as chromatic aberration in a simple lens.

In one possible technique for overcoming this problem, proposed by Kaminow et al.,[5] fibers are manufactured using triple dopants to produce their gradation. The particular combination proposed was P_2O_5, GeO_2, and B_2O_3 deposited from the vapor phases along with SiO_2. These authors have shown that in theory, by carefully controlling the spatial disposition of these three components, it is possible to make a fiber having a profile close to ideal from the visible to well into the infrared (beyond 1100 nm). If proved in production, such techniques would allow the installation of wideband graded fiber networks that either could be operated simultaneously at multiple wavelengths or, perhaps more realistically, could initially be operated at 840 nm with GaAs sources and later refitted with longer wavelength sources of more advanced specification.

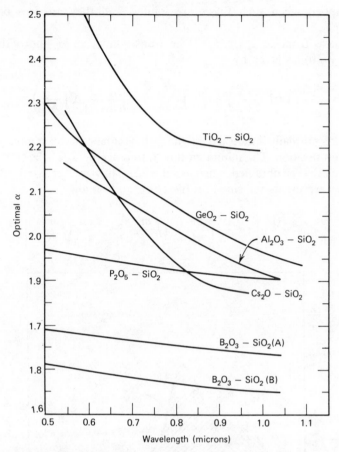

Figure 6.3. Curves of the optimum value of the profile parameter, α, versus wavelength for a number of combinations of materials, showing the effects of profile dispersion. Reproduced with permission from H. M. Presby and I. P. Kaminow, *Appl. Opt.*, **15**, 3029 (1976).

It is interesting to note that this solution is an exact analogue of the one used to correct chromatic aberration in a simple lens, where several simple lenses with different dispersion materials are combined to form corrected doublets or triplets.

6.3.3 Optical Equalization in a Graded Fiber Sequence

A reasonably common occurrence in operational systems is the jointing of graded fibers having profiles that differ randomly about the optimum, since in production the optimum will be sought and errors are as likely on one side as on the other. It is therefore interesting to contemplate what may happen when such fibers are jointed and they are free of mode coupling.

We have from equation 6.3.7 the transit time, τ, for the m'th mode (setting $y = 0$), as given by

$$\tau_{m'} = \frac{N_1 L}{c} \left[1 + \left(\frac{\alpha - 2}{\alpha + 2} \right) \Delta \left(\frac{m'}{M'} \right)^{\alpha/(\alpha+2)} + \frac{3\alpha - 2}{2(\alpha + 2)} \Delta^2 \left(\frac{m'}{M'} \right)^{2\alpha/(\alpha+2)} \right]$$

We are normally interested in the behavior about the minimum time dispersion position. The minimum time difference between the slowest and fastest modes is obtained when $\alpha = 2 - 2\Delta$. Thus we set $\alpha = 2 - 2\Delta + \delta\alpha$, where $\delta\alpha$ represents the small profile error, and obtain

$$\tau_{m'} \approx \frac{N_1 L}{c} \left[1 + \frac{\Delta^2}{2} \left(\frac{m'}{M'} - \sqrt{\frac{m'}{M'}} \right) + \frac{\delta\alpha \, \Delta}{4} \sqrt{\frac{m'}{M'}} \right] \qquad (6.3.17)$$

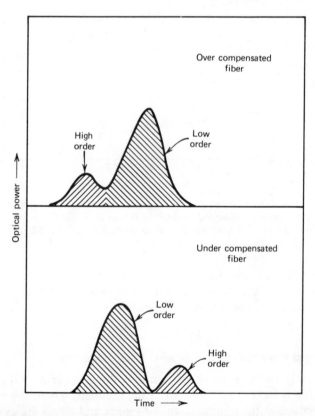

Figure 6.4. Pulse response of over- and undercompensated graded fibers. The detailed pulse shape can vary greatly.

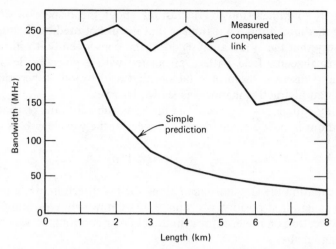

Figure 6.5. The 3 dB bandwidth of an experimental fiber link 5 km long, measured on installed cable using alternate over- and undercompensation for the 1 km component sections. Reproduced with permission from M. Eve, *Opt. Quant. Electron.*, **10**, 41 (1978).

This shows that the change in transit time for any given mode, m', is linear in $\delta\alpha$. It follows that, if we joint two fibers that are closely similar but are characterized by profile errors of $+$ and $-(\delta\alpha)$ in series, provided that they are free of mode coupling we can expect the errors to cancel, and the two low bandwidth fibers will lead to a single high bandwidth fiber of double length. Such an effect seems extraordinary in terms of normal transmission media but is observed in cable designs in which the mode coupling is low enough to allow mode power conservation over a distance of several kilometers.[6] (See also Section 13.6.4 for a discussion of the behavior of jointed graded fibers.) In Figure 6.4 we show the pulse responses of over- and undercompensated fibers (low and high α values, respectively) as they would appear in impulse testing on an oscilloscope. The potential for equalization is clear. Figure 6.5 shows the measured 3 dB electrical bandwidth (see Appendix 5) of a 6 km cable link,[6] using alternate over and undercompensated fibers.

6.4.   RAY PROPAGATION IN A GRADED-INDEX MEDIUM

In Section 6.3 we analyzed the propagation of electromagnetic waves in the generalized graded-index medium, using the WKB method. This approach is valid only under the same conditions that we would have to impose if we were to make a ray analysis, namely, slowly varying refractive index and a large core diameter compared with the wavelength. It is reasonable to expect, therefore, that a ray analysis should be capable of leading us to rather similar conclusions about the dispersion of such a guiding medium, although it is clear that it will not predict discrete modes in the way that the WKB analysis does, and consequently it will not allow us to visualize the field pattern associated with a given mode. But it is possible to discover the profile factors that lead to low dispersion and to do so in an instructive manner, as shown by Eve.[7]

The ray equation of a generalized medium, subject to the restriction that the medium is slowly changing with respect to the wavelength, is given by

$$\frac{d}{ds}\left(n\frac{\partial \mathbf{r}}{\partial s}\right) = \mathbf{grad}\,(n) \tag{6.4.1}$$

In this equation the length is taken along the ray direction, while the vector $\mathbf{r}$ is the generalized position coordinate. If we now split equation 6.4.1 into its three vector components for cylindrical polar coordinates, we obtain the following results:

$$\frac{d}{ds}\left(n\frac{dr}{ds}\right) - nr\left(\frac{d\phi}{ds}\right)^2 = \frac{dn}{dr} \tag{6.4.2}$$

$$n\frac{dr}{ds}\frac{d\phi}{ds} + \frac{d}{ds}\left(nr\frac{d\phi}{ds}\right) = 0 \tag{6.4.3}$$

$$\frac{d}{ds}\left(n\frac{dz}{ds}\right) = 0 \tag{6.4.4}$$

for the r, ϕ, and z components, respectively. Multiplying equation 6.4.3 by r allows us to reduce it to the simple form

$$\frac{d}{ds}\left(nr^2\frac{d\phi}{ds}\right) = 0 \tag{6.4.5}$$

If we now consider the case of interest, namely, ray propagation in a graded-index fiber, we define angles and coordinates as shown in Figure 6.6. The ray enters the fiber at position (r_0, ϕ_0, z_0), making angles α, β, γ with the x, y, z coordinates defined at that point, so that its direction cosines are $\cos\alpha$, $\cos\beta$, and $\cos\gamma$. Then, by some manipulation of equations 6.4.2 to 6.4.5, we can obtain the following results, using in the process the fact that the index n is a function of r only:

$$n\left(\frac{\partial z}{\partial s}\right) = n_0\left(\frac{\partial z}{\partial s}\right)_0 = n_0\cos\gamma_0 \tag{6.4.6}$$

$$nr^2\left(\frac{\partial \phi}{\partial s}\right) = n_0 r_0^2\left(\frac{\partial \phi}{\partial s}\right)_0 = n_0 r_0(\cos\beta_0\cos\phi_0 - \cos\alpha_0\sin\phi_0) \tag{6.4.7}$$

$$r^2\left(\frac{\partial \phi}{\partial z}\right) = r_0\sec\gamma_0(\cos\beta_0\cos\phi_0 - \cos\alpha_0\sin\phi_0) \tag{6.4.8}$$

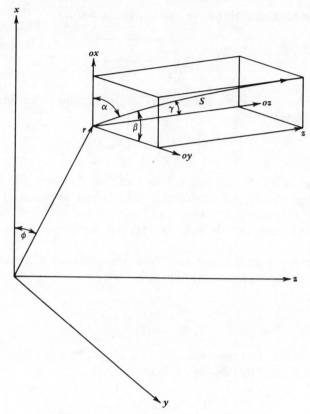

Figure 6.6. Notation used in ray tracing analysis.

The motion of the ray in the medium can now be described in terms of two new variables, defined as follows:

$$\bar{E} = n\left(\frac{\partial z}{\partial s}\right) \qquad \text{(analogous to energy)} \qquad (6.4.9)$$

$$\bar{I} = nr^2\left(\frac{\partial \phi}{\partial s}\right) \qquad \text{(analogous to angular momentum)} \qquad (6.4.10)$$

which lead us to our final general equation governing the ray motion:

$$\frac{\partial^2 r}{\partial z^2} - \frac{\bar{I}^2}{r^3} = \frac{1}{2E^2}\frac{\partial(n^2)}{\partial r} \qquad (6.4.11)$$

Once a ray is launched into the fiber, the parameters $\bar{E}$ and $\bar{I}$ are fixed by the conditions of its launching and its subsequent motion is full determined.

From the equations above, we can derive the result

$$\frac{\partial r}{\partial z} = \sqrt{\frac{n^2}{\bar{E}^2} - \frac{\bar{I}^2}{r^2} - 1}$$

(6.4.12)

If we look for the turning points of r with respect to z, we find, setting $\partial r/\partial z = 0$, that

$$n(r)^2 - \frac{\bar{I}^2 \bar{E}^2}{r^2} - \bar{E}^2 = 0$$

(6.4.13)

Referring to Figures 6.7 and 6.8, we see that there are two solutions, given by R_{max} and R_{min}. If we define z_0 to be located so that $r(z_0) = R_{min}$, it is clear that r is symmetric about z_0 and also around z_1, the position of the first R_{max}. It follows that the function $r(z)$ is periodic in z with a period of $2(z_1 - z_0)$.

By a similar argument we can show the presence of periodicity in ϕ-dependent motion. From equation 6.4.10 we have the result

$$\frac{\partial \phi}{\partial s} = \frac{\bar{I}}{n(r)r^2}$$

(6.4.14)

We already know that the value of $r(z)$ oscillates between R_{min} and R_{max}. It therefore follows that $\partial \phi/\partial s$ oscillates between

$$\left(\frac{\partial \phi}{\partial s}\right)_{min} = \frac{\bar{I}}{n(R_{min}) \cdot R_{min}^2}$$

(6.4.15)

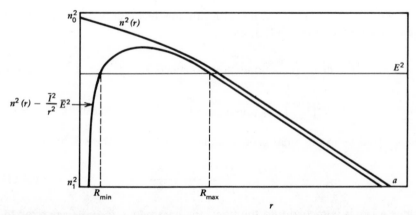

Figure 6.7. Locations of the values of R_{min} and R_{max}. Reproduced with permission from M. Eve, *Opt. Quant. Electron.*, **8**, 285 (1976).

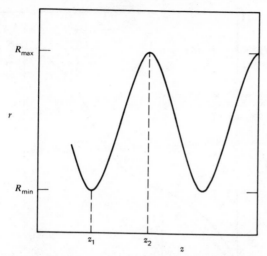

Figure 6.8. Generalized ray path in r/z plane. Reproduced with permission from M. Eve, *Opt. Quant. Electron.*, **8**, 285 (1976).

and

$$\left(\frac{\partial \phi}{\partial s}\right)_{max} = \frac{\bar{I}}{n(R_{max}) \cdot R_{max}^2} \qquad (6.4.16)$$

It follows that we can visualize the development of the ray path in terms of r and ϕ with respect to z in the manner shown in Figure 6.9, with the r component oscillating back and forth between the minimum and maximum values and the ϕ component increasing steadily but having an oscillatory motion superimposed. This is entirely in agreement with the general picture that we obtained from the plane-wave-like WKB analysis.

The value of the wavelength of oscillation in the r plane along the z axis is given by (from equation 6.4.12)

$$2(z_1 - z_0) = 2 \int_{R_{min}}^{R_{max}} \frac{dr}{\sqrt{(n^2/\bar{E}^2) - (\bar{I}^2/r^2) - 1}} = P(\bar{E}, \bar{I}) \qquad (6.4.17)$$

Since ϕ is also pseudoperiodic, we can observe that, over the same interval $P(\bar{E}, \bar{I})$, the value of ϕ changes thus: $\phi(z_0) \rightarrow \phi(z_1) = \phi(z_0) + \Phi(\bar{E}, \bar{I})$, where

$$\Phi(\bar{E}, \bar{I}) = \int_{z_0}^{z_0+P} \frac{\partial \phi}{\partial z} \delta z = 2 \int_{R_{min}}^{R_{max}} \frac{\bar{I} \, dr}{r^2 \sqrt{(n^2/\bar{E}^2) - (\bar{I}^2/r^2) - 1}}$$

$$(6.4.18)$$

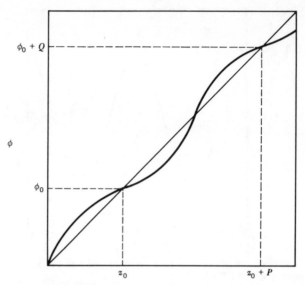

Figure 6.9. Generalized ray path in ϕ/z plane. Reproduced with permission from M. Eve, *Opt. Quant. Electron.*, **8**, 285 (1976).

The functions P and Φ are very important for the study of time dispersion of the ray paths, since they characterize the repeat distances for a given ray. This is clear when we apply to the problem Fermat's principle, which states, "If all rays starting from a point A and traveling via the medium arrive at the same point B, the optical paths from A to B are the same length." It follows that, if we launch a bundle of rays from some generalized point at the end of the fiber with values of $\bar{E}$ and $\bar{I}$ within the ranges $\bar{E}_1 < \bar{E} < \bar{E}_2$ and $\bar{I}_1 < \bar{I} < \bar{I}_2$, then, for the minimum time dispersion between these rays, we require the optical paths of the rays to be equal to some other point in the fiber. It is convenient to consider that point spaced at a distance $P(\bar{E}, \bar{I})$ and $\Phi(\bar{E}, \bar{I})$ away. Then we can say that the first conditions for low time dispersion will be given by

$$\left[\frac{\partial P(\bar{E}, \bar{I})}{\partial \bar{E}}\right]_{\substack{\bar{E}a \\ \bar{I}a}} = 0 \qquad \left[\frac{\partial \Phi(\bar{E}, \bar{I})}{\partial \bar{E}}\right]_{\substack{\bar{E}a \\ \bar{I}a}} = 0$$

$$\left[\frac{\partial P(\bar{E}, \bar{I})}{\partial \bar{I}}\right]_{\substack{\bar{E}a \\ \bar{I}a}} = 0 \qquad \left[\frac{\partial \Phi(\bar{E}, \bar{I})}{\partial \bar{I}}\right]_{\substack{\bar{E}a \\ \bar{I}a}} = 0$$

where $\bar{E}_a$ and $\bar{I}_a$ are average values, defined as

$$\bar{E}_a = \frac{\bar{E}_1 + \bar{E}_2}{2}, \qquad \bar{I}_a = \frac{\bar{I}_1 + \bar{I}_2}{2}$$

Ideally we would like the higher derivatives of $\bar{E}$ and $\bar{I}$ to be zero also, but that is unlikely to be possible.

We are now in a position to examine the performance of given theoretical profiles. If we take the alpha profile already discussed in the WKB analysis, we have

$$n(r) = n_0 \left[1 - 2\Delta \left(\frac{r}{a} \right)^\alpha \right]^{1/2} \qquad r < a$$

$$n(r) = n_0 (1 - 2\Delta)^{1/2} \qquad r \geqslant a$$

Then writing the expressions for $\bar{E}$ and $\bar{I}$ from equations 6.4.17 and 6.4.18 gives the following results:

$$P(\bar{E}, \bar{I}) = \frac{2\bar{E}}{n_0} \int_{R_{\min}}^{R_{\max}} \left[1 - 2\Delta \left(\frac{r}{a} \right)^\alpha - \frac{\bar{I}^2 \bar{E}^2}{n_0^2 r^2} - \frac{\bar{E}^2}{n_0^2} \right]^{-1/2} \tag{6.4.19}$$

$$\Phi(\bar{E}, \bar{I}) = \frac{2\bar{E}\bar{I}}{n_0} \left(\frac{r}{r_0} \right) \int_{R_{\min}}^{R_{\max}} \left\{ r^2 \left[1 - 2\Delta \left(\frac{r}{a} \right)^\alpha - \frac{\bar{I}^2 \bar{E}^2}{n_0^2 r_0^2} - \frac{\bar{E}^2}{n_0^2} \right]^{1/2} \right\}^{-1} \tag{6.4.20}$$

By definition, $R_{\max} < a$ for a fully guided mode (nonleaky). Thus we take $\bar{E}_a = \frac{1}{2}(1 + \sqrt{1 - 2\Delta}) \simeq (1 - \Delta/2)$ and $\bar{I}_a = 0$, since rays are symmetrically spread about the axis. We now have to solve

$$\left[\frac{\partial P(\bar{E}, 0)}{\partial \bar{E}} \right]_{\bar{E}_a} = 0 \qquad \left[\frac{\partial \Phi(\bar{E}, 0)}{\partial \bar{E}} \right]_{\bar{E}_a} = 0$$

$$\left[\frac{\partial P(\bar{E}_a, \bar{I})}{\partial \bar{I}} \right]_0 = 0 \qquad \left[\frac{\partial \Phi(\bar{E}_a, \bar{I})}{\partial \bar{I}} \right]_0 = 0$$

We can now write equation 6.4.19 in the form

$$P(\bar{E}, \bar{I}) = \frac{2\bar{E}a}{\sqrt{1 - \bar{E}^2}} \left(\frac{1 - \bar{E}^2}{2\Delta} \right)^{1/\alpha} \int_0^1 \frac{du}{\sqrt{1 - u^\alpha}} \tag{6.4.21}$$

and set

$$c(\alpha) = \int_0^1 (1 - u^\alpha)^{-1/2} du$$

We now have

$$\left[\frac{\partial P(\overline{E},\overline{I})}{\partial \overline{E}}\right]_{\overline{E}_a} = \frac{2ac(\alpha)}{(2\Delta)^{1/\alpha}}\Delta^{1/\alpha-3/2}\left[1-\frac{2(1-\Delta)}{\alpha}\right] \qquad (6.4.22)$$

so that $[\partial P/\partial \overline{E}]_{\overline{E}_a}=0$ implies $\alpha_{min}=2-2\Delta$.

Referring to equation 6.3.13, we see that our WKB analysis predicted $\alpha_{min}=2-\frac{12}{5}\Delta$. The discrepancy arises because that result was based upon minimizing the rms pulse width, whereas this analysis has minimized the meridional ray dispersion. Gloge and Marcatilli[2] derived the result $\alpha_{min}=2-2\Delta$ from a WKB analysis under the assumptions used here. As for the other conditions that we wish to satisfy, since $\Phi(\overline{E},0)=0$, then $\partial(\overline{E},0)/\partial \overline{E}=0$ for all $\overline{E}$, and since $P(\overline{E}_a,\overline{I})=P(\overline{E}_a,-\overline{I})$, then $[\partial P(\overline{E}_a,\overline{I})/\partial \overline{I}]_0=0$.

There remains the condition $[\partial \Phi(\overline{E}_a,\overline{I})/\partial \overline{I}]_0=0$ that we would like to satisfy. This is expected to be small since (for very small $\overline{I}$) the function $\Phi(\overline{E}_a,\overline{I})\simeq \pi[1+0(P^2E^2)]$. Thus we may expect the derivative with respect to $\overline{I}$ to be very close to zero. It follows that the alpha profile is an ideal profile since all four first derivatives are zero.

For comparison, we note that the visually very similar parabolic profile described by

$$n(r)=n_0(1-Ar^2)$$

leads to a value of $\partial P/\partial \overline{E}=2\pi/n_0\sqrt{A}$ with the remaining three derivatives zero. Eve[7] also discussed other profiles in his paper.

In summary, then, we can say that the ray model has led us to draw conclusions similar to those obtained with the WKB approach. It has the virtue of making clearer the different components leading to the observed pulse dispersion in a given profile, by virtue of the separation of the conditions to be satisfied, but it does not allow us to form a picture of the modal field patterns.

6.5. EXCITATION OF A MULTIMODE GRADED CORE FIBER BY A LAMBERTIAN (DIFFUSE) SOURCE

We have already discussed the plane wave approximation as a means of visualizing the modes in a generalized circular fiber. It is a simple matter to extend that analysis to obtain general results which include not only the purely guided mode but also the weakly leaky modes, and to use these results to estimate the excitation solid angles associated with both at any given point on the fiber end face. In so doing, we can estimate the power that would be coupled from a diffuse source into either or both groups of modes for a smoothly varying multimode fiber.

We have already shown that if the refractive index of the fiber is given by

$$n(r) = n_2 \qquad r > a$$
$$n_1 \geqslant n(r) \geqslant n_2 \qquad r < a$$

then the components of the **k** vector at the point (r_0, ϕ_0, z_0) are given by

$$^z k(r_0) = {}^z k \tag{6.5.1}$$

$$^\phi k(\phi_0) = \pm \frac{l}{r_0} \tag{6.5.2}$$

$$^r k(r_0) = \sqrt{n^2(r)k_0 - {}^z k^2 - l^2/r_0^2} \tag{6.5.3}$$

and, by reference to Figure 6.6, the direction cosines for the plane wave directions are given by

$$\cos \gamma_0 = \frac{{}^z k}{n(r_0)k_0} \tag{6.5.4}$$

$$\cos \alpha_0 = \frac{l}{r_0 n(r_0)k_0} \tag{6.5.5}$$

The magnitudes of these components are usefully illustrated in Figure 6.10, which shows that, for a generalized graded fiber, $^r k$ may be real only over a small range of radius in the core of the fiber. However, the possibility also exists for it to become real again as $r \to \infty$, the case discussed under leaky modes in Section 5.4.

The range of values of $^z k$ corresponding to tunneling leaky modes is given by

$$n_2^2 k_0^2 - \frac{l^2}{a^2} \leqslant {}^z k^2 < n_2^2 k_0^2 \tag{6.5.6}$$

This result can be rewritten in terms of the direction cosines from equations 6.5.4 and 6.5.5 to yield the result

$$n_2^2 - \left(\frac{r_0}{a}\right)^2 n^2(r_0) \cos^2 \alpha_0 \leqslant n^2(r_0) \cos^2 \gamma_0 < n_2^2 \tag{6.5.7}$$

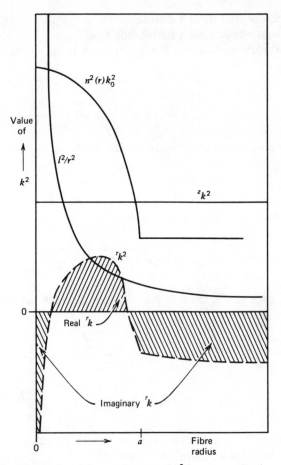

Figure 6.10. Magnitudes of the components of k^2 for a generalized graded fiber.

The angles α_0 and γ_0 are measured internally, relative to the fiber axis. Experimentally we are more interested in measuring angles external to the fiber, and this is easily done using the following relations:

$$\sin I = n(r_0) \sin \gamma_0 \qquad (6.5.8)$$

$$\cos \psi = \frac{\cos \alpha_0}{\sin \gamma_0} \qquad (6.5.9)$$

We can now insert these results into equation 6.5.7 to obtain the range of external angles corresponding to the excitation of tunneling leaky modes, in the form

$$\frac{n^2(r_0) - n_2^2}{1 - (r_0/a)^2 \cos^2\psi} \geqslant \sin^2 I > n^2(r_0) - n_2^2 \qquad (6.5.10)$$

Examination of this result shows that, for $\psi = \pi/2$, there is no range for I. Thus only guided modes are excited. This is to be expected since this condition corresponds to the case of $l = 0$. The maximum range of I is obtained for $\psi = 0$, corresponding to the maximum allowed values of l. The angular variation of the excitation range for guided and leaky modes is illustrated in Figure 6.11, where for any particular point, r_0, on the fiber face a solid cone of excitation angle $I = \sin^{-1}\sqrt{n^2(r_0) - n_2^2}$ exists with an elliptical section around it within which leaky modes will be excited.

From this analysis it follows, by integrating over the solid angles defined by equation 6.5.10, that the power accepted as a function of radius in the fiber is given by[8]

$$\frac{P(r)}{P(0)} = \left[\frac{n^2(r) - n_2^2}{n^2(0) - n_2^2}\right]\frac{1}{\sqrt{1 - (r/a)^2}} \qquad (6.5.11)$$

The second term under the square root sign represents the correction in launched power to be applied because of the presence of leaky modes. The

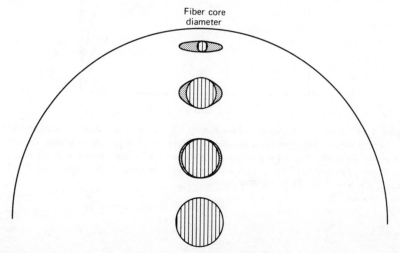

Figure 6.11. Ranges of angular excitation for a graded fiber versus position on fiber end (after Ref. 8).

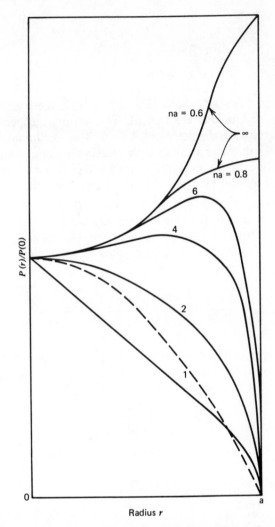

Figure 6.12. Plots of the near field intensity, including leaky mode contributions for Lambertian source excitation for a number of different profile parameter values. The dotted line shows the profile shape for a parabolic ($\alpha = 2$) profile for comparison with the near field ($\alpha = 2$) plot. Reproduced with permission from M. J. Adams, D. N. Payne, and F. M. E. Sladen, *Electron. Lett.*, **11**, 238 (1975).

results of evaluating equation 6.5.11 for a number of different alpha profiles $[n(r) = n(0)(1 - 2\Delta(r/a)^\alpha]^{1/2}$ are shown in Figure 6.12; included also is the profile for $\alpha = 2$, shown dotted for comparison. Note that, in evaluating the case of the step-index fiber, it is necessary to ensure that the acceptance angle given by equation 6.5.8 and 6.5.9 does not exceed $\pi/2$, a result which the theory can predict but which is not useful.

The results are significant in several ways. First, the relationship for the launched power from a diffuse source (equal power per mode or per solid angle) given by equation 6.5.11 can be used as the basis for a measurement of the index profile, by experimentally measuring the curves of Figure 6.12. However, great care must be taken to correct for the effects of differential attenuation in the transmission of the leaky modes.

Second, the results show that the power distribution in a fiber from a diffuse source cannot simply be assumed to be uniform in angle and position. This affects both insertion loss measurements and calculations of system performance.

Finally, we should note that in a graded fiber power launched at a given angle to the axis in one part of the face of the fiber couples to guided modes only, while at another part (larger radius) it can couple to guided and leaky modes. Experimentally, this means that one cannot launch the guided modes of a graded fiber without simultaneously exciting leaky modes if a simple aperture limited optical system is used Thus measurements of graded-index fibers can be considerably complicated in comparison to step-index fibers.

REFERENCES

1. W. Streiffer and C. N. Kurtz, *J. Opt. Soc. Am.*, **57**, 779 (1967).

2. D. Gloge and E. A. J. Marcatilli, *Bell Syst. Tech. J.*, **52**, 1563 (1973).

3. S. D. Personick, *Bell Syst. Tech. J.*, **50**, 843 (1971).

4. R. Olshansky and D. B. Keck, *Appl. Opt.*, **15**, 483 (1976).

5. I. P. Kaminow, H. M. Presby, J. B. McChesney, and P. B. O'Connor, Postdeadline Paper, *Optical Fiber Transmission, II*. Williamsburg, Va.: OSA/IEEE, February 1977.

6. a. M. Eve, *Electron Lett.*, **13**, 315 (1977).

   b. M. Eve, *Opt. Quant. Electron.*, **10**, 41 (1978).

7. M. Eve, *Opt. Quant. Electron.*, **8**, 285 (1976).

8. M. J. Adams, D. N. Payne, and F. M. E. Sladen, *Electron Lett.*, **11**, 238 (1975).

9. H. M. Presby and I. P. Kaminow, *Appl. Opt.*, **15**, 3029 (1976).

10. J. Arnaud, *Opt. Quatn. Electron.*, **9**, 111 (1977).

7

Mode Coupling in a
Step-Index Fiber

7.1. INTRODUCTION

Mode coupling in multimode fibers has become a subject of considerable interest since it affects the transmission properties in several subtle
and sometimes unexpected ways. In this chapter we examine mode coupling in just two situations which are of practical interest. The coupling
can be studied using a full mode solution for the fiber. However, this leads,
in general, to rather complex mathematics, and a simpler approach has
attracted widespread interest. Before discussing this simpler approach, we
must first clarify our ideas on a point of detail concerning mode nomencla-
ture.

Without resorting to any mathematics, we can discern two limiting cases
of mode mixing in a multimode fiber and deduce their effects. In the first
case we imagine a pure single mode being launched in the fiber, most
conveniently the LP_{01}. If we assume that there is no mode mixing, it will
propagate as the LP_{01} throughout the fiber, and the bandwidth and other
characteristics will be those of the single-mode fiber.

In the same situation, but with some mode mixing, energy will be
coupled from the LP_{01} mode into higher order modes having lower group
velocities, and additional pulse dispersion will inevitably result.

The second extreme case is to imagine the same multimode fiber, with
all modes uniformly excited. Without mode mixing we will obviously
observe the maximum pulse spreading, given by the pulse width expression
derived as equation 6.3.10. If we now start to introduce some mode mixing,
we can see that any particular photon, starting its journey in the mth
mode, will be randomly scattered to other modes. If there are many such
scattering events in the length of fiber under examination, the photon will,
on the average, have spent some time in many different modes, and will
have traveled many short distances at different group velocities. We may
guess, then, that under these conditions the observed pulse width will be

reduced because of the velocity averaging that will result; and it has been shown that under such conditions the pulse spreading builds up, not as L, the distance traveled, but as $\sqrt{L}$, so that useful gains can be achieved. In reality, however, we will usually observe propagation that is a combination of all these situations, leading to a very complex problem to analyze mathematically. In this chapter we restrict ourselves to examining the conditions under which coupling can occur, and leave for further reading the study of how a particular model situation evolves and what time response is associated with it.

7.2. MODE NUMBERS

It is often convenient to refer to modes in multimode dielectric guides by a single integer and also to be able to estimate quickly how many modes are likely to propagate within a given guide. With this in mind, we present the following approximations, which allow just such estimates to be made.

7.2.1 Single-Mode Number Notation

We know that the LP_{lm} mode can propagate provided that the value of V for the guide is such that $U<V$, where U is defined by the mth root of the equation

$$J_{l-1}(U)=0$$

Now, if we examine the asymptotic expansion for the Bessel function, $J_l(z)$, for large arguments, we find that

$$J_l(z)=\sqrt{\frac{2}{\pi z}}\ \cos\left(z-\frac{l\pi}{2}-\frac{\pi}{4}\right) \tag{7.2.1}$$

We can now see that, for there to be a zero in the field for the LP_{lm} mode at $r=a$, we must satisfy the condition that

$$U-\frac{l\pi}{2}-\frac{\pi}{4}=\left(m+\tfrac{1}{2}\right)\pi \tag{7.2.2}$$

or, approximately,

$$U=\frac{\pi}{2}(2m+l) \tag{7.2.3}$$

It is now an obvious step to associate a single mode number with all the modes characterized by the integer

$$M=2m+l \tag{7.2.4}$$

with cutoff values in the region of

$$U = \frac{M\pi}{2} \tag{7.2.5}$$

It is now evident that for guides characterized by the parameter V values of l_{max} and m_{max} will be given by

$$l_{max} = \frac{2V}{\pi} \tag{7.2.6}$$

$$m_{max} = \frac{V}{\pi} \tag{7.2.7}$$

and the total number of integer groups will be given by

$$I = \frac{1}{2}\left[\frac{2V^2}{\pi^2}\right] = \frac{V^2}{\pi^2} \tag{7.2.8}$$

However, since each LP mode integer group is associated with four discrete but degenerate modes, the total number of modes in the guide is given by

$$I_{total} = \frac{4V^2}{\pi^2} = \tfrac{1}{4}M_{max}^2 \tag{7.2.9}$$

This result can also be obtained in a quite different way. If we examine the acceptance angle for a dielectric guide, we find that it is closely related to the numerical aperture (NA), defined as follows:

$$\text{NA} = \sin\theta_1 = \sqrt{n_1^2 - n_2^2} = \sqrt{2n_1^2\Delta} \tag{7.2.10}$$

The solid acceptance angle for the same guide (in air) is $\pi\theta_1^2$, assuming small angles (see Figure 7.1). The product of solid angle ψ and cross

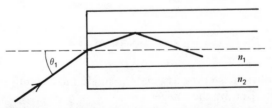

Figure 7.1. Schematic of a step-index fiber defining the ray angle, θ_1, and the numerical aperture, $\sin\theta_1$.

sectional core area A is then given by

$$\psi A = \pi^2 a^2 \left(n_1^2 - n_2^2 \right) \tag{7.2.11}$$

The solid angle for a single electromagnetic radiation mode (black-body mode or mode emanating from a laser or waveguide) is λ^2/A, where A is the area of the aperture that the mode is leaving or entering. This is a plane polarized mode and thus can have two polarization orientations. The total number of modes the guide can support, couple to, and radiate into is therefore

$$I_{\text{total}} = \frac{2\pi^2 a^2}{\lambda^2} \left(n_1^2 - n_2^2 \right) \tag{7.2.12}$$

But since the V parameter is given by the relation

$$V = \frac{2\pi a}{\lambda} \left(n_1^2 - n_2^2 \right)^{1/2} \tag{7.2.13}$$

we have

$$I_{\text{total}} = \frac{V^2}{2} \simeq \frac{4V^2}{\pi^2} \tag{7.2.14}$$

Obviously both relations (equations 7.2.9 and 7.2.14) are approximate, but either gives a useful guide to the number of modes carried by a multimode guide.

If we now return to the single quantum number characterizing an LP mode, we see that, apart from convenience in notation and in associating modes characterized by the relation

$$M = 2m + l \tag{7.2.15}$$

it has further significance. The value of $^z k$ for the mode is characterized by the relation

$$^z k = \left(n_1^2 k_0^2 - \frac{U^2}{a^2} \right)^{1/2}$$

$$= \left(n_1^2 k_0^2 - \frac{M^2 \pi^2}{4a^2} \right)^{1/2} \tag{7.2.16}$$

The transverse component of this wave, which is conserved when the wave enters free space at the end of the waveguide, is given by

$$^Tk = \frac{M\pi}{2a} \tag{7.2.17}$$

Thus, in the far field, the wave will radiate into a cone of semiangle

$$\theta = \frac{\pi M}{2ak_0} = \frac{M\lambda}{4a} \tag{7.2.18}$$

The far-field angle for radiation leaving the guide is simply related to the single-mode number M, so that experimental studies framed in terms of modes characterized by the integer M allow relatively easy measurement of the power associated with modes having that number. In practice, it will probably not be possible to detect the power associated with the modes characterized by the integer M alone. We should also note that, for a given value of M, the number of associated modes is given by the nearest integer below or equal to $M/2$.

7.2.2 General Relations for Mode Numbers in Power Law Profile Guides and Relationship between Linearly Polarized and Wentzel-Kramers-Brillouin Modes

The results presented above were derived for step-index multimode fibers. However, it is common to work with graded-index guides, and some similar results are easily established for them. In Section 6.2 we discussed the power law profile guides, characterized by the parameter α, and obtained an expression for ^zk for the m'th modes of the structure, in the form (equation 6.2.16)

$$^zk^2 = n_1^2 k_0^2 \left[1 - 2\Delta \left(\frac{m'}{M'} \right)^{\alpha/(\alpha+2)} \right] \tag{7.2.19}$$

We can relate this mode numbering system to the LP mode system by using a result derived by Streiffer and Kurtz.[1] For an alpha profile guide they derived the following result for an LP_{lm} mode:

$$^zk^2 = n_1^2 k_0^2 \left\{ 1 - \left[\frac{\sqrt{\pi}\,(2m+l+1)\Gamma(\frac{3}{2}+1/\alpha)}{n_1 k_0 a \Gamma(1+1/\alpha)} \right]^{2\alpha/(\alpha+2)} (2\Delta)^{2/(\alpha+2)} \right\} \tag{7.2.20}$$

Equating these two results, we obtain the general result

$$m' = \frac{\alpha\pi}{2(\alpha+2)}\left[\frac{(2m+l+1)\Gamma\left(\frac{3}{2}+1/\alpha\right)}{\Gamma(1+1/\alpha)}\right]^2 \qquad (7.2.21)$$

Two special cases of this general relation now follow directly. For $\alpha = 2$, the parabolic guide, we find that

$$m' \simeq (2m+l+1)^2 \qquad (7.2.22)$$

while for $\alpha = \infty$, the step-index guide, we have the result

$$m' \simeq \frac{\pi^2}{8}(2m+l+1)^2 \qquad (7.2.23)$$

In either case we see that it is a good approximation to use the simple approximation of $m' = (2m+l+1)^2$. This result also gives us another way of relating the simple single-quantum-number method of labeling to the double-quantum-number LP system and to our single quantum number M, derived in Section 7.2.1, since $m' \simeq M^2$.

7.3.   MODE COUPLING IN MULTIMODE FIBERS

An understanding of the properties and effects of mode coupling in multimode fibers is important since waves propagate for enormous distances, relative to a wavelength, and mode mixing is almost certain to occur over such distances. Indeed, it is remarkable that individual modes can propagate for distances of the order of 1 km in a glass fiber without large energy transfers to adjacent modes, even when the fiber is of exceptionally good quality and is not strained or bent by its surroundings. We should note that 1 km is of the order of 10^9 wavelengths, indicating quite exceptional mechanical uniformity. However, slight perturbations of the fiber lay or small imperfections introduced at the pulling stage can easily disturb this situation.

We first note that the ϕ and z dependences of the propagating fields associated with the jth mode are of the form

$$E_j(\phi,z) = E_j \exp \pm (il_j\phi) \cdot \exp(-ik_j z) \qquad (7.3.1)$$

It is faily evident that to couple the jth mode to the j'th mode requires an irregularity of the general form

$$\psi = \psi_0 \exp \pm i(l_j - l_{j'})\phi \cdot \exp\left[-i(k_j - k_{j'})z\right] \qquad (7.3.2)$$

Irregularity in the fiber induced during fiber pulling is most likely to be due to pulling speed fluctuation or to meniscus fluctuation, either of which will tend to lead to fluctuations in the fiber diameter but not in its circular symmetry. Such fluctuations will thus be of the category described by setting $l_j - l_{j'} = 0$ in the above.

Once the fiber has been made, it is extremely easy to impose small lateral deformations on it, so that instead of lying in a straight path it lies along a sinusoidal path about the central straight axis. Such deformations have the general symmetry of a $\cos\phi$ variation, corresponding to $l_j - l_{j'} = 1$. It is this type of fluctuation that appears to be most common in optical fibers, as one would expect and as is borne out by measurements of mode coupling coefficients and the effects predicted by a simple theory based upon such a postulate. See Appendices 1 and 2.

By reference to Section 7.2.1 on mode numbering, we recall that for the LP_{lm} mode we can write approximately

$$^z k = n_1 k_0 \left[1 - \frac{\pi^2}{8 n_1^2 k_0^2 a^2} (2m + l)^2 \right]^{1/2} \qquad (7.3.3)$$

Modes having the same value of $(2m + l)$ will be nearly degenerate in $^z k$, and mode pairs for which $\Delta l = 1$ and $\Delta m = 0$ will be closely spaced in $^z k$ and thus will need only small values of $\Delta^z k$ or long fluctuation wavelengths to couple them. The possibility also arises of coupling between modes for which $\Delta m = \pm 1$ and $\Delta l = \mp 1$. Such modes will show only weak coupling because the overlap of the radial (r-dependent) wavefunction will tend to average to zero. A more precise evaluation of these coupling parameters is given in Appendicies 1 and 2. However, without resort to further mathematics, we can easily perceive that it is likely to be a good approximation to say that mode coupling will occur only between pairs of modes for which the reduced mode number, M, differs by unity, that is,

$$\Delta M = \Delta(2m + l) = \pm 1 \qquad (7.3.4)$$

This observation immediately suggests the application of a formalism based upon M to analyze the problem, and the most useful results obtained to date have been obtained by that approach. Our analysis follows loosely that of Gloge.[2]

7.4. MODE DIFFUSION ANALYSIS OF COUPLING

We will now set up the differential equations governing power flow between the modes of the Mth group and those of the $M-1$ and $M+1$ groups. Since in the Mth group of modes there are $M/2$ discrete LP mode patterns, each consisting of two degenerate modes with two discrete polarizations, the number of modes within the Mth group is $2M$. Thus we can say that

$$M\frac{\partial P_M}{\partial z} = -M\alpha_M P_M + M d_M(P_{M+1} - P_M)$$

$$+(M-1)d_{M-1}(P_{M-1} - P_M) \qquad (7.4.1)$$

where for the Mth mode group, α_M is the power attenuation coefficient, P_M is the power at z and d_M is the power coupling coefficient to the $M+1$ mode group. Now, setting

$$\frac{P_{M+1} - P_M}{\theta_{M+1} - \theta_M} = \frac{dP_M}{d\theta} \qquad (7.4.2)$$

$$\theta_M = \frac{M\lambda}{4a} \qquad \text{(equation 7.2.18)} \qquad (7.4.3)$$

$$\theta_{M+1} - \theta_M = \Delta\theta = \frac{\lambda}{4a} \qquad (7.4.4)$$

we can now write equation 7.4.1 in the form

$$\frac{\partial P_M}{\partial z} = -\alpha_M P_M$$

$$+ \frac{\Delta\theta}{\theta_M}\left(\theta_M d_M \frac{\partial P_M}{\partial\theta} - \theta_{M-1} d_{M-1} \frac{\partial P_{M-1}}{\partial\theta}\right) \qquad (7.4.5)$$

or

$$\frac{\partial P(\theta)}{\partial z} = -\alpha(\theta)P(\theta) + \frac{(\Delta\theta)^2}{\theta}\frac{\partial}{\partial\theta}\left[\theta d(\theta)\frac{\partial P(\theta)}{\partial\theta}\right] \qquad (7.4.6)$$

We expect the loss coefficient α to consist of two terms, one that is mode or θ independent and one that depends upon the mode number and will be largely radiative in nature. Thus we set $\alpha = \alpha_0 + \alpha_1(\theta)$. Since the term α_0 will be the same for all modes, it is easily accounted for at the close of the analysis, if we wish, by simply multiplying all amplitudes by $\exp(-\alpha_0 z)$, and for this reason it will be neglected now. The mode-dependent loss is expected to be of the general form $A\theta^2$ since the power density at the core-cladding interface varies roughly as U^2 and hence as θ^2.

The mode coupling coefficients are also likely to be θ dependent to some degree, but for the purposes of this analysis will be assumed to be predominantly constant and set to d_0.

Given these approximations, then, equation 7.4.6 reduces to the form

$$\frac{\partial P}{\partial z} = -A\theta^2 P + \frac{D}{\theta}\frac{\partial}{\partial \theta}\left(\theta\frac{\partial P}{\partial \theta}\right) \tag{7.4.7}$$

where

$$D = (\Delta\theta)^2 d(\theta) = (\Delta\theta)^2 d_0 \tag{7.4.8}$$

This equation can now be recognized as a standard form of the diffusion equation, and we state that the process of energy redistribution within the multimode guide will be a diffusion-like process. It is of course assumed here, without explicitly stating so, that the number of modes is large enough for them to be treated as a continuum characterized by the continuous variable θ.

Equation 7.4.8 has been solved for a number of different situations with several different objects in view. The simplest solution to find is that for the steady state mode distribution. Evidently, this will be of the form $P(z,\theta) = P(\theta)\exp-(\gamma z)$ since, by definition, in the steady state all modes will attenuate equally with distance, even though power diffuses between them.

Thus, substituting for $P(\theta)$, we find that

$$(A\theta^2 - \gamma)P(\theta) = \frac{D}{\theta}\frac{\partial}{\partial \theta}\left[\theta\frac{\partial P(\theta)}{\partial \theta}\right] \tag{7.4.9}$$

Solving this equation yields the steady state power profile, $P(\theta)$, as a function of θ for the parameters D, A, and γ. The solutions take the form of Laguerre-Gaussian polynomials. The minimum values of γ are obtained for the situation

$$P(\theta) = P(0)\exp-\left(\frac{\theta^2}{\Theta_\infty^2}\right) \tag{7.4.10}$$

where

$$\Theta_\infty = \left(\frac{4D}{A}\right)^{1/4} \tag{7.4.11}$$

and

$$\gamma = 2\sqrt{AD} \tag{7.4.12}$$

This distribution gives the best compromise for the power distribution among the guided modes, assuming the presence of the $A\theta^2$ loss law. It predicts, in the steady state, the Gaussian power distribution in the far field. In practice such a distribution is frequently seen in high order, multimode step-index fibers.

Gloge[2] has further analyzed the buildup of the stable distribution above from an initial excited distribution when the initial distribution is Gaussian but has a different decay parameter from Θ_∞. Gambling et al.[3] have used the mode coupling diffusion equation to analyze the case of an excitation corresponding to an incident laser beam, impinging on the fiber not at normal incidence, as in Gloge's case, but at an angle ψ. In this case they show that by studying the far-field radiation pattern after a length of fiber, L, one can deduce numerical values for the coupling parameter d_0 by noting when the observed far-field radiation pattern changes from a hollow doughnut appearance to a filled circle (as the excitation angle is decreased).

We should note carefully, however, that the result of this type of mode coupling has been to lead to a steady diffusion of energy, not only between guided modes, but also between guided modes and cladding modes. In other words, a loss γ has been introduced over and above any loss already present in the fiber from absorption or other causes. However, such a fiber will show an improved pulse response because of the mixing of energy between modes, so that the additional loss brings some reward.

7.5. PULSE RESPONSE OF A MODE COUPLED FIBER

We have shown simply by a consideration of the steady state power distribution in the mode coupled step-index fiber that a power distribution Gaussian with respect to θ is obtained and that, associated with this, is an additional loss γ_∞ (km^{-1}) or 4.35 γ_∞ (dB/km). However, the analysis has told us nothing at all about the detailed time dependence of the fiber power when subjected to a pulse input, although by an intuitive argument we have already derived some general results.

However, equation 7.4.7 can easily be converted to time-dependent form by noting that

$$\frac{dP}{dz} = \left(\frac{\partial P}{\partial z}\right) + \left(\frac{\partial P}{\partial t}\right)\left(\frac{\partial t}{\partial z}\right) \tag{7.5.1}$$

By simple geometry (path length) we note that

$$\frac{\partial z}{\partial t} = \frac{c\cos\theta}{n} \approx \frac{c}{n(1+\theta^2/2)} \tag{7.5.2}$$

We then have by substitution the result

$$\frac{\partial P}{\partial z} = -A\theta^2 P - \frac{n}{2c}\theta^2\frac{\partial P}{\partial t} + \frac{1}{\theta}\frac{\partial}{\partial\theta}\left(\theta D\frac{\partial P}{\partial\theta}\right) \tag{7.5.3}$$

We now proceed by substituting for $P(\theta,z,t)$ its Laplace transform:

$$Q(\theta,z,s) = \int_0^\infty \exp(-st)P(\theta,z,t)\,dt \qquad (7.5.4)$$

Thus, multiplying equation 7.5.3 by $\exp(-st)$ and integrating over time from zero to infinity, we obtain the following result:

$$\frac{\partial Q}{\partial z} = -A\theta^2 Q - \frac{n\theta^2}{2c}[sQ - P(\theta,z,0)] + \frac{1}{\theta}\frac{\partial}{\partial\theta}\left(\theta D\frac{\partial Q}{\partial\theta}\right) \qquad (7.5.5)$$

Here we have used the result that

$$\int_{0}^\infty \exp(-st)\left(\frac{\partial P}{\partial t}\right)dt = -P(t=0) + sQ$$

Fortunately, since we are interested in solutions at $z \neq 0$, we can set $P(t=0)=0$ because it is so for all z other than $z=0$.

Equation 7.5.5 can be simplified by writing it in the form

$$\frac{\partial Q}{\partial t} = -A\sigma^2\theta^2 Q + \frac{1}{\theta}\frac{\partial}{\partial\theta}\left(\theta D\frac{\partial Q}{\partial\theta}\right) \qquad (7.5.6)$$

where

$$\sigma = \sqrt{1 + ns/2cA}$$

This result is now identical with that of equation 7.4.7 except that we must replace the A of equation 7.4.7 with $B = \sigma A$. With that observation in mind, we propose a solution of the form

$$Q(\theta,z,s) = Q(z,s)\exp-\left[\frac{\theta^2}{\Theta^2(z,s)}\right] \qquad (7.5.7)$$

If we substitute this in equation 7.5.5, we obtain the following result:

$$\frac{\partial Q(z,s)}{\partial z} + Q(z,s)\left(\frac{2\theta^2}{\Theta^3}\right)\left(\frac{\partial\Theta}{\partial z}\right)$$

$$= -A\sigma^2\theta^2 Q(z,s) + DQ(z,s)\left(\frac{-4}{\Theta^2} + \frac{4\theta^2}{\Theta^4}\right) \qquad (7.5.8)$$

Since this equation must hold for all values of θ, we can separate it into two separate equations as follows:

$$\frac{\partial\Theta}{\partial z} = \frac{-A\sigma^2\Theta^3}{2} + \frac{2D}{\Theta} \qquad (7.5.9)$$

$$\frac{\partial Q(z,s)}{\partial z} = \frac{-4D}{\Theta^2}Q(z,s) \qquad (7.5.10)$$

Evidently, when z is very large, $\partial\Theta/\partial z$ must tend to zero, so that we have

$$(\Theta'_\infty)^4 = \frac{4D}{A\sigma^2} = \frac{\Theta^4_\infty}{\sigma^2} \qquad (7.5.11)$$

while from equation 7.5.10 we can deduce that since, for very large z, $\Theta(z,s)$ is constant at Θ'_∞, $Q(z,s)$ will take the form

$$Q(z,s) = Q(s)\exp - (\gamma'_\infty z) \qquad (7.5.12)$$

where

$$\gamma'_\infty = 2\sqrt{\sigma AD} = \sqrt{\sigma}\,\gamma_\infty \qquad (7.5.13)$$

This tells us that in the time-dependent solution the far-field angles after a long fiber length and the measured loss will be slightly different in the case of continuous wave (CW) excitation ($s=0$ and hence $\sigma=1$) and in the case of pulse excitation and $\sigma\neq0$. Physically this discrepancy arises because the time spread of energy is related to the spatial spread of energy, large angle, high order modes becoming spatially separated along the z axis from low order modes. However, to retain uniformity of notation with the time-independent analysis and with Gloge's original work, we will use Θ_∞ and γ_∞ and insert the σ where appropriate.

To find the time- and length-dependent solution, we now return to equation 7.5.9 and note that, by setting $\psi=\Theta^2$, it can be written in the form

$$\frac{\partial\psi}{\partial z} = -4D\left(\frac{\sigma^2\psi^2}{\Theta^4_\infty + 1}\right) \qquad (7.5.14)$$

Noting that

$$\int (a^2 + b^2x^2)^{-1} dx = \frac{1}{ab}\arctan\left(\frac{bx}{a}\right)$$

we can arrive, after some algebra, at the desired result for $\Theta(z,s)$:

$$\Theta^2(z,s) = \frac{\Theta^2_\infty}{\sigma}\left[\frac{\sigma\Theta^2_0 + \Theta^2_\infty\tanh(\sigma\gamma_\infty z)}{\Theta^2_\infty + \sigma\Theta^2_0\tanh(\sigma\gamma_\infty z)}\right] \qquad (7.5.15)$$

In deriving this result, we have evaluated the constant of integration at $z=0$, postulating an initial power distribution among modes of the form

$$Q(\theta,z,s) = f(z,s)\exp\left(\frac{-\theta^2}{\Theta^2}\right) \qquad (7.5.16)$$

Equation 7.5.16 now describes the change in angular power distribution with distance along the fiber from the initially postulated distribution to the distribution when $z = \infty$, which is Gaussian with angular width: $\Theta'_\infty = \Theta_\infty / \sqrt{\sigma}$. Returning to equation 7.5.9, we can use equations 7.5.15 and 7.5.16 to rewrite it in the form

$$\frac{\partial f(z,s)}{\partial z} = \frac{-4D\sigma f(z,s)}{\Theta_\infty^2} \frac{\Theta_\infty^2 \cosh(\sigma\gamma_\infty z) + \sigma\Theta_0^2 \sinh(\sigma\gamma_\infty z)}{\sigma\Theta_0^2 \cosh(\sigma\gamma_\infty z) + \Theta_\infty^2 \sinh(\sigma\gamma_\infty z)} \quad (7.5.17)$$

This equation can be greatly simplified by writing it in the form

$$\frac{\partial f}{\partial z} = \left(\frac{-4D}{\Theta_\infty^2 \gamma_\infty}\right) \frac{f}{g} \frac{dg}{dz} \quad (7.5.18)$$

where

$$g = \sigma\Theta_0^2 \cosh(\sigma\gamma_\infty z) + \Theta_\infty^2 \sinh(\sigma\gamma_\infty z)$$

But $\Theta_\infty^2 \gamma_\infty = 4D$, so that we have

$$\frac{\partial f}{f} = -\frac{\partial g}{g}$$

or

$$\log f = -\log g + \text{constant}$$

It follows that, by simple substitution and the use of the boundary conditions for $z = 0$,

$$f(z,s) = \frac{f(0,s)\sigma\Theta_0^2}{\Theta_\infty^2 \sinh(\sigma\gamma_\infty z) + \sigma\Theta_0^2 \cosh(\sigma\gamma_\infty z)} \quad (7.5.19)$$

Clearly, if we take the case of CW excitation and $s = 0$, and launch a distribution of angular width Θ_∞, it will propagate unchanged (under the terms of this model), while any other distribution will flow smoothly into the distribution in question, given sufficient fiber length. To discover how the pulse shape develops, we must evaluate $f(z,s)$ for some chosen cases of interest. Several cases are analytically soluble. Those for which $\sigma\gamma_\infty z$ is either very large or very small allow simple approximations to be used in the place of the hyperbolic functions.

Thus, for the case in which we can approximate $\sinh x$ and $\tanh x$ by x and $\cosh x$ by 1 (i.e., small $\sigma \gamma_\infty z$), if we launch an angular distribution characteristic of the CW steady state ($\sigma = 0$), we find that

$$Q(\theta, z, s) = \frac{f(0, s)}{1 + \gamma_\infty z} \exp\left[-\theta^2 \left(\frac{1}{\Theta_\infty^2} + \frac{nzs}{2c} \right) \right] \qquad (7.5.20)$$

This can be transformed to give an expression for $P(\theta, z, t)$ of the form

$$P(\theta, z, t) = F\left(0, t - \frac{n\theta^2 z}{2c}\right) \exp -\left(\frac{\theta^2}{\Theta_\infty^2} \right)(1 + \gamma_\infty z)^{-1} \qquad (7.5.21)$$

This expression tells us the pulse shape for no mode coupling for the conditions of this model. The angular power distribution is conserved, but the power is attenuated by the term $(1 + \gamma_\infty z)^{-1}$, while the power propagating at the angle θ has been time delayed by an amount $(n\theta^2 z/2c)$. Exactly the same result would have been obtained by a simple ray analysis showing rays traveling at an angle, and calculating the additional path distance traveled. To see what pulse shape is produced we must postulate some starting pulse form, described by $F(0, t)$. If we choose to launch a delta function pulse, the Laplace transform of F is given by $f(0, s) = 1$, and we find that the power output as a function of time, after integrating over the angle θ, is given by

$$P(z, t) = \frac{2c\pi}{nz(1 + \gamma_\infty z)} \exp\left(-\frac{2ct}{n\Theta_\infty^2 z} \right) \qquad (7.5.22)$$

Obviously this will be valid only if the small argument expansion of the hyperbolic functions is valid. It should also be noted that the particular pulse shape predicted in this case arises not from mode coupling in the fiber, but from the assumption that a Gaussian distribution of power was launched among the guided modes and conserved for the measured length. Obviously this result is therefore extremely sensitive to changes in launching. Moreover, it is likely to be very difficult to reproduce experimentally unless one is launching into a fiber in which heavy mode coupling is deliberately introduced in the first few centimeters of length. Experimentally, this will have a negligible effect upon the temporal distribution of power (in the region of coupling), but it will rapidly generate a power distribution close to the stable distribution but very near the launch end.

The second analytically soluble case, that of large argument in the hyperbolic functions (i.e., large $\sigma \gamma_\infty z$) leads to the following expression for the pulse shape:

$$P(z,t) = \Theta_\infty^2 \sqrt{\frac{\pi}{Tt}} \left(\frac{t}{\gamma_\infty z T} + \frac{1}{2} \right)^{-1} \exp\left(-\frac{\gamma_\infty^2 z^2 T}{4t} - \frac{t}{T} \right) \quad (7.5.23)$$

where

$$T = \frac{n}{2cA} = \frac{n}{2c} \frac{\Theta_\infty^2}{\gamma_\infty}$$

The pulse shape predicted by this result is shown in Figure 7.2 for various values of $\gamma_\infty z$ as calculated by Gloge.[2]

Following a rather different argument but using the same notation, Gloge derived expressions for the mean pulse delay and the mean pulse width as a function of propagation distance, z, under the assumption that the launched power distribution is characterized by $\Theta_0 = \Theta_\infty$ as before. He found that

$$\delta_P = \frac{T}{2} \left\{ \gamma_\infty z + \frac{1}{2} \left[1 - \exp(-2\gamma_\infty z) \right] \right\} \quad (7.5.24)$$

$$\tau_P = \frac{T}{2} \left\{ \gamma_\infty z \left[1 - 2\exp(-2\gamma_\infty z) \right] + \frac{3}{4} \right.$$

$$\left. - \exp(-2\gamma_\infty z) + \frac{1}{4} \exp(-4\gamma_\infty z) \right\}^{1/2} \quad (7.5.25)$$

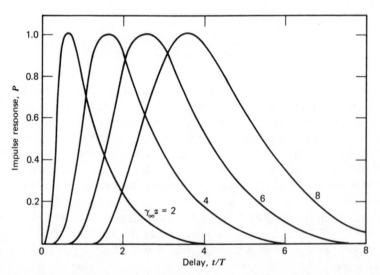

Figure 7.2. Pulse shapes for various degrees of mode coupling after unit distance in a step-index fiber. Reprinted with permission from D. Gloge, *Bell Syst. Tech. J.*, **52**, 801 (1973), copyright 1973, The American Telephone and Telegraph Company.

Here δ_p is the mean time delay of the pulse, and τ_p is the mean pulse width. For small $\gamma_\infty z$ the pulse width is given by

$$\tau = T\gamma_\infty z \qquad (7.5.26)$$

while for large values of $\gamma_\infty z$ it develops as

$$\tau = \frac{T}{2}\sqrt{\gamma_\infty z} \qquad (7.5.27)$$

showing that, under conditions in which mode coupling predominates, the pulse width develops as the square root of the distance. If we approximate the true pulse width versus distance curve by two straight line approximations (given by equations 7.5.26 and 7.5.27, we find that the turnover point from linear to square root dependence is given by $z_T = 1/4\gamma_\infty$. We note, therefore, that if the additional loss of a fiber due to mode coupling can be measured, the pulse width versus length can be fully predicted, assuming that the above theoretical model provides a good fit to practice. The additional loss (in dB/km) is given by $h = 4.35\gamma_\infty$, so that the turning point from linear to square root pulse broadening occurs at

$$z_T = \frac{1}{4\gamma_\infty} \approx \frac{1.1}{h} \text{ km} \qquad (7.5.28)$$

In evaluating the formulas above, the parameter T is critical. It was defined as

$$T = \frac{n}{2cA} = \frac{n}{2c}\frac{\Theta_\infty^2}{\gamma_\infty} = \frac{4.35n}{2c}\frac{\Theta_\infty^2}{h}$$

Both h and Θ_∞ must be found experimentally since they cannot be calculated without access to the intimate details of the mode coupling perturbation, about which nothing of any detail has been said. However, with this approach a good first order agreement is found between the measured pulse widths, the fiber lengths, the additional loss, and the far-field angle of radiation from a fiber. Detailed investigation, however, raises problems. It is generally not the case that the angle Θ_∞ is very much less than the acceptance angle of the fiber, with the power distribution among modes falling sharply to the highest order modes and the angle Θ_∞ being perhaps half of the acceptance angle. This means that in detail this model is too crude. However, it has the immense virtue of being mathematically soluble and valuable for allowing one to develop some picture of the power flow under these conditions.

Before leaving the Gloge model for mode coupled propagation, we will note several other results which, despite the limitations of the approach mentioned above, help to explain the effects that we have been discussing. Returning to equations 7.5.7, 7.5.15, and 7.5.19 and setting the launched distribution to the value

$$\Theta_0 = \frac{\Theta_\infty}{\sigma} \tag{7.5.29}$$

we obtain the following particularly simple relations:

$$Q(\theta, z, s) = f(z, s) \exp\left(-\frac{\sigma \theta^2}{\Theta_\infty^2}\right) \tag{7.5.30}$$

$$f(z, s) = f(0, s) \exp(-\sigma \gamma_\infty z) \tag{7.5.31}$$

$$f(0, s) = 1 \qquad \text{for the delta function}$$

so that

$$Q(\theta, z, s) = \exp\left[-\sigma\left(\frac{\theta^2}{\Theta_\infty^2} + \gamma_\infty z\right)\right] \tag{7.5.32}$$

By using the substitution $s' = 1 + Ts = \sigma^2$, we can reduce equation 7.5.32 to a standard form having the Laplace inverse transform:

$$P(\theta, z, t) = \left(\frac{g}{2}\right)\sqrt{\frac{T}{\pi t^3}} \ \exp\left(-\frac{t}{T} - \frac{g^2 T}{4t}\right) \tag{7.5.33}$$

where

$$g = \frac{\theta^2}{\Theta_\infty^2} + \gamma_\infty z$$

Integrating this result with respect to θ to find the total power in the pulse as a function of time yields the result

$$P(z, t) = \int 2\pi P(\theta, z, t)\theta \, d\theta$$

$$= \Theta_\infty^2 \sqrt{\frac{\pi}{tT}} \ \exp\left(-\frac{t}{T} - \frac{\gamma_\infty^2 z^2 T}{4t}\right) \tag{7.5.34}$$

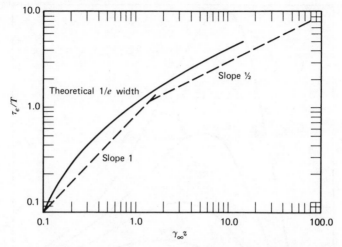

Figure 7.3. The $1/e$ pulse widths for pulse propagation in a fiber having a value of $\gamma_\infty = 1$ versus distance, z, showing the change from linear to square root length dependence.

This result yields the same solutions as equation 7.5.23 in the region of $t/T < 1$, which is where the main pulse energy occurs, but predicts slightly different results for the tails, which remain launching dependent. However, it does have the attraction that it can be used to describe the pulse from the point of launch, through the linear growth region, and into the square root growth region.[4] In Figure 7.3 we plot the $1/e$ widths for the pulses predicted by the result, equation 7.5.34, versus distance, clearly showing the turnover. The steeper than linear rise in the very short length region is associated with the use of the $1/e$ width rather than an average width. The significance is appreciated by reference to Figure 7.2, in which the actual pulse shapes are plotted, normalized to give comparable peak heights.

In Figure 7.4 we show the values of $P(\theta, z, t)$ calculated from equation 7.5.33 for different values of θ^2/Θ_∞^2 and with $\gamma_\infty z = 4$. We see that the gradient of power is very much greater across the leading edge of the pulse than across the trailing edge. This is made clearer in Figure 7.5, which shows the values of $P(\theta, z, t)$ plotted versus θ^2/Θ_∞^2 for different values of t/T. Also shown on the same figure is the angular falloff associated with the stable distribution, $\exp -(\theta^2/\Theta_\infty^2)$. Clearly, at all points in the pulse there is a negative gradient of P with respect to θ, leading at all points to a net outward diffusion of power from the pulse. However, if we remove the mean outward flow pattern and concentrate only on the variation of the flow patterns across the pulse, we can see that a circulating power flow is present. This is illustrated in Figure 7.6. It operates as follows.

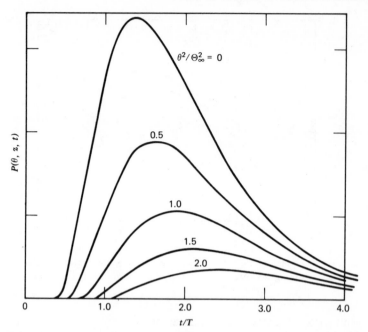

Figure 7.4. Values of the power in a step-index fiber versus time for $\gamma_\infty z = 4$, and for different values of the propagation angle. Reproduced with permission from J. E. Midwinter, *Opt. Quant. Electron.*, **8**, 531 (1976).

Power traveling in the low angle modes travels faster and tends to forge ahead of the main pulse. As a result, it soon finds that the adjoining modes at higher angles have less power than would otherwise be expected, leading to a preferential diffusion to them. This diffused power now travels more slowly and tends to fall back into the main body of the pulse. At the same time, the energy in high angle modes tends to fall behind the main pulse, finds less power than expected in the low angle modes, and preferentially diffuses into them, thus tending to catch up with the main body of the pulse. Thus preferential diffusion always tends to clip the leading and trailing edges of the propagating pulse, thus shortening it. The particular characteristic of the stable pulse distribution, with width growing as $\sqrt{L}$, is not just that overall it maintains the "right" distribution of power between angles, but also that the pulse of power in each mode is correctly phased relative to its neighbors in time and space, so that the preferential pulse clipping takes place.

We see from this picture of pulse shortening that the linear growth region of pulse width with distance after launching is obviously not fundamental but arises only while the power redistributes in time into the

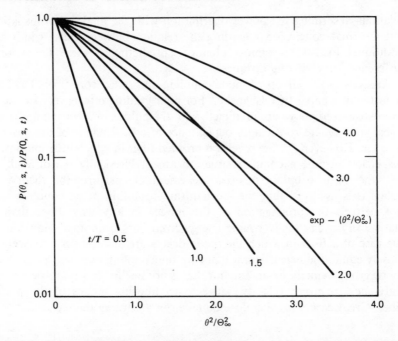

Figure 7.5. Values of the power at angle θ versus θ for $\gamma_\infty z = 4$ for various points in the time pulse, showing the gradient of power varying with mode or angle as the pulse propagates. Reproduced with permission from J. E. Midwinter, *Opt. Quant. Electron.*, **8**, 531 (1976).

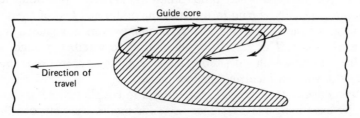

Figure 7.6. Schematic of the circulating pulse power distribution under conditions of mode coupling.

distribution described above.[4] However, to launch the above time-space distribution would be exceedingly difficult, using a simple source and lenses.

A different model describing power flow has been developed by Marcuse[5] in which he assumes coupling between modes, described by a diffusion equation, but applies the boundary condition that the power at the fiber boundary is identically zero. This leads to Bessel function relations describing the power distribution with the angle of propagation, but the final conclusion regarding the relation between pulse width and additional loss, h, is almost identical, predicting a mere factor of 2 difference for very long distances.

Analysis of the effects of mode coupling on graded core fibers has been reported in a paper by Olshansky.[6] For poor grading effects very similar to those discussed above are predicted, with the pulse spread changing from a linear to a square root length dependence after a distance of the order of $1/h$ km. However, for the precision profiled fiber in which little multimode dispersion occurs, a complex situation arises. Physically the difficulty is that the mode coupling theories are essentially perturbation theories in which it is assumed that the mechanism promoting mode coupling does not perturb the propagation of the modes in any way other than to statistically occasionally couple energy from one to another. However, in the case of a fiber in which the mode delays are already very precisely or closely equal, the perturbation inducing the coupling may well perturb the mode group velocity by an amount that is not negligible with respect to the difference in group velocity between neighboring modes of the perfect fiber. Great care must therefore be taken in analyzing these cases.

7.6. CONTROLLED MODE COUPLING

The preceding analysis has shown that even randomly induced mode coupling can lead to a beneficial effect in a step-index fiber, since the pulse spreading is reduced when propagation occurs over very long lengths. However, a penalty is paid in terms of the additional loss of energy from the fiber, arising out of the energy coupling from the guided into the unguided modes. This raises the question as to whether it might not be possible to induce the desirable mode coupling to increase the fiber bandwidth without incurring the loss penalty.

If we consider the change in ^zk required to couple two modes together, then, by examining the difference in ^zk for modes characterized by m, l, and $m+\Delta m$, $l+\Delta l$, we find, using equation 7.3.3, that

$$\Delta(^zk)_{\Delta m \Delta l} \simeq \frac{M\pi^2}{8k_0 a^2}(2\Delta m + \Delta l)\left(1 + \frac{M^2\pi^2}{8nk_0^2 a^2}\right)^{-1/2} \qquad (7.6.1)$$

where $M = 2m + l$ for the LP_{lm} mode. This shows that the spacing between mode groups, M, increases as M increases. Thus, if we limit the perturbation of the fiber to the $\cos\phi$ type already discussed, so that we are involved only with $\Delta M = \Delta(2m + l) = \pm 1$, it is conceptually possible to prevent coupling to the highest order modes of the fiber by restricting the spectrum of $\Delta^z k$ values to all values up to but not including the one for the cutoff mode, M_c. Thus all frequencies can be present up to

$$\Delta(^z k)_{co} = \frac{M_c \pi^2}{4k_0 a^2} \sqrt{1 + \frac{M_c^2 \pi^2}{8n k_0^2 a^2}}$$

If such a perturbation spectrum could be achieved by modulating the fiber lay along the z direction with sufficient accuracy, all modes of the fiber would be intercoupled up to but excluding the modes closest to cutoff. If they were not coupled to the other guided modes, energy should not be coupled through them to the nonguided cladding modes, and no additional loss should be experienced, while the bandwidth should be increased because of the interchange of energy between guided modes. It must be said, however, that experimentally it looks extremely difficult to achieve such a precisely controlled fiber environment, and at the time of writing no reports of its experimental testing are known.

REFERENCES

1. W. Streiffer and C. N. Kurtz, *J. Opt. Soc. Am.*, **57**, 779 (1967).
2. D. Gloge, *Bell Syst. Tech. J.*, **52**, 801 (1973).
3. W. A. Gambling, D. N. Payne, and H. Matsumura, *Appl. Opt.*, **14**, 1538 (1975).
4. J. E. Midwinter, *Opt. Quant. Electron.*, **8**, 531 (1976).
5. D. Marcuse, *Theory of Dielectric Optical Waveguides*. New York: Academic, 1974.
6. R. Olshansky, *Appl. Opt.*, **14**, 935 (1975).

8

Materials for Optical Fibers

8.1. INTRODUCTION

We saw in the foregoing theoretical analyses that to produce a guiding fiber a composite structure must be made consisting of a core surrounded by a cladding material of lower refractive index. The cladding helps to confine the electromagnetic energy largely to the core region of the structure by total reflection at the core-cladding interface. There is some penetration into the cladding, as discussed under the heading "Goos-Haenchen Shift" and elsewhere in the earlier chapters. However, by making the cladding sufficiently thick, it is possible to produce a fiber in which the field of the guided modes is negligible at the outer surface of the cladding, so that propagation is largely unaffected by the surrounding material, such as a plastic extrusion for cable makeup. (This is true only if the coating or surrounding does not physically distort the fiber. Bending and similar effects give rise to scatter losses.) Cladding thicknesses of 20 to 50 microns are typical for optical fibers, with core diameters ranging from 3 up to 80 microns.

The choice of materials to be used in the fabrication of such fibers is influenced by the need to satisfy simultaneously many requirements. Obviously the material must be formable into a fine filament, transparent and available with two different refractive indices for core and cladding, respectively. These requirements alone more or less limit the field to plastics or glasses, although a liquid has been used to form the core of a fiber drawn from a hollow tube of glass. Many plastics are excluded from further consideration because the presence of hydrogen in their structures gives rise to very high losses and because their molecular size leads to large scattering losses. And within the infinite number of possible glasses, most are ruled out by other considerations. To appreciate this situation more fully, we need to examine in more detail the physical mechanisms involved, particularly those controlling optical loss since most optical communications systems require fibers of exceptionally low attenuation at the optical carrier frequency. Usually less than 20 dB/km is sought.

128

8.2. GLASS MATERIALS

Glasses are formed from fused mixtures of metal oxides, sulfides, or selenides. Because they are fused mixtures rather than fixed compounds with crystal structures, their compositions are infinitely variable within certain regions of their respective phase diagrams, and large numbers of different glasses are manufactured by industry. Most of these fall into the category of oxide glasses, since these are the optically transparent ones, the sulfide and selenide glasses being used in the infrared region from approximately 0.6 micron to 14 microns or more.

Of the oxide glasses, by far the most common are silica (SiO_2); sodium calcium silicates, frequently used for plate and window glass; sodium borosilicates, often used for oven ware and chemical apparatus; and lead silicates, which are the crystal glasses having relatively high refractive indices and thus appearing "shiny."

Typical starting materials for these glasses are sodium and calcium carbonates, boric oxide (B_2O_3) or boric acid, silica (sand), and lead oxide. For the optical fiber materials, new conditions arise, namely, the need for very low optical loss, which means very high chemical purity, so that the materials sources used are usually different from those serving large scale industry.

This has led to a great deal of interest in glasses which can be prepared directly from the vapor phase rather than by fusing mixtures of oxide powders. Such glasses are usually very high in silica content and are produced by the reaction with oxygen of silane (SiH_4) or silicon tetrachloride ($SiCl_4$), with small amounts of dopants present in the gas stream to modify the refractive index of the material so produced. In this chapter we will not discuss the preparation of these materials since they are usually produced directly as composite preforms and are discussed as such in Chapter 9.

The structure of glasses is noticeably different from that of the solids usually encountered in the electronics industry, namely, crystals. In the latter the individual atoms are well defined in space according to very precise and repeated patterns, lying in exact three-dimensional lattices. Glass, on the other hand, consists of a loosely connected network formed by groups, which can be added to or modified by other components.

For example, the addition of sodium (a network modifier) tends to break up the SiO_2 network, as shown in Figure 8.1, in a sodium silicate glass. The result is less strongly bound than pure silica, and the melting temperature is thereby lowered. Since B_2O_3 is also a network former like SiO_2, a series of glasses such as sodium borates exist, paralleling sodium silicates but having very much lower melting temperatures since the B-O bond is much weaker than the Si-O bond. In Figure 8.2 we show the main network forming and network modifying glass components.

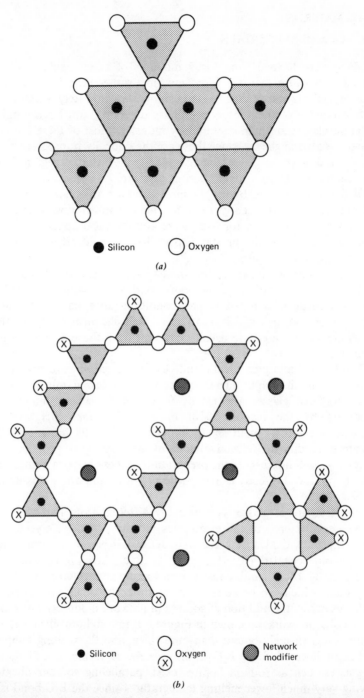

Figure 8.1. Networks involving SiO_2 groups, shown schematically in two dimensions. (a) A regular SiO_2 lattice (two-dimensional quartz crystal). (b) The effect on the lattice of the addition of a network modifier.

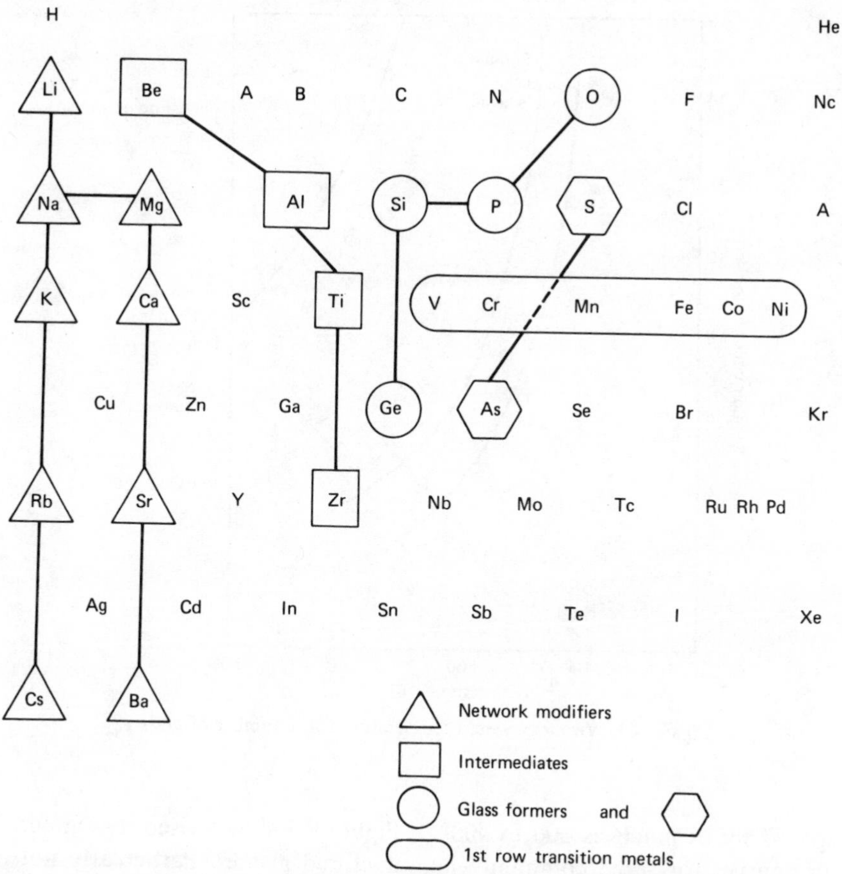

Figure 8.2. Periodic table of the elements, highlighting the major glass forming elements with network modifiers and intermediate elements. Also shown are the first row transition metals which give rise to strong glass coloration or fiber attenuation by absorption....

Another result of the randomly connected network structure of glasses is that there is not a precise melting temperature, as in the case of many crystalline substances. Instead, one finds that the viscosity of the glass varies smoothly over many orders of magnitude through a temperature range of many hundreds of degrees. It is thus a matter of definition as to where melting occurs, and in practice a number of temperatures are defined. These are the strain temperature ($\eta = 10^{14.5}$ poise), and anneal temperature ($\eta = 10^{13.5}$ poise), the softening temperature ($\eta = 10^{7.5}$ poise), and the working temperature are roughly $\eta = 10^4$ poise. These may be

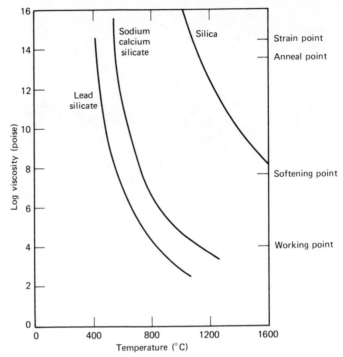

Figure 8.3. Viscosity versus temperature for a number of glasses.

spaced by as much as 300 to 400°C. Figure 8.3 shows viscosity-temperature curves for some commonly manufactured glasses. Particularly noteworthy is the fact that the viscosity of a pure silica glass is much higher
than that of multicomponent glasses at the same temperature.

Just as the viscosity of a glass is a function of its composition, so too are
the refractive index and the thermal expansion coefficient. Since there is an
infinity of glass compositions, full data are not available for all glasses; in
fact, comprehensive data are restricted to a few glass groups that have
been extensively studied because of large scale commercial applications.
The sodium calcium silicate (NCS) group is one such, and Figure 8.4
shows the curves of constant expansion coefficient and constant refractive
index superimposed upon the phase diagram.[1] This highlights an attractive
feature of these glasses; in addition to there actually being published data,
Figure 8.4 shows that the refractive index and expansion lines run almost
at right angles, allowing the index to be varied for a core-cladding pair
while maintaining the same expansion coefficients to ensure a low strain
fiber. However, as shown by the viscosity-temperature curves in Figure 8.3,

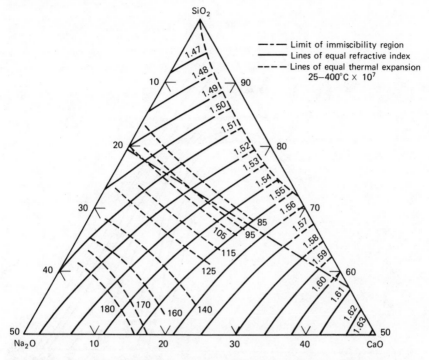

Figure 8.4. Curves of refractive index and thermal expansion coefficient versus composition for the sodium calcium silicate system. Reproduced with permission of B. Scott and H. Rawson, University of Sheffield, and using data taken from Ref. 3.

this glass system has to be worked at relatively high temperatures, and this fact has presented serious practical impediments to its use.

A second glass system of great current interest is the sodium borosilicate (NBS) group of glasses. The phase diagram[2] for this system shown in Figure 8.5 illustrates several points. There is a large region of the phase diagram, toward the sodium-rich region, in which glasses do not form at all but crystalline phases will separate out.[2] Then in addition, in the very low sodium region and for a wide range of silica/boric oxide ratios, there is a phase separating region. Here the glass formed is unstable and has a strong tendency to separate into two more stable glasses, one with high silica content and one with high boric oxide content. In such glasses the two phase regions grow slowly, and the size of the microdroplets of each glass depends very much on the composition of the starting mixture, the temperature, and the time it is held there. Thus, in a fiber puller, when the glass is very rapidly cooled from molten to solid state, phase separation may not be visible to the eye, whereas if the same composition were cooled

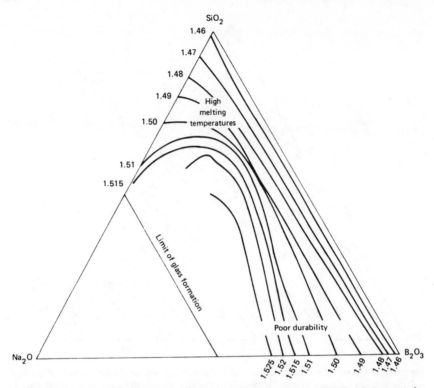

Figure 8.5. Curves of (*a*) refractive index, (*b*) thermal expansion, and (*c*) the glass properties versus composition for the sodium boro-silicate system. Reproduced with permission from (*a*) G. R. Newns, unpublished work, (*c*) T. Inoue, K. Koizumi, and Y. Ikeda, *Proc. IEE*, **123**, 577 (1976). Curve (*b*) plotted from data in Ref. 3.

slowly, separation would occur and the glass would appear milky white from the light scattered by the phase (refractive index) boundaries. Obviously, these composition regions should be avoided. Thus, in the NBSs, a limited range of acceptable compositions exists. Within this region the published data on refractive index and expansion coefficient are somewhat sketchy and may well not be entirely reliable. However, the data that are available[3, 14] are shown in Figure 8.5, which suggests that the freedom to modify the index, while holding the expansion coefficient constant, that existed in the NCSs is far more limited in these glasses. However, curves for viscosity versus temperature for a number of NBS glasses show that very much lower working temperatures are possible in this system, a factor of importance in making low loss fibers with them.

Another group of glasses that may be of interest in fiber manufacture is the lead silicates. These are of potential interest since they allow large refractive index differences to be obtained between core and cladding

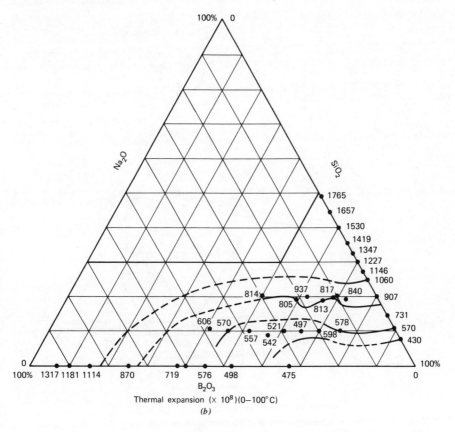

Figure 8.5. (b)

glasses. For example, a pair with indices of 1.5 and 1.65 showing a 10% index difference can probably be obtained by using the highly polarizable lead ions to dope the higher index material. Fibers with such large index difference will be characterized by a higher level of Rayleigh scattering from the lead (of the order of 10 dB/km, compared to 1 dB/km for silica at 900 nm) and have not yet been produced with very low loss (i.e., under 15 or 20 dB/km). But for many short link systems applications these disadvantages are more than offset by the large acceptance angle for such a high index difference fiber.

All these three groups of glasses are of interest because they can be made by nearly conventional glass melting techniques in large quantities at low cost. They are formed by mixing the appropriate powders and fusing them in a crucible to form a glass. But there is another group of materials, as already noted, all based upon pure silica with small additions of one or

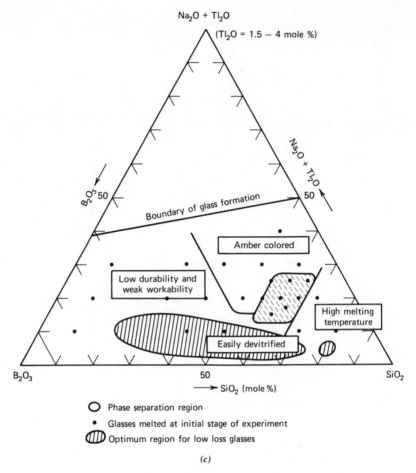

Figure 8.5. (c)

more other oxides. These materials are worked at much higher temperatures (1500 to 2000°C, compared to 850 to 1100°C). The disincentive to their use offered by the higher working temperatures is largely offset by the fact that they can be deposited in the form required directly from the vapor phase, thus minimizing the need to handle or work them. The result is that these materials are of great interest for fibers, particularly where exceptionally low loss is required (see Section 9.4).

We now turn our attention to the optical loss mechanisms common in glasses in the visible region of the spectrum. These fall into two general categories, radiative and absorptive. We discuss them separately since they arise from quite different sources.

8.3. ABSORPTIVE LOSSES IN GLASSES

Optical fiber systems choose to operate in one of two wavelength bands, the GaAs device region from 800 to 900 nm or the Nd laser region at 1.06 microns, with a remote possibility of an extension to perhaps 1.2 or 1.3 microns using other semiconductor sources. This choice is dictated by good physical reasons.

Wavelengths shorter than about 600 to 650 nm are ruled out because Rayleigh scattering in the glass becomes severe and impurity absorption spilling over the band edge absorption is a problem. Beyond about 1.3 microns the first overtone of the OH stretching vibration appears at 1.4 microns, giving heavy absorption. Some spillover from multiphonon bands from the glass constituents probably adds to this, and detectors become far less efficient because the quantum energy $h\nu$ rapidly approaches kT under conditions of room temperature operation. Thus we are concerned only with mechanisms that contribute to loss in the narrow wavelength band of 600 to about 1300 nm (1.3 microns).

8.3.1 Band Edge Absorption

Five absorption mechanisms are of concern in a glass that is considered for use in optical fibers. The first two are associated with the basic glass constituents themselves, typically a combination of oxides of silicon, sodium, boron, calcium, germanium, and so on. The glass has a band edge absorption somewhere in the ultraviolet region of the spectrum. Such absorption is extremely intense; and although the wavelength of interest for the operation of a system is a considerable distance away, there has been serious debate as to whether the tail of the band edge could provide a significant loss mechanism. Current evidence suggests (and in some cases proves) that it does not.

For example, we may note a paper by Pinnow et al.[4] in which the authors attempt to establish the ultimate lower limit of absorption in both pure silica and a soda lime silicate glass. They report the results of measurements of the ultraviolet absorption edge and note that they can be fitted to an exponentially decreasing loss with wavelength. They show for silica that points measured in the region of 0.5 to 1.0 micron in fiber and points measured in the region of 0.2 to 0.3 micron in bulk samples fall on the same curve. From this they conclude that the lower absorption limit for silica at 0.8 micron is about 2 dB/km and at 0.5 micron is about 10 dB/km. However, recent results with germanium silicate fibers and phosphosilicate fibers have shown that both of these figures are wrong, since the sum of absorption and scatter together in the fiber at these wavelengths is less than the absorption estimated.

We draw two conclusions. In the region of 0.8 to 1.1 microns, the band edge absorption component is almost certainly less than 1 dB/km in the silica and glass materials most commonly studied. Precise measurements of the intrinsic band edge are exceedingly difficult to make because impurity absorptions usually mask it. However, absorption from it is unlikely to be a problem unless a system is to be operated at a much shorter wavelength, and this in itself is most unlikely since Rayleigh scattering and impurity absorption then become more serious problems. This more optimistic view is further supported by the work of Bagley and French,[5] who estimated an upper limit of absorption due to band edge spillover in the GaAs device region for silica of <0.01 dB/km, using extrapolation of band edge measurements made much further into the vacuum ultraviolet region than those of Pinnow et al.[4]

8.3.2 Infrared Absorption from Vibrations of Glass Constituents

The individual constituents of a glass are bonded together chemically. Thermal energy maintains them in a constant state of random motion so that any particular bond, say a SiO bond in the glass, is continually vibrating. Associated with the bond is an electric dipole whose strength is modulated by the vibration, and this allows interaction to occur between the electric vector of an electromagnetic field and the bond, resulting in transfer of energy from the field to the structure, manifesting itself as absorption. These absorption mechanisms are very strong, as would be expected from the very large number of bonds present. Each bond oscillates at a characteristic frequency, and at that frequency a line absorption is seen. However, these lines have finite width and spillover to the neighboring wavelengths. But the evidence suggests at present that these absorptions, like the band edge absorption, are of no significance for the optical fiber system and become serious only beyond about 1.5 micron wavelength.

We list in Table 8.1 the characteristic frequencies for the stretching vibrations associated with some of the common bonds present in optical glasses, and note that they all fall well beyond the O-H bond frequencies.

TABLE 8.1 CHARACTERISTIC STRETCHING FREQUENCIES OF BONDS MET IN FIBER GLASSES

Bond	Wavelength (microns)	Frequency (cm^{-1})
Si-O	9.0	1100
B-O	7.3	1370
P-O	8.0	1250
Ge-O	11.0	910

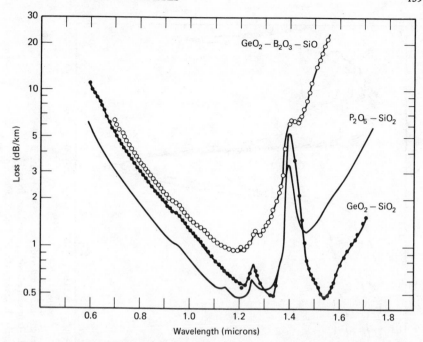

Figure 8.6. Losses versus wavelength of some low water content fibers showing the matrix influence on the infrared transmission. Reproduced with permission from H. Osonai, T. Shioda, T. Moriyama, S. Araki, M. Horiguchi, T. Izawa, and H. Takata, *Electron. Lett.*, **12**, 550 (1976).

Figure 8.6 shows the losses of some fibers with very low water contents made with different constituents in the core by the chemical vapor deposition technique[6], and in Figure 8.7 the infrared spectra of some of these same materials, showing the relative size and position of the infrared absorptions, are presented. The work suggests that GeO_2 doping for the core is the most favorable because of the longer wavelength at which the GeO_2 stretching vibration occurs, and this leads to an estimate[6] for the lowest loss for this type of fiber, at about 1.5 microns, of about 0.3 dB/km, as shown in Figure 8.8. Evidently, to obtain figures even remotely approaching these, very low water contents must be achieved.

8.3.3 Transition Metal Ion Contamination

The remaining absorption mechanisms are all related to impurities or defects in the glass and are therefore not intrinsic. The mechanism that has given the greatest trouble to date is absorption by traces of the transition metals in the glass as impurities. In Figure 8.2 we showed the periodic table and indicated the transition metal elements causing the most trouble,

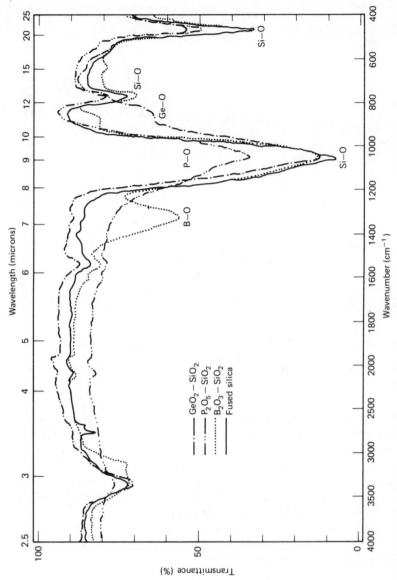

Figure 8.7. Infrared transmission spectra of the materials associated with the fibers of Figure 8.6, showing the various absorption bands involved. Reproduced with permission from H. Osonai, T. Shioda, T. Moriyama, S. Araki, M. Horiguchi, T. Izawa, and H. Takata, *Electron. Lett.*, **12**, 550 (1976).

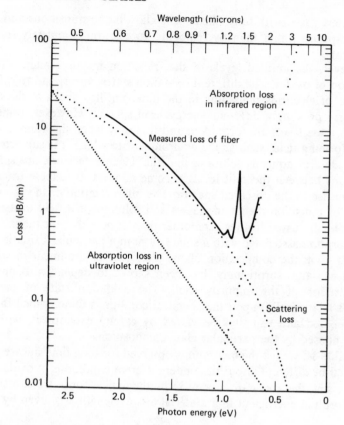

Figure 8.8. Predicted losses arising from both the ultraviolet edge absorption and the infrared absorption for a GeO_2-SiO_2 fiber. Reproduced with permission from M. Horiguchi, *Electron. Lett.*, **12**, 311 (1976).

namely, the elements appearing in the first row. These all give rise to broad, intense absorptions, mostly entering the wavelength band of interest for our system. The later transition metal elements occur at very low levels and also generally have narrow line absorptions that are less troublesome. In practice they are not usually observed at all in glasses produced for optical communications.

The reason for these absorptions by the transition metals is that the latter have incompletely filled inner electron shells. Transitions between levels of the unfilled shells give rise to their characteristic absorptions, just as the sodium or lithium atom shows a hydrogen-like spectrum from its single electron in its unfilled outer shell. The chemical bonding process involved in forming, say, the oxide with metals normally pairs the outermost electron(s) so that both atoms achieve an electron status akin to that of the inert gases, with fully filled shells allowing no transitions and hence no absorptions (other than by removing an electron completely from the structure).

However, the unfilled levels of the transition metals remain after the formation of oxides. The different oxidation states correspond to different numbers of electrons remaining in the inner unfilled shells of the atoms, with correspondingly different energy level spectra. However, transitions between these levels are forbidden, and it is only when the atom is placed in a deforming field, such as the environment to which it is subjected in a silicate glass or aqueous solution, that the degenerate levels are split and transitions between the split levels become allowed. It is these transitions that give rise to the observed spectra. The interpretation of the absorption spectra of transition metals in glasses and solutions is a field in itself and will not be discussed further here, other than to note that the magnitude of the absorption associated with a single atom in a particular oxidation state depends upon the composition of the glass (the environment in which it finds itself) and, surprisingly, has been shown to depend also upon the concentration of the impurity itself. The oxidation state of the atom depends (statistically) upon the chemical conditions under which the glass was prepared and can thus be varied by careful experiment, within the limits imposed by the particular glass composition.

In Table 8.2 we list the colorations observed for soda lime silicate glasses produced by different impurities under different conditions of oxidation.

Expressing the above discussion in mathematical form, we can say that the absorption coefficient in a material at wavelength λ is given by

$$\alpha(\lambda) = \sum_{\text{ions}} C_M \left(\frac{C_M^{n+}}{C_M} \right) \epsilon_{M^{n+}} S(\lambda)$$

Here the concentration of the ion M in the material is C_M, and the proportion in the $n+$ state is given by (C_M^{n+} / C_M), the redox ratio. Also $\epsilon_{M^{n+}}$ is the extinction coefficient for the ion, the absorption at the peak of the absorption line for unity concentration, while $S(\lambda)$ describes the shape of the absorption spectrum and is a factor $\leqslant 1$. The value of $\epsilon_{M^{n+}}$ is

TABLE 8.2   COLORATIONS OBSERVED BY EYE IN HEAVILY DOPED GLASSES

Dopant	Oxidized		Reduced	
Fe	Yellow to brown	Fe^{3+}	Green to blue	Fe^{2+}
Cu	Blue	Cu^{2+}	—	
Mn	Purple	Mn^{3+}	—	
Ni	Brown to green	Ni^{2+}	—	
Co	Purple to blue	Co^{2+}	—	
Cr	Green	Cr^{6+}	Yellow	Cr^{3+}

dependent upon melting conditions, glass composition, and also impurity concentration. In Figure 8.9 we show some absorption curves for the most serious impurities in a sodium calcium silicate glass.[7] Since the glass was melted in an air atmosphere, the impurities present will tend to be present in the oxidized states, where a potential exists. The losses are plotted in units of decibels per kilometer per part per million. It is immediately apparent that with this glass system, if a loss by absorption of 10 dB/km at 900 nm was to be obtained, all the impurities and, in particular, iron and copper would have to be held to an exceptionally low level, typically a few parts per billion. The NCS glasses were extensively studied by numerous groups for use as optical fiber materials, partly because more data are probably available on this glass system in the published literature than on almost any other system. However, for a variety of reasons, interest has tended to center in recent years on the sodium borosilicate glasses.

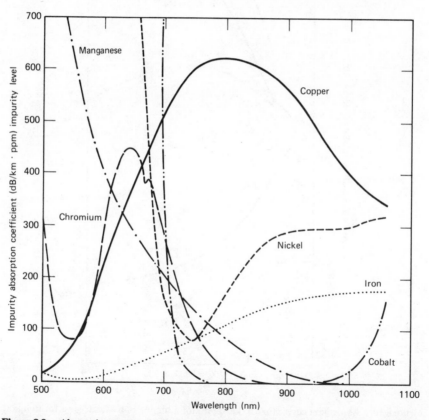

Figure 8.9. Absorption spectra for various transition metal ions in a sodium calcium silicate glass. Reproduced with permission from K. J. Beales, J. E. Midwinter, G. R. Newns, and C. R. Day, *Post Off. Elec. Eng. J.*, **67**, 80 (1974).

One of several reasons for this emphasis is illustrated in Figure 8.10, which shows the absorption curves for iron and copper in a typical glass of the NBS system.[8] Loss curves are given for the glass when melted in oxidizing and in reducing atmospheres. These make apparent two features: the loss for iron (in dB/(km·ppm)) is less than for the NCS glass; and by controlling the melting conditions, the relative absorption due to iron and copper can be varied over a great range. It is relatively easy to see that, for a given concentration of each impurity, there will be an optimum degree of

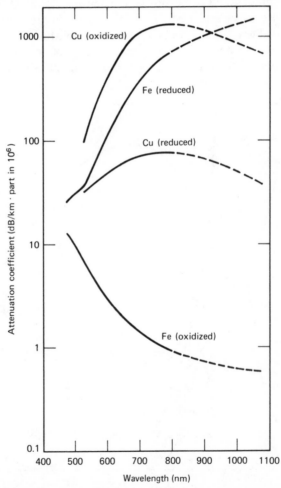

Figure 8.10. Absorption curves (in dB/km·ppm) for a sodium borosilicate glass for iron and copper when melted under reducing and oxidizing conditions. Reproduced with permission from K. J. Beales, C. R. Day, W. J. Duncan, J. E. Midwinter and G. R. Newns, *Proc. IEE,* **123,** 591 (1976).

oxidation or reduction for the minimum loss in the material, at a chosen wavelength. The production of this minimum loss and the stabilization of the glass in that state are the subjects of intensive research and development, and no details have been published as yet of the techniques involved. However, an indication of the success of this approach[14] is given in Figure 8.11, which shows, for the same NBS glass composition and the same impurity contents (of iron and copper, the most serious contaminants), the loss of the glass when melted under reducing, oxidizing, and balanced conditions. This figure also makes clear that one can make a useful estimate of the balance obtained between oxidation and reduction by looking at the form of the absorption curve. If the glass is too reduced, the absorption curve tends to rise significantly at long wavelengths; if the opposite is true, it tends to rise more at short wavelengths (~800 nm and less). The flat curve in the region of 800 to 900 nm is close to the optimum for that range with the balance of impurities commonly found. An analysis of the impurities in a typical 20 dB/km NBS glass is given in Table 8.3.

The transition metal ion impurities in pure silica give rise to very similar absorption spectra. Curves directly comparable to those for the NCS and NBS glasses are given for silica in Figure 8.12, taken from the paper by Schultz.[9] Particularly striking is the very low level ascribed to copper in silica when compared to NBS and NCS glasses, while the role of vanadium shows up as being of much greater significance.

In all cases control of the optical loss in the glass occurs, once the impurity level is fixed, through the balancing of a chemical reaction which is driven by the partial pressure of the oxygen in the melt. The reaction is summarized by the following equations, which indicate the reversible nature of the reaction:

$$M^{n+} + \tfrac{1}{2}O \rightleftharpoons M^{(n+1)+} + \tfrac{1}{2}O^{2-}$$

TABLE 8.3 ANALYSIS OF IMPURITY CONTENT OF A GLASS WITH LOSS IN 20 TO 30 DB/KM RANGE[a]

Impurity	Content (parts in 10^6)
Fe	0.18–0.24
Cu	<0.01
Ni	<0.01
Mn	<0.01
Co	<0.01
Cr	<0.05

[a]Ref. 15.

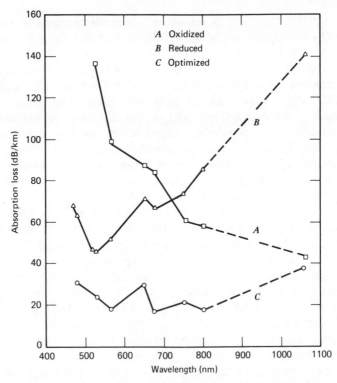

Figure 8.11. Attenuation curves for a glass containing iron and copper in the ratios that might occur from impurity contamination, showing the loss varying with melting conditions. Reproduced with permission of G. R. Newns, unpublished work.

In switching from a strongly reducing melting condition to a strongly oxidizing one, the partial pressure of oxygen may be varied by as much as 16 orders of magnitude. The effect of the reaction is to convert one ion species to another, such as ferrous to ferric, cuprous to cupric, or vice versa. The measured absorption due to a given concentration of a particular element thus changes as the proportion of one oxidation state changes in relation to the other. Prediction of the optical loss in a glass for a given concentration of total element is complicated by the fact that, in addition to the need to know the relative proportions of the different oxidation states present, the absorption due to a single atom, say of iron, depends also upon the glass composition, the melting temperature and time, the presence or absence of additives, and the total concentration of the contaminant species. Thus, for a particular glass composition and set of melting conditions, it has been found that the absorption per ion of iron varies with the composition and concentration of this element according to

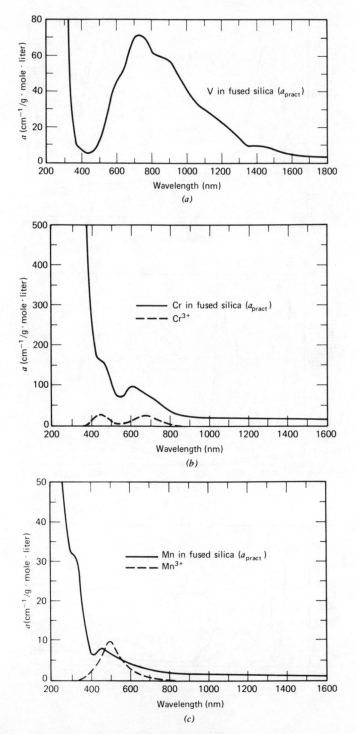

Figure 8.12. Absorption spectra for the transition metals in silica. Copyright, The American Ceramic Society, Columbus, Ohio, 1974. Reproduced with permission from P. C. Schultz, *J. Am. Ceram. Soc.*, **57**, 309 (1974).

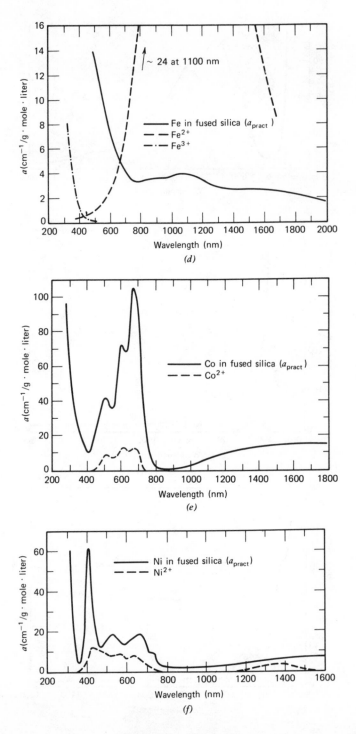

Figure 8.12. (*Continued*)

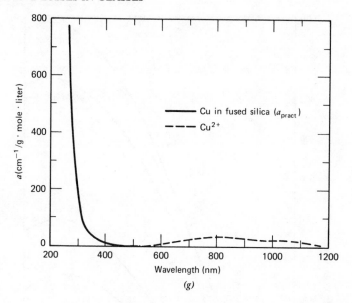

Figure 8.12. (*Continued*)

the curves of Figures 8.13 and 8.14. The result is that published data on the absorption coefficients for contaminants in glasses are useful as a guide to what can happen and to the order of magnitude of effects likely to be observed, but beyond that they are highly particular to the glass composition and apparatus used for the preparation.

Experimentally, the oxidation conditions prevailing in the melt can be controlled in a number of ways. Traditionally, oxides of arsenic and antimony have been added to glasses to "decolorize" them, and this technique appears to be still widely used. However, an alternative approach stems from the use of gas bubbling to stir the melt. The bubbling gas can be used to modify the oxidation state in a controlled manner, by bubbling with oxygen to push the melt toward a more oxidized state or by bubbling with reducing gases such as CO, mixed with CO_2, to produce a more reduced melt. This technique is particularly attractive for experimental studies of the oxidation/reduction process since the same melt can be varied in time and samples taken as it varies. In addition, the need for very pure additives is removed since the gases can be produced free of transition metal contamination with relative ease.

Figure 8.15 shows a breakdown of the loss mechanisms in a fiber made from a sodium borosilicate glass.[8,18] In this case the impurity levels of iron and copper have been held so low that the optimum transmission is obtained by reducing the glass to the point at which the copper contributes almost nothing to the total loss, but in its place the iron makes a small

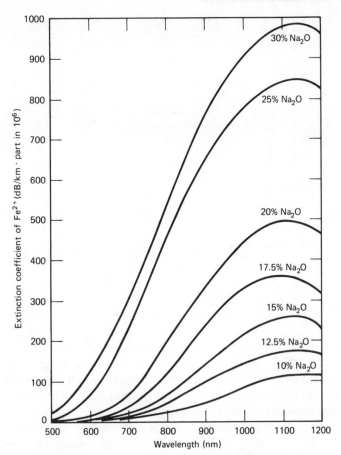

Figure 8.13. Variation of the Fe^{2+} extinction coefficient with glass composition. Reproduced with permission from K. J. Beales, C. R. Day, W. J. Duncan, J. E. Midwinter, and G. R. Newns, *Proc. IEE.*, **123**, 591 (1976).

contribution and chromium and nickel appear as significant secondary impurities. The loss in the glass has been varied by chemical means after starting materials and handling techniques were improved to the point that substantial decreases in impurity in the finished glass appear unlikely. The production of glass with essentially zero impurity level is at present possible only by formation directly from the vapor phase.

8.3.4 Hydroxyl Ion Overtone Absorption

A problem with most optical fibers is to reduce the hydroxyl ion (OH) content to a sufficiently low level, usually a few parts per million. The OH ion is undesirable because overtones of the fundamental stretching vibra-

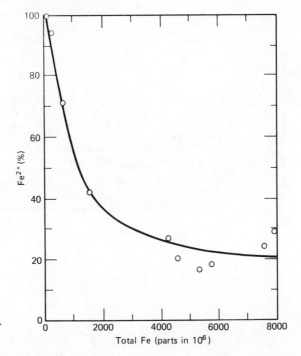

Figure 8.14. Percentage of Fe^{2+} versus total concentration of iron in the melt of a sodium borosilicate glass. Reproduced with permission from K. J. Beales, C. R. Day, W. J. Duncan, J. E. Midwinter, and G. R. Newns, *Proc. IEE.*, **123**, 591 (1976).

tion (centered around 2.8 microns) give rise to absorptions at 1.4 microns and 970 and 750 nm, and thus interfere with the transmission band of interest in glass. It need hardly be pointed out that OH, in the form of water or water vapor, is commonly available and will contaminate any material in a laboratory unless special precautions are taken to exclude it.

The OH stretching vibration is fundamentally that of a simple harmonic oscillator, although clearly with some anharmonicity present, as shown by the presence of harmonics. The line shape of the harmonic oscillator is Lorentzian and takes the form

$$g(\omega) = \frac{A}{(\omega - \omega_0)^2 + (\Delta\omega)^2}$$

A feature of such a line shape is that, in comparison to the Gaussian line, the wings spread much further; or, alternatively, at any distance from the line center the absorption contribution is greater for the Lorentzian line. This leads to the situation in which one must seriously question how much

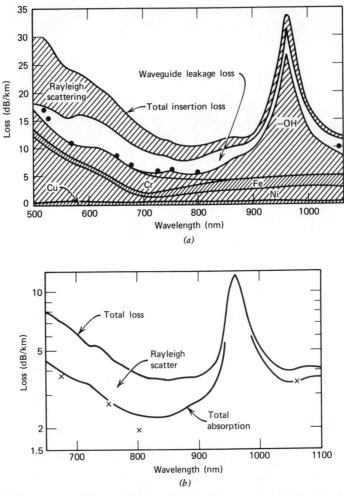

Figure 8.15. Analysis of the attenuation of (a) a sodium borosilicate fiber made from glass having 5 dB/km absorption and (b) a good fiber made by the same process. Reproduced with permission from (a) K. J. Beales, C. R. Day, W. J. Duncan, J. E. Midwinter, and G. R. Newns, *Proc. IEE.*, **123**, 591 (1976) and (b) K. J. Beales, C. R. Day, W. J. Duncan, and G. R. Newns, *Electron. Lett.*, **13**, 755 (1977).

absorption in the wavelength band of interest (800 to 900 nm) is due to the fundamental and first overtones, as well as the second and third overtones, which straddle the region. Experimentally, the estimation of these contributions is extremely difficult unless a glass can be made in which it can be confidently assumed that all the absorption is due to the OH vibrations alone.

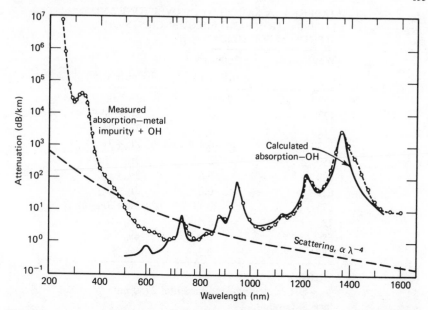

Figure 8.16. Loss breakdown for a silica fiber, showing the estimated components arising from the OH overtone and combination bands, the transition metals, and the Rayleigh scattering. The overtone and combination bands involved are listed in Table 8.4. Reproduced with permission from D. B. Keck, R. D. Maurer, and P. C. Schultz, *Appl. Phys. Lett.*, **22**, 307 (1973).

Several attempts have been made to carry out such an assignment. In Figure 8.16 and Table 8.4 we reproduce the results of such an assignment by the workers at the Corning Glass Works,[10] in which they broke down, into the constituents, the loss of a silica fiber. The component due to OH overtone vibrations is clearly shown, although the exact assumptions used in making the assignment were not specified.

A similar exercise carried out for sodium borosilicate glasses[8] showed two features rather clearly. The individual absorption lines which are clearly visible in the fiber measurements are not Lorentzian lines. This is illustrated in Figure 8.17, in which a Lorentzian line is superimposed upon the measured line for the second overtone, making clear the marked asymmetry in the measured line.[11] A second fact follows rapidly, namely, that if one seeks a best fit, using a single Lorentzian curve, to each of the observed lines, the projected contribution due to the sum of the wing components of the fundamental, first, second, and third overtones is far greater at 800 nm than the total absorption of the fiber.

Two conclusions seem inescapable. The first is that the lines observed are in fact composites resulting from the overlap of many lines, all formed from slightly different center oscillator frequencies. This is consistent with

TABLE 8.4   ASSIGNMENT OF OVERTONE AND
COMBINATION BAND FREQUENCIES IN OH
SPECTRUM FOR SILICA[a]

Wavelength (nm)	Intensity (dB/km)	Identification
1370	2900	$2\nu_0$[b]
1230	150	$2\nu_0 + \nu_1$[c]
1125	3.4	$2\nu_0 + 2\nu_1$
1030	0.4	$2\nu_0 + 3\nu_1$
950	72	$3\nu_0$
880	6.6	$3\nu_0 + \nu_1$
825	0.8	$3\nu_0 + 2\nu_1$
775	0.1	$3\nu_0 + 3\nu_1$
725	6.4	$4\nu_0$
685	0.9	$4\nu_0 + \nu_1$
585	0.5	$5\nu_0$

[a]Ref. 10.
[b]ν_0 = OH stretching→fundamental 2730 nm.
[c]ν_1 = Si-O bond.

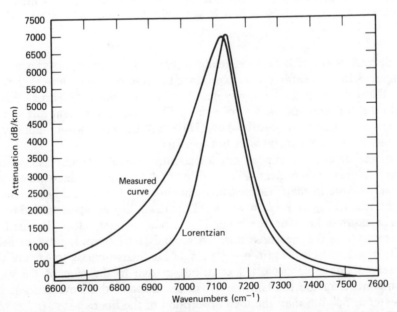

Figure 8.17. Detailed form of a measured OH overtone line centered around 1.4 micron wavelength in a sodium borosilicate glass fiber showing the asymmetry present when compared to a Lorentzian line.

the individual OH ions finding themselves in slightly varying surroundings within the glass matrix.

The second conclusion, which then follows, is that we can learn little or nothing about the true linewidth of individual oscillators by examining the measured lines in the regions of their peaks. However, if we hold to the assumption that the lines are fundamentally Lorentzian in character, we can take the linewidths as adjustable parameters and use additional data on the estimated absorption due to OH at particular wavelengths to obtain best guess values and hence to predict the form of the complete spectrum.

The results of such an analysis process are shown in Figure 8.18, in which use has been made of the remaining loss at 1060 nm after the known absorption from transition metal impurity has been subtracted, and the estimated curve for the OH component alone has been adjusted to give the observed overtone and fundamental peak heights and the estimated contribution at 1060 nm. This curve has been found to be consistent with extensive data available on the absorption at 800 nm, insofar as the

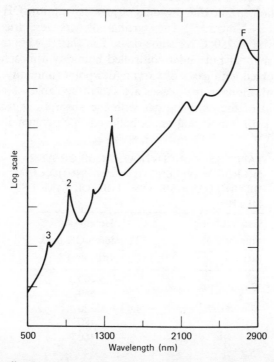

Figure 8.18. Predicted total absorption versus wavelength from OH in a sodium borosilicate fiber after adjusting the oscillator linewidth to fit the experimental data. Reproduced with permission from K. J. Beales, C. R. Day, W. J. Duncan, J. E. Midwinter, and G. R. Newns, *Proc IEE.*, **123**, 591 (1976).

measured total absorption at that wavelength has been found to vary as predicted with changes in OH concentration, which can be measured by the changes in the peak heights of the overtones. Notice also that this estimated curve does not include all the combination band lines which arise from combination oscillations of the OH overtone modulated by the Si-O bond frequency. These are clearly visible in the curves of Figure 8.16, although away from the peak positions they make little contribution to the absorption and often can be neglected.

It is important to stress that in both of the above cases the scientific justification for the curve fitting exercise described is by no means complete or particularly well proved. However, it appears from extensive measurements of fibers covering a large dynamic range that the general picture presented is right. Further study of the precise details is needed before a fully verified description can be given. It also seems evident from careful study of the fiber transmission curves that the linewidth of the O-H oscillator is significantly greater in the multicomponent borosilicates than in pure silica, so that the contribution away from the peak is proportionately greater in the former.

The production from oxide powders of glasses with low OH content has been described by Newns.[12, 16] The starting powders were dried by heating in a vacuum oven at 250°C for some days. The glasses were then prepared in the normal manner but under controlled humidity atmosphere, and the melts were bubbled with gases of varying dewpoint (humidity). The resulting OH concentrations in the glasses are shown in Table 8.5 as a function of the dewpoint of the bubbling gas with the absorptions associated. An absorption of 1 dB/km at 900 nm is believed to correspond roughly to 1

TABLE 8.5 CONCENTRATION OF OH AT 2.8 MICRON WAVELENGTH AS A FUNCTION OF DRYING GAS BUBBLING TIME FOR VARIOUS MELTS[a]

Gas bubbling time (400 g) (hr)	Height of 2.8 micron OH peak (dB/km)
0	4600
1	550
2	220
3	180
4	160
5	140
20	87
After 20 hr with 8 g melt→21 dB/km	

[a]Ref. 16.

ppm weight of water in the glass. The results clearly show that melt bubbling with very dry gas greatly reduced the loss from OH absorption (from 580 to 16 dB/km estimated at 960 nm).

8.3.5 Metallic Inclusions in Glass

Glasses that are melted in platinum crucibles, or crucibles of similar materials, in the presence of oxygen will dissolve some of the metal during the period in which the glass is held molten. Typical platinum levels arising from a 24 hr melting period for a sodium calcium silicate glass are in the region of a few parts per million by weight. Provided that the glass is melted under oxidizing conditions, it appears that this metal enters the glass in the form of oxide and, as such, does not give rise to any measurable absorption in the 800 to 900 nm region. However, the same glass reduced will contain the platinum as microinclusions of free metal, giving the glass a distinct coloration and very high loss. It follows that, if reduced glasses are to be produced, melting in platinum or similar crucibles is generally to be avoided.

The amount of dissolution is dependent upon the melt temperature. Hence it appears that no problems are encountered in holding molten reduced sodium borosilicate glass in a platinum double crucible, since runs extending for many days have been reported in which the loss of the fiber did not alter measurably. In this case the crucible temperature is in the region of 850°C, compared to something of the order of 1100 to 1300°C for melting and refining NCS glasses.

8.4. RAYLEIGH SCATTERING

The phenomenon of Rayleigh scattering is well known to all of us as the mechanism responsible for blue sky. It is the scattering of light from microirregularities in the dielectric medium through which the electromagnetic wave is propagating. The physical scale of the irregularities is of the order of one-tenth wavelength or less, so that each irregularity acts as a point source for scattered radiation. The resulting radiation pattern from the induced dipole is doughnut shaped, being uniform in the plane perpendicular to the dipole and varying as $\sin\phi$ in the plane containing the dipole, where ϕ is the angle between the observation direction and the dipole axis.

In the sky, Rayleigh scattering arises from the minute density fluctuations in the atmospheric gas caused by the constant thermal fluctuation of the medium, while the $\sin\phi$ component causes it to be highly polarized at ground level. In glass, Rayleigh scattering can arise from two separate effects, density and composition fluctuations.

We have already seen that glasses are disordered structures, loosely connected in a largely random sequence. Evidently, in such a structure there are local regions in which the average density is higher than in other regions. The dense and less dense regions can be traced back to thermally driven fluctuations in density arising from the Brownian movement of the liquid glass before it froze. The magnitude of the fluctuation is thus expected to be related to the freezing temperature—the higher the temperature, the greater the density fluctuation. One might therefore expect that a high melting material such as silica would show a higher Rayleigh scattering loss than a lead glass of lower melting temperature. But this is not so for another reason.

In a glass containing several oxide constituents, not only can the overall density fluctuate with position, but so also can the composition. Density fluctuations are subject to mechanical restoring forces, tending to reduce them against the driving force of the kinetic energy. Composition fluctuations are subject to the restoring force of the total free energy of the glass system, which will also seek to minimize itself. If the particular composition of glass melted does not choose to bond together and form a favorable state of minimum free energy, large fluctuations of the glass composition can occur locally without leading to an unstable state.

The magnitude of the density fluctuation scattering is given by the following expression:[20]

$$\alpha_{\text{scat},\rho} = \frac{8\pi^3}{3\lambda^4}(n^8 p^2)(kT_f)\beta_T$$

Here p is the photoelastic coefficient for the glass; T_f is the fictive temperature, defined as the temperature at which it becomes possible for the glass to reach a state of thermal equilibrium and closely related to the anneal temperature; k is Boltzmann's constant; and β_T is the isothermal compressibility. Evidently, the energy kT_f is the driving energy for the fluctuation giving rise to the loss mechanism described by $\alpha_{\text{scat},\rho}$.

The expression for the scattering due to composition is more complex, taking the form[20]

$$\alpha_{\text{scat},c} = \frac{32\pi^3 n^2}{3\lambda^4 \rho N_A} \sum_{j=1}^{m}$$

$$\left[\left(\frac{\partial n}{\partial x_j}\right)_{T_f, x_i \neq x_j} + \left(\frac{\partial n}{\partial \rho}\right)_{T_f, x_i}\left(\frac{\partial \rho}{\partial x_j}\right)_{P, T_f, x_i \neq x_j} \right]^2 M_j x_j$$

In this formula, M_j and x_j are, respectively, the molecular weight and the weight fraction of the jth modifier, and N_A is Avagadro's number. The partial derivatives of the refractive index and the density, ρ, may be determined from experimental data so that the summation can be performed.

Schroeder[21] has derived an expression for the mean composition fluctuation in the glass, which takes the form

$$\langle \Delta C^2 \rangle_{T_f} = \frac{\rho_0}{C_0}\left(\frac{kT_f}{V}\right)\left(\frac{\partial \mu}{\partial C}\right)^{-1}_{C_0, T_f}$$

where C is the mole fraction of one of the constituents in the composition C_0, V is the molar volume, and μ is the chemical potential difference between the major and minor constituents of the (binary) glass. Evidently the composition fluctuation is minimized if $\partial \mu / \partial C$ is large and the fictive temperature is small.

Since the refractive index of a glass is a sensitive function of composition, it is frequently the case that the composition fluctuation swamps the density fluctuation component. Thus a low melting glass may well have a much higher scatter loss than silica, and in general this is bound to be the case. However, some particular glass compositions have been found to lead to exceptionally low total Rayleigh scattering losses. The potassium silicates[13] are one such group which show a minimum scattering loss at one composition, as shown in Figure 8.19. In Table 8.6 we list the Rayleigh scattering at 900 nm wavelength for a number of glasses that are of interest for optical fiber systems.

TABLE 8.6 RAYLEIGH SCATTERING LOSSES
FOR A NUMBER OF FIBER GLASSES (QUOTED
FOR 900 NM)

Glass	Loss (dB/km)
Silica	~1.2
Sodium borosilicate (Ref. 14)	~1.5
Phosphosilicate (0.7% index difference from silica) (Ref. 10, Chapter 9)	~1.6
Germania silicate (0.18% index difference from silica) (Ref. 17)	~1.2
Selfoc (Ref. 3, Chapter 9)	2.0

The experimental measurement of these loss mechanisms is important because it represents the irreducible minimum loss for a fiber made out of

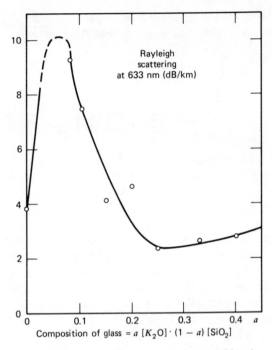

Figure 8.19. Rayleigh scattering loss at 633 nm versus composition for a sodium silicate glass. Plotted using data from Ref. 13.

a particular glass. Since the Rayleigh scattering varies as λ^{-4}, it is useful to measure it as a function of wavelength as a check to separate it from other mechanisms. This is desirable because it is easy to incur additional scatter losses in glasses, particularly in bulk glasses, such as scatter from strain, inhomogeneity from imperfect glass stirring, or particulate inclusions from refractory material or undissolved starting materials. In general, all of these will lead to a scatter loss component that is essentially independent of wavelength, since the scattering object will be large compared to the optical wavelength.

REFERENCES

1. B. Scott and H. Rawson, University of Sheffield, Private communication.
2. T. Inoue, K. Koizumi, and Y. Ikeda, *Proc. IEE*, **123**, 577 (1976).
3. G. W. Morey, *The Properties of Glass*, 2nd ed. New York: Reinhold, 1954.
4. D. A. Pinnow, T. C. Rich, F. W. Ostermeyer, and M. DiDomenica, *Appl. Phys. Lett.*, **22**, 527 (1973).
5. B. G. Bagley and W. French, *J. Am. Ceram. Soc.*, **52**, 701 (1973).
6. a. M. Horiguchi, *Electron Lett.*, **12**, 311 (1976).

b. H. Osonai, T. Shioda, T. Moriyama, S. Araki, M. Horiguchi, T. Izawa, and H. Takata, *Electron Lett.*, **12**, 550 (1976).

7. K. J. Beales, J. E. Midwinter, G. R. Newns, and C R. Day, *Post Off. Elec Eng. J*, **67**, 80 (1974).

8. K. J. Beales, C. R. Day, W. J. Duncan, J. E. Midwinter, and G. R. Newns, *Proc. IEE*, **123**, 591 (1976).

9. P. C. Schultz, *J. Am. Ceram. Soc.*, **57**, 309 (1974).

10. D. B. Keck, R. D. Maurer, and P. C. Schultz, *Appl. Phys. Lett.*, **22**, 307 (1973).

11. J. E. Midwinter, Unpublished work.

12. G. R. Newns, *2nd European Conference on Optical Communications, Paris*, 1976. Paris: SEE.

13. J. Schroeder, R. Mohr, P. B. Macedo, and C. J. Montrose, *J. Am. Ceram. Soc.*, **56**, 510 (1973).

14. G. R. Newns, Unpublished work.

15. C. E. E. Stewart, D. Tyldesley, B. Scott, H. Rawson, and G. R. Newns, *Electron Lett.*, **9**, 482 (1973).

16. K. J. Beales, C. R. Day, W. J. Duncan, and G. R. Newns, *3rd European Conference on Optical Communication, Munich, September 1977*. Berlin: VDE.

17. M. Kawachi, A. Kawara, and T. Miyashita. *Electron Lett.*, **13**, 442 (1977).

18. K. J. Beales, C. R. Day, W. J. Duncan, and G. R. Newns, *Electron Lett.*, **13**, 755 (1977).

19. For general reading on glasses and their properties see C. L. Babcock, *Silicate Glass Technology Methods*. New York: Wiley, 1977.

20. D. A. Pinnow, T. C. Rich, F. W. Ostermayer, Jr., and M. DiDomenico, Jr., *Appl. Phys. Lett.*, **22**, 527 (1973).

21. J. Schroeder, R. Mohr, P. B. Macedo, and C. J. Montrose, *J. Am. Ceram. Soc.*, **56**, 510 (1973).

9

Fiber Preparation

9.1. INTRODUCTION

In Chapter 8 on materials for fibers, we discussed some of the factors affecting the choice of materials for fiber preparation, particularly the features which affect fiber loss. In this chapter we consider the methods by which the different materials can be formed into fibers with the desired properties.

Several discernible stages emerge in the techniques that have to date produced low loss fiber in any quantity. In one process involving multi-component low melting temperature glasses the glass is prepared in a large melt and subsequently formed into fiber in a second, and discrete state. A second process is usually based upon silica and utilizes a preform containing both core and cladding materials directly formed from the vapor phase and subsequently pulled down into fiber. These stages are discussed in separate sections under the appropriate headings, so that to follow through a single process several sections need to be read together. A third process involves pulling a rod preform of single composition and coating the fiber obtained with plastic to provide a cladding.

9.2. GLASS MELTING

Several techniques have been described for melting glasses that are specifically intended for use in the preparation of optical fibers. Each is based upon the use of powders which are premixed, heated in a crucible until they fuse, and then agitated to produce a homogeneous mix. The heating may be applied to the crucible through black-body radiative coupling from the walls of an electrically heated furnace; it may be generated by the coupling of radio-frequency (RF) radiation to the crucible when the crucible is made of a conducting material such as platinum; or, alternatively, RF energy may be coupled directly to the melt glass, provided that is is preheated to a temperature at which it begins to conduct. In the last case, the crucible remains relatively cold, thus helping to reduce problems of crucible contamination. In all cases it is necessary to use for the melting an enclosure which excludes contamination from the laboratory environment. Typically, silica liners are used to provide such isolation.

The simplest technique—and one which appears to be capable of excellent results—is to melt the glass powders in a silica crucible within a silica lined electric furnace.[1] The apparatus is cheap, and a clean environment is readily achieved. If atmosphere control is desired, the same system can be readily sealed against the laboratory atmosphere. Silica crucibles are easily obtained in varying degrees of purity and in a range of sizes, although the fact that a crucible can generally be used only once increases to some extent the operating costs of this approach. Reuse is impossible because the glass constituents attack and dissolve the walls of the crucible during the melting procedure. It is generally impossible to cool a silica crucible to room temperature without its shattering after glass has been melted in it, since the expansion coefficients of the glass skin remaining and of the silica are widely different. The same electric furnace, operated with a platinum crucible, allows multiple melts to be made with a single crucible. However, as we have already noted, attempts to produce reduced glasses in such a system may lead to the precipitation of free platinum in the glass.

In the electric furnace, energy is transferred from the furnace wall to the crucible by black-body radiation. A great deal of early melting work was done using platinum crucibles contained within silica enclosures for cleanliness, but with the heat applied through RF energy coupled from an encircling coil, as shown in Figure 9.1. This technique has the attraction that there is no refractory insulating material anywhere near the melt.[2] However, precise temperature control is generally more difficult, the thermal time constant of the apparatus being much shorter, and the apparatus is far more expensive. It also requires a conducting crucible, ruling out the use of silica.

A modification of this technique, designed to avoid crucible contamination more or less completely, uses a higher frequency of RF radiation to couple directly to the glass material (typically 5 MHz versus 100 kHz). The apparatus for this is shown in Figure 9.2. The glass powders are first fused and made molten by radiative (black-body) heating from a carbon susceptor heated by the RF radiation and placed under the crucible. Once the glass has fused and begun to conduct (and couple to the RF field directly), the susceptor is lowered away and the glass becomes directly heated by the RF radiation for the remainder of the melt time. The attraction of this technique is that, once the start-up period is passed, the crucible remains relatively cold, the RF energy being coupled to a toroidal shaped region within the molten glass itself.

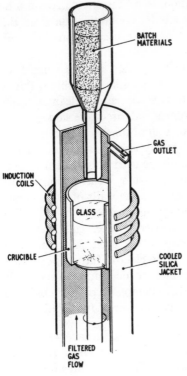

BATCH MATERIALS

GAS OUTLET

INDUCTION COILS

GLASS

CRUCIBLE

COOLED SILICA JACKET

FILTERED GAS FLOW

Figure 9.1. Radio frequency heating of a platinum crucible for glass melting in a clean enclosure. Reproduced with permission from K. J. Beales, J. E. Midwinter, G. R. Newns, and C. R. Day, *Post Off. Elec. Eng. J.*, **67**, 80 (1974).

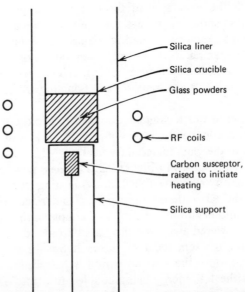

Silica liner

Silica crucible

Glass powders

RF coils

Carbon susceptor, raised to initiate heating

Silica support

Figure 9.2. Apparatus for direct heating of the glass material, showing the carbon susceptor used to initiate the heating process from the high frequency RF field. After Ref. 2.

Because the crucible remains cold, both attack of and contamination by the crucible are minimized, so that it is likely that the purest possible glasses will be made from powders in this way. However, against this advantage, the technique suffers from the defect that the nonuniform heating and cooling present lead to large variations in temperature and viscosity throughout the melt, making homogenization of the glass very difficult. At present, insufficient data appear to have been published to allow a balanced appraisal of the different melting techniques. The lowest absorption glass reported to date has been made using an electric furnace with a silica liner and crucible.

In each of the melting techniques, after the initial heating stage in which the powders fuse and carbon dioxide is liberated from the carbonates to leave oxides, several further stages follow in which the crude glass is improved to the point that it is usable for fiber manufacture. The glass must be homogenized and fined. For homogenization some form of stirring must be carried out to ensure a uniform composition, although the powders are already very well mixed. This stirring can be done either by the use of a metal (platinum) or ceramic stirrer, or by bubbling gas through the melt so that the rising bubbles agitate it. Both techniques have been reported as suitable for the production of low loss glass. This stage is then followed by a period of refining, in which remaining gas bubbles are allowed to rise to the surface. Once the glass has fully fined and homogenized, it is ready for transfer to the fiber pulling stage. Control of the oxidation state of impurities in the glass and of the OH content must be completed before this stage is reached.

The traditional technique for achieving this control is to add chemicals to the glass which will either preferentially reduce or oxidize the melt, the oxides of arsenic and antimony being commonly used in the glass industry. An alternative approach has been described which can be combined with the homogenization stage if gas bubbling is used. This involves using either a reducing CO and CO_2 mixture or an oxidizing (O_2) gas mixture so that, while the glass is being homogenized, the redox ratio for the transition metal ions is also being controlled. A further attraction of this approach is that the OH content of the glass can also be reduced in the same operation by using gases of very low OH content.

The simplest and most convenient technique for removing the ultrapure glass from the crucible for storage before loading in a fiber pulling apparatus seems to be to pull it vertically from the melt into rods. The apparatus for doing this is shown schematically in Figure 9.3. Rods are formed by dipping a seed rod into the melt and pulling it slowly upwards, through the cooled ring. The treacle-like glass pulls up after the seed and solidifies to form a cane, typically 5 to 10 mm in diameter and a few meters long. In this form the glass is readily stored in a clean environment

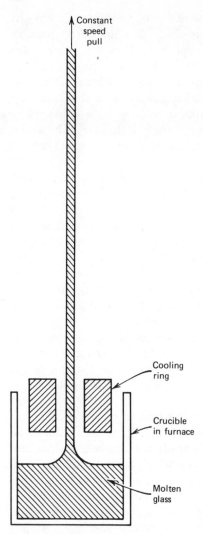

Constant
speed
pull

Cooling
ring

Crucible
in furnace

Molten
glass

Figure 9.3. Apparatus for pulling glass rods from a melt.

by sealing it in silica tubes, and either it can be fed back into a double-crucible fiber puller or the rods can be used for measurement purposes.

9.3. FIBER PULLING BY DOUBLE-CRUCIBLE APPARATUS

9.3.1 General Features

The production of fiber using a double-crucible apparatus is conceptually extremely simple.[1,3] The basic apparatus is illustrated in Figure 9.4. Two concentric crucibles are held with their axes vertical. Each crucible

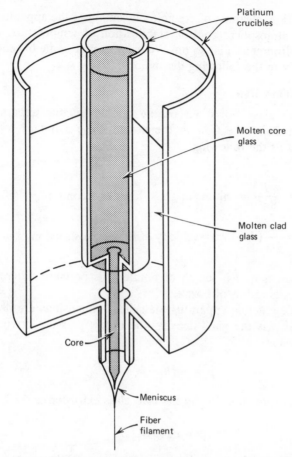

Figure 9.4. Apparatus for double crucible fiber pulling. Reproduced with permission from K. J. Beales, J. E. Midwinter, G. R. Newns, and C. R. Day, *Post Off. Elec. Eng. J.*, **67**, 80 (1974).

has in its base a central circular nozzle, the inner being carefully aligned to be concentric with the outer and perhaps 1 cm above it. The inner crucible is filled with core composition glass, and the outer with cladding composition glass. With the glasses molten, the core glass flows through the inner nozzle into the cladding stream and is subsequently carried out of the cladding outer nozzle surrounded by cladding glass, so that a composite glass flow is produced. The molten glass exudes into the space outside the outer crucible, and a filament is pulled from the exudant to form a cored glass fiber.

In practice, the production of low loss fiber by this process requires a great deal of careful attention to detail at every stage of the process; the glass preparation, cleaning of crucibles, assembly of the apparatus, loading of the glass, atmosphere control, stabilization of temperature, control of meniscus environment, and so on. We will examine some of these features more carefully in the following discussion.

9.3.2 Glass Flow Rate

The molten glass flows out of the double-crucible apparatus under conditions of Poiseuille flow in the nozzles. The standard formulas for such flow tell us the following:

$$Q = \frac{\pi p R^4}{8\eta l} = \text{volume of glass flow per second} \tag{9.3.1}$$

$$V(r) = \frac{p}{4\eta l}(R^2 - r^2) = \text{velocity of flow across section of nozzle} \tag{9.3.2}$$

where η is viscosity, l is length of nozzle, p is pressure difference across nozzle, and R is radius of nozzle.

In a double-crucible apparatus the salient dimensions are illustrated in Figure 9.5. If ρ is the glass density, the excess pressure driving the core glass extrusion is

$$p_1 = \rho g(h_1 - h_2 + l_1 + l_3) \tag{9.3.3}$$

and the excess pressure driving the clad glass extrusion is

$$p_2 = \rho g(h_2) \tag{9.3.4}$$

The volume flows of core and composite glasses are therefore

$$Q_1 = \frac{\pi p_1 R_1^4}{8\eta_1 l_1} \tag{9.3.5}$$

$$Q_2 = \frac{\pi p_2 R_2^4}{8\eta_2 l_2} \tag{9.3.6}$$

We assume that the viscosities and densities of the core and cladding glasses are very closely similar. Strictly, in the evaluation of Q_2 some account should be taken of the fact that the stream not only is composed of cladding material but also contains core material. In practice, these glasses are likely to be chosen to have very similar mechanical properties.

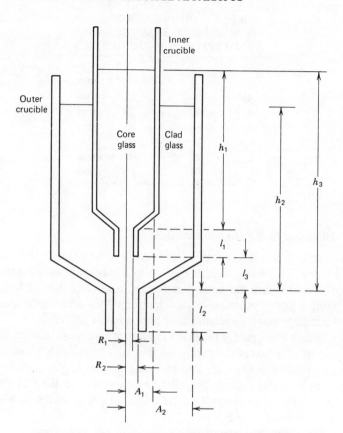

Figure 9.5. Dimensions of a double crucible for analysis.

The core/cladding ratio (in terms of radii) must be given by

$$\frac{a_{\text{core}}}{a_{\text{clad}}} = \sqrt{\frac{Q_1}{Q_2}} \qquad (9.3.7)$$

Typical values of the parameters in the above formulas as used in the author's laboratory are given in Table 9.1. These are for sodium borosilicate glasses and yield fibers with core diameters of the order of 15 to 20 microns and overall diameters of about 70 to 100 microns.

TABLE 9.1 DIMENSIONS OF A
TYPICAL DOUBLE CRUCIBLE
(WITH REFERENCE TO FIGURE 9.5)

R_1	0.5	–	1.5	mm
R_2	2	–	3	mm
A_1	5	–	15	mm
A_2	40	–	50	mm
h_1	30	–	80	mm
h_2	30	–	80	mm
h_3	40	–	100	mm
l_3	10	–	30	mm

9.3.3 Diffusion in the Double Crucible

The double-crucible apparatus has been used to produce fiber having a graded index. This has been done by using a glass pair which interdiffuse. In particular, thallium has been described as a suitable dopant for the core glass, being a highly polarizable ion which rapidly increases the glass index of refraction for small concentrations and is easily diffusible.

When such a glass pair is used, the core glass enters the molten cladding glass as soon as it leaves the core glass nozzle. The diffusible species starts to exchange between the core and cladding glasses as the stream of core glass moves with the surrounding cladding glass toward the exit nozzle. It is usually assumed that the bulk of the diffusion occurs in the parallel glass flow region, labeled l_3 in Figure 9.5.

In circular symmetry the diffusion equation takes the form

$$\frac{\partial^2 N}{\partial r^2} + \frac{1}{r}\frac{\partial N}{\partial r} = \frac{1}{D}\frac{\partial N}{\partial t} \qquad (9.3.8)$$

where N represents the concentration of the diffusible species.

If one assumes that the cladding is of infinite thickness, as it would effectively be in the l_3 region, the boundary conditions to be applied in seeking a solution of equation 9.3.8 are, at $t=0$,

$$N = N_0 \qquad 0 \leqslant r < a$$

$$N = 0 \qquad r \geqslant a$$

Equation 9.3.8 has standard solutions, which can be derived from Carslaw

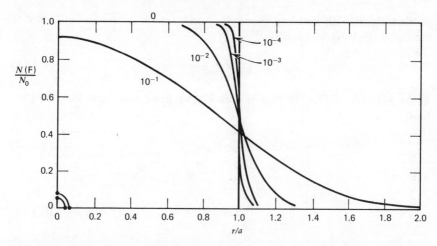

Figure 9.6. Index profiles predicted theoretically for double crucible fiber by solving the diffusion equation for various values of Dt/a^2. Reproduced with permission from K. B. Chan, P. J. B. Clarricoats, R. B. Dyott, G. R. Newns, and M. A. Savva, *Electron. Lett.*, **6**, 748 (1970).

and Jaeger,[4] assuming the same diffusion properties in core and cladding:

$$\frac{N(r/a)}{N_0} = \int_0^\infty \exp\left[-\left(\frac{Dt}{a^2}\right)u^2\right]J_0\left(\frac{ur}{a}\right)J_1(u)\,du \qquad (9.3.9)$$

where D is the diffusion constant. This equation has been evaluated by computer[5] for the range of values of $F = r/a$ of interest and for a number of different values of the diffusion parameter, Dt/a^2, as shown in Figure 9.6.

The values of $N(F)/N_0$ plotted in Figure 9.6 can be related to the refractive index of the guide by the relationship

$$n(r) = (n_1 - n_2)\left[\frac{N(F)}{N_0}\right] + n_2 \qquad (9.3.10)$$

where n_1 and n_2 are the refractive indices of the glasses before diffusion has occurred, and a linear dependence on concentration, which is likely to be a good approximation for small changes, is assumed.

We should note that, if the apparatus is designed so that the diffusion occurs in the parallel flow region, l_3, the diffusion profile obtained is not easily altered over a very great range. The profile depends upon the numerical value of the function Dt/a^2. If we assume that we are to produce fiber with winding speed V_w and core radius a_f, the quantity of core glass flowing is given by

$$Q_1 = V_w \pi a_f^2 \qquad (9.3.11)$$

If the velocity of flow in region l_3 is V_3, we must also have the relations that

$$Q_1 = V_3 \pi a^2 \qquad (9.3.12)$$

and

$$t = \frac{l_3}{V_3} \qquad (9.3.13)$$

It follows that the parameter Dt/a^2 can be expressed as

$$\frac{Dt}{a^2} = \frac{D\pi l_3}{Q_1} \qquad (9.3.14)$$

Changing the nozzle size, r_1, will not at first affect the value of the diffusion profile obtained, since the critical parameters are l_3 and D. Since the latter is controlled by the diffusible species that are available, l_3 remains as the major controlling factor except insofar as variations in temperature can be used to vary D. Unfortunately, temperature variation also changes the viscosity, η, and hence the flow rate Q, in such a way that they tend to cancel each other.

The index profiles obtained in practice using the double crucible are significantly different from those shown in Figure 9.6, and are much closer to those described by the alpha profile theory. This situation probably arises from several effects. The theory outlined above contains two gross simplifications. The first simplification assumes that the diffusion coefficient for the same species in the core and cladding glasses is the same (i.e., D is the same for both glasses). In practice this is unlikely to be exactly so. If, for example, the diffusion coefficient for the cladding glass is much larger than that for the core glass, the effect will be to greatly sharpen the core-cladding boundary by fixing the concentration of impurity at a nearly constant value throughout the cladding.

The second approximation is to assume that all the diffusion takes place in a region of laminar glass flow. In practice some part of the diffusion will almost inevitably take place in a region of Poiseuille flow as the composite glass stream flows out through the exit (cladding) nozzle. This will perturb the profile obtained, although apparently no calculations have been published that show the extent or nature of this change.

A further factor not taken into account in the analysis given above is the possibility that after infinite diffusion time the glass indices of core and cladding might form a step-index distribution. Thus the glasses used for core and cladding not only may be different insofar as one contains a diffusible species that the other does not, but may also differ in nondiffusible species, giving an underlying step-index distribution. This will tend to sharpen the wings of the diffused distribution described in Figure 9.6 and bring it much closer to an alpha profile.

Finally, we should note that the profile obtained in practice may also be modified by the presence of concentration-dependent diffusion, in which the diffusion constant of the desired diffusible species may vary with the local concentration of that species, being higher at low than at high concentrations. Of these many effects, there appears to be no published work which allows one to interpret the actual mechanisms in particular diffused fibers, since such vital information as the compositions of the glasses used is always omitted.

In Figure 9.7 we show a measured profile for a Selfoc fiber[6] to which has been added the profile for $\alpha = 2.4$. This value of α was deduced from propagation studies as being necessary to explain the observed delay dispersion, assuming that the fiber propagation was governed by such a profile law. One can see that an excellent agreement has been obtained. However, whether the profile obtained with the double-crucible diffusion

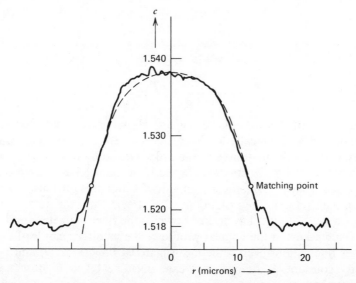

Figure 9.7. A measured profile for a diffused double crucible fiber of the Selfoc type. Reproduced with permission from J. Guttman, H. J. Heyke, and O. Krumholtz, *Opt. Quant. Electron.*, **7**, 305 (1975).

process can be further modified to that of the optimum profile for very low dispersion remains to be seen.

9.3.4 Dimensional Stability

We take as a general target for our fiber a dimensional stability figure of 1% in diameter, both for core and for cladding. This is imposed by the fact that most jointing techniques aim to use the outer surface of the fiber to achieve alignment of the core, and to obtain the very low coupling losses desired, tight mechanical tolerances must be maintained at the fiber production stage.

If we refer to equations 9.3.5 and 9.3.6, we see that the temperature enters into the flow rates, Q_1 and Q_2, through the dependence of the viscosity of the glasses on temperature. A 1% fluctuation in diameter corresponds to a 2% fluctuation in area and hence in flow rate. Thus we require

$$\frac{\delta\eta}{\eta} \leqslant 0.02 \qquad (9.3.15)$$

The functional relationship between viscosity and temperature for most glasses is of the form

$$\eta(T) = \eta_0 \exp\left(\frac{A}{T}\right) \qquad (9.3.16)$$

Hence

$$\frac{\delta\eta}{\eta} = A\,\delta T \qquad (9.3.17)$$

An evaluation of this formula for typical glasses shows that the value of the parameter A is typically 3×10^{-3} °K^{-1}, whence the maximum allowable temperature fluctuation is of the order of 2 to 6 °K.

A second source of fluctuations in the fiber diameter is associated with the meniscus shape. A significant amount of material is obviously stored in the meniscus between the exuding parallel stream from the nozzle and the point at which the glass has necked down to the fiber diameter and frozen. Small fluctuations in the angular profile of the meniscus can lead to local diameter fluctuations even if the mean material flow rates, Q_1 and Q_2, are perfectly constant. It follows that care must be taken to obtain stable conditions here.

Long term dimensional stability of the fiber is always a factor of importance in a fiber pulling process. The high silica fibers pulled from preforms generally have their core/cladding ratios determined at the preform manufacture stage. However, in the case of the double-crucible fiber it is clear that the dimensions are governed by the instantaneous flow rates during the run. Since the glass levels fall as the run proceeds, it is of importance to see whether there are preferable situations in which the changing glass head levels will not, to a first approximation, affect the core/cladding ratio. The overall diameter can then be stabilized by controlling the winding speed. (In principle, glass could be fed into the crucibles to maintain constant glass heads at all times. In practice it seems more likely that replenishment of glass at regular time intervals would be used.)

If we refer to Figure 9.5 for dimensions, we can write the following relations concerning the glass flow rates:

$$Q_1 = \frac{\pi \rho_1 R_1^4 g}{8 \eta_1 l_1} (h_3 - h_2) \quad = P_1 (h_3 - h_2) \tag{9.3.18}$$

$$Q_2 = \frac{\pi \rho_2 R_2^4 g}{8 \eta_2 l_2} h_2 \quad\quad = P_2 h_2 \tag{9.3.19}$$

For a constant core/cladding ratio we require that $Q_1 = MQ_2$, where M is a constant factor. It also follows that, if the factor M is to remain invariant in time to the first order, the following relation must also apply:

$$\frac{\partial Q_1}{\partial t} = M \frac{\partial Q_2}{\partial t} \tag{9.3.20}$$

But we have

$$\frac{\partial Q_1}{\partial t} = P_1 \left(\frac{\partial h_3}{\partial t} - \frac{\partial h_2}{\partial t} \right) \tag{9.3.21}$$

$$\frac{\partial Q_2}{\partial t} = P_2 \frac{\partial h_2}{\partial t} \tag{9.3.22}$$

and

$$\frac{\partial h_3}{\partial t} = \frac{Q_1}{\pi R_1^2} \tag{9.3.23}$$

$$\frac{\partial h_2}{\partial t} = \frac{Q_2 - Q_1}{\pi (R_2^2 - R_1^2)} \tag{9.3.24}$$

Substituting these relations and rearranging terms leads us to the result

$$\left(\frac{R_2}{R_1}\right)^2 = \left[\frac{P_2}{P_1}(1-M) + \frac{1}{M}\right] \tag{9.3.25}$$

Now, we should note that P_1/P_2 is itself a function of R_1/R_2, as follows:

$$\frac{P_2}{P_1} = \left(\frac{\rho_2\eta_1 l_1}{\rho_1\eta_2 l_2}\right)\left(\frac{R_2}{R_1}\right)^4 = F_{12}\left(\frac{R_2}{R_1}\right)^4 \tag{9.3.26}$$

Equation 9.3.25 reduces to the form

$$M\left(\frac{R_2}{R_1}\right)^4 F_{12}(1-M) - \left(\frac{R_2}{R_1}\right)^2 + \frac{1}{M} = 0 \tag{9.3.27}$$

having as solution

$$\left(\frac{R_2}{R_1}\right)_s^2 = \left\{1 \pm \left[1 - \frac{4F_{12}(1-M)}{M}\right]^{1/2}\right\}[2F_{12}(1-M)]^{-1} \tag{9.3.28}$$

so that for real solutions we must have

$$F_{12} \leqslant \frac{1}{4}\frac{M}{1-M} \tag{9.3.29}$$

Referring to equations 9.3.18 and 9.3.19, we see that a simple relationship exists between h_3 and h_1 as follows:

$$\left(\frac{h_3}{h_2}\right)_s = \left(1 + M\frac{P_2}{P_1}\right)_s = 1 + MF_s\left(\frac{R_2}{R_1}\right)_s^4 \tag{9.3.30}$$

Our theory has thus predicted conditions for stable flow with a constant core/cladding ratio, provided that we choose the initial levels, h_3 and h_1, correctly. Under such conditions, if we wished to obtain constant fiber diameter as well, we would have to servocontrol the pulling speed to account for the slowly falling glass head and the consequent decrease in both Q_1 and Q_2. At this time no experimental results appear to have been published which would make it possible to check the predictions of the above analysis.

9.3.5 Atmosphere Control

The gaseous atmosphere surrounding the double crucible must be carefully controlled for several reasons. The most obvious is to exclude particulate or other contamination from entering the highly pure glass.

A second and less obvious reason is to exclude oxygen, which can be electrolytically transported through the glass and liberated on the inner crucible wall in the region of the core glass nozzle. The driving force for this transport process is the electrolytic cell formed by the confluence of two dissimilar glasses in the molten state in the region of a shorting electrode. This transport can be disastrous for the fiber so formed since the gas liberated in the nozzle region is easily carried by the glass flow into the core-cladding interface to appear as uniformly distributed gas microbubbles at regular intervals along it. The result can be very high scatter loss from the fiber, together with a substantial amount of mode coupling. Thus it is important to exclude oxygen from the furnace atmosphere.

Finally, a third reason for controlling the atmosphere is to reduce its OH content for the production of fiber having low absorption approaching the region of 960 nm, where the second overtone occurs. No figures appear to have been published on the extent of OH pickup through this cause, although the author's laboratory[7] has published the overall results of an early study to reduce OH throughout the glass melting-fiber pulling process (see Figure 9.8).

9.3.6 Crucible Loading

There appear to be three separate approaches to loading a double-crucible apparatus with glass. In each the objective is to fill the crucibles with glass, ultimately in the molten state, without introducing any contamination such as transition metals or water, and without trapping gas bubbles. One approach is to cut blocks of glass to fit the crucibles as closely as possible, perhaps polishing their surfaces to reduce the possibility of particulate pickup, but at the risk of attracting impurities from the polishing process, and to subsequently melt these blocks slowly in the crucibles, allowing plenty of time for gas bubble to escape. The use of large, well fitting blocks minimizes the surface area available to trap gas in the first place.

The second approach is to attempt to fill the crucible directly with molten glass. In practice this appears to be difficult because, at the temperatures involved, the glass is too viscous to pour easily into a relatively small enclosure. Again, there is a high probability of trapping gas bubbles in such a pouring process, particularly when refilling.

A third approach is to use glass in the form of cane, that is to say, rod a few millimeters in diameter. Such cane is readily produced by pulling upwards (with a seed rod) from a crucible of molten glass. Once in cane form, the glass can be fed into a heated double crucible without entrapping gas bubbles provided that the feed speed is carefully controlled to allow time for the cane tip to reach the environment temperature before entering

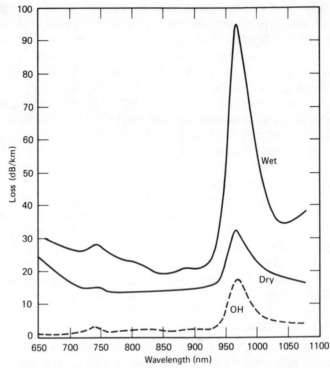

Figure 9.8. Effect of careful OH control by the use of a very dry environment on a sodium borosilicate glass fiber. Reproduced with permission from K. J. Beales, W. J. Duncan, P. L. Dunn, G. R. Newns, *Optical Fibre Communication, IEE Conf. Publ.* 132, London, 1975, p. 32.

the melt. It is important to guide the cane in order to prevent the tip from folding over. Great care must also be taken to maintain cleanliness of the cane surface between pulling and loading, since the large surface/volume ratio produced is obviously susceptible to recontamination.

The third approach has been described by the author's laboratory and has been shown to be capable of use for loading during a fiber pulling run without altering the loss of the fiber produced, thus making possible truly continuous running with the double-crucible apparatus.

9.4. PREFORM METHODS

9.4.1 Preform Production by Chemical Vapor Deposition

Preforms consisting of a cladding surrounding a higher index core material have been made by a variety of techniques in which the starting material is derived directly from the vapor phase. These techniques all use pure silica (SiO_2) as a base material and then add to it small amounts of dopants to change its refractive index sufficiently to allow a waveguide to be formed. The techniques that have evolved seem to fall into three classes.

The first is the "soot" process, described and developed by the Corning Company,[8] in which the reaction between the gaseous components occurs in a flame to form a fine particulate deposit of SiO_2 or doped SiO_2. This white powder, with very small grain size, deposits onto a suitable surface and is subsequently sintered or compacted into an optical quality material (see Figure 9.9a).

In the second class of techniques, the reaction is arranged to occur at a surface, typically the inner surface of a silica tube. The layer produced is usually made to be glassy in form, rather than sootlike, to avoid problems of gas entrapment in the later stages of collapsing and pulling the preform. This process we will refer to as chemical vapor deposition (CVD) (see Figure 9.9b).

The CVD process, in which the reaction takes place at a relatively low temperature at the surface of the silica tube, involves a heterogeneous reaction, stimulated by the surface. The reaction can be made to proceed faster at higher temperature, in which case it occurs in the gas phase, a homogeneous reaction. In this case the resulting product is more akin to soot, and the final result depends very much upon the detailed conditions, so that it appears to be a matter of opinion as to where the soot process ends and the CVD process begins.

Finally, a third possibility is to produce bulk material by flame hydrolysis, both in doped and in undoped form, and to subsequently cut a rod of the core material and surround it by a tube of the cladding material. Such a process enables the independent laboratory to make use of the bulk manufacturing capability of an established manufacturer but appears to offer few other advantages.

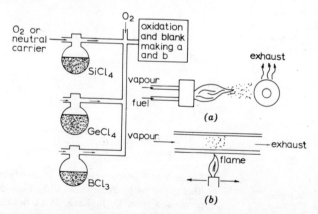

Figure 9.9. Two processes for making silica fiber: (a) the soot process and (b) the CVD process. Reproduced with permission from R. D. Maurer, *Proc. IEE.*, **123**, 581 (1976).

9.4.2 Soot Process

This process has the distinction of being the one used to produce the first 20 dB/km fiber in the world. It consists of hydrolyzing a mixture of $SiCl_4$ and O_2 in a gas burner to produce a spray of ultrafine SiO_2 particles, for cladding material. In the first experiments the gas stream was doped with $TiCl_4$ to produce a titania doped core material. The stream was directed down the core of a silica tube, as shown in Figure 9.10, where a layer of the soot was built up. If the tube was then heated and collapsed, the doped silica layer adhering on the inside would form a doped core inside the collapsed tube. Fiber was then pulled from the preform so produced. One major disadvantage of this approach was the fact that the titanium could exist in the Ti^{4+} and the Ti^{3+} states. Reduction from the $4+$ state to the $3+$ state tended to occur at the fiber pulling stage, and it was necessary to reverse this process by annealing the fiber in an oxygen

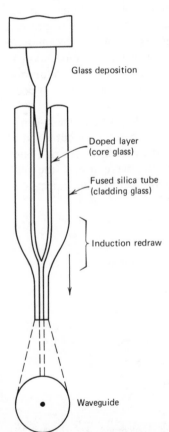

Figure 9.10. Process described in the first low loss fiber patent for producing silica fiber. Reproduced with permission from D. B. Keck, P. C. Schultz, and F. Zimar, U.S. Patent 3,737,292.

environment after pulling. This led to a fiber that was excessively fragile. It was also extremely difficult to produce a fiber with a large core by this process.

These problems appear to have been overcome by introducing a modified process as illustrated in Figure 9.9a. In this process the flame hydrolysis stage is essentially the same as before, but the soot is now deposited on the outside of a silica mandrel.[9] A typical sequence might involve a number of layers of doped silica to subsequently form the core, followed by a number of layers of pure silica to provide a cladding. The soot layers so formed are then sintered into a solid glassy mass and the mandrel removed, presumably by core drilling and polishing, to produce once again a preform which can be collapsed and pulled into fiber. Some of the engineering problems of this process are apparent, namely, the difficulty of handling the sintered preform and removing the mandrel without producing any contamination. However, one must conclude that these have been largely overcome since fibers with losses lower than 4 dB/km at 800 nm have been routinely produced by the Corning Glass Works, seemingly by this process.

A further change introduced was to shift from $TiCl_4$ dopant to $GeCl_4$. This avoids the problems associated with the oxidation/reduction of the titanium. However, it probably introduces other difficulties since the germanium is quite volatile, and at the temperature needed to collapse and manipulate preforms, typically 1800 to 2000°C, significant loss of the dopant by evaporation can occur with a resultant change in the fiber index profile. But against this disadvantage the mandrel process seems to be able to produce fiber having cores in the range of 50 to 70 microns with numerical apertures in the range from 0.15 to 0.23. Such fibers are excellent for use with LED sources, where a multimode fiber is necessary if enough power is to be captured.

A further very great attraction of the soot process is the potential it offers for very rapid deposition of material, since the reaction rate depends upon the rate of flow of gaseous reactants through the burner. The process also allows the production of graded core fibers in which the index profile is very well controlled. Such fibers are produced by depositing many layers of soot, the composition being carefully changed from layer to layer so that the resulting preform has exactly the right profile. This attraction is shared by all the CVD processes, although the problems of precision control of the index profile are likely to vary in detail.

9.4.3 Chemical Vapor Deposition Process

The CVD process[9] as illustrated in Figure 9.9b can have a number of gas feeds, of which only a few would be used at any one time. The major difference compared to the soot system is that the reaction of the chemicals involved no longer takes place in a flame at the entrance to the apparatus; rather, the unreacted gases are mixed together and flow to the deposition region, where the reaction is initiated by the heated surface and surroundings. In the first work using this process, the gases reacted were silane (SiH_4) and Germane (GeH_4). The reaction temperatures were low, so that the silica support tube remained rigid, but the reaction rate was very low, typically taking 1 to 2 days for a preform yielding a 1 km length of fiber.

Later modifications of the process have involved the use of chlorides (as shown in Figure 9.9b) with the oxygen gas stream. Much higher concentrations of the reacting species have become possible, although the reaction now occurs at a higher temperature, in the range of 1300 to 1600°C, so that problems arose from sagging of the support tube. This was overcome by rotating it on a glass lathe. The net result was that the reaction and deposition rates were greatly speeded up. The reaction now occurs partially on the surface of the preform but also partially in the gas phase, generating a fine soot deposit. However, because of the high perform temperature in the reaction zone, the soot fuses with the preform surface as it is deposited, yielding a glasslike layer. In this way deposition rates have apparently been increased by several orders of magnitude, although the distinction between this modified CVD process and the Corning soot process becomes even less clear cut. The switch to chlorides greatly reduces the problems of OH contaminates in the fiber.

The choice of dopants in both the CVD and the soot process is wide, and the precise advantages or disadvantages in the production of each do not appear to have been stated. We have already noted that titanium is unattractive because of the necessity of controlling its oxidation state. A popular dopant is germanium but two problems appear to be associated with its use. The first, as already noted, is its volatility, leading to a loss of the dopant at the preform collapse stage, which in turn gives rise to a fiber having a doughnut or toroidal core cross section instead of a cylindrical section. The second arises because germanium doped silica has a lower viscosity than undoped silica, so that, when a preform with a germanium doped layer inside a pure silica support tube is collapsed, the germanium tends to become liquid while the support is still substantially rigid. This makes precision dimensional control extremely difficult.

Another dopant that has been successfully used in place of germanium is P_2O_5,[10] produced from the gas phase $POCl_3$. This appears to have advantages over germanium in both the above areas, suffering less from volatilization and apparently less also from differential viscosity problems, although few firm data are available on these matters. Large core fibers (50 to 70 micron diameter) are produced rapidly, and low loss levels have been achieved with this materials system.

Another interesting variant concerns the use of BCl_3 as the dopant gas[11] to generate a deposit of B_2O_3 mixed with SiO_2. This technique was developed after it was shown that a bulk glass of SiO_2 with B_2O_3 rapidly quenched had a refractive index significantly lower than that of pure silica.[12] It follows that such a composition deposited from the vapor phase can lead to a suitable cladding material when used with a pure silica core. Several techniques based upon this observation have been described, in which both the SiO_2-B_2O_3 cladding and the SiO_2 core have been deposited from the vapor phase and the cladding only has been deposited, the resulting lined tube being collapsed onto a rod of pure silica, which subsequently became the core of the fiber.

The fibers show a greater index difference between the core and cladding than would be expected on the basis of bulk measurements of refractive index on glasses of the same composition. This arises because pure silica has a very low coefficient of expansion, so that the borosilicate glass in the fiber is held in a state of mechanical extension, and thus exhibits a lower density than the bulk equivalent material. This extension or strain can be annealed out in the fiber, in which case a fiber of very low index difference is obtained. The rate of annealing with temperature has been studied by Camlibel et al.[13] and is shown in Figures 9.11 and 9.12, which illustrate the fact that, if operation at an elevated temperature is not required, this fiber is likely to be perfectly satisfactory.

9.4.4 Graded Core Preforms

We should note that increasing interest centers upon the deposition of more than one dopant for a given fiber and also upon depositing multiple layers, the concentration in each layer being carefully monitored and varied in a predetermined manner. In the former category a typical deposition might be of B_2O_3-SiO_2 to form a cladding and GeO_2-SiO_2 to form the core. Such a pair could be chosen to enable the viscosities of the two layers to be more equally matched, thus facilitating the collapse stage. Also, it is found that the presence of some of the dopants allows greatly increased gas flow and reaction rates to be used, a factor of considerable commercial importance in implementing a process.

In the deposition of multiple layers with different dopant concentrations, the apparatus used is essentially as shown in Figure 9.9a or 9.9b except that the flow valves are program controlled. Typically, the hot zone is traversed back and forth along the tube, holding a constant gas composition for each traverse. After a traverse is completed, the hot zone returns to the start of the tube, the composition is adjusted, and another layer is

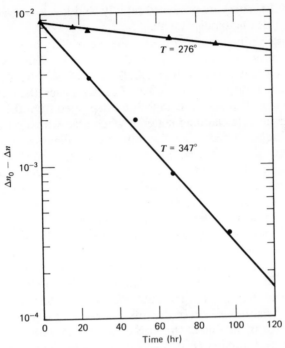

Figure 9.11. Measured values of index difference versus temperature for a silica core fiber with borosilicate cladding. Reproduced with permission from I. Camlibel, D. A. Pinnow, and F. W. Dabby, *Appl. Phys. Lett.*, **26**, 185 (1975).

deposited, until perhaps 100 layers are put down. In this way the index profile obtained in the fiber can be controlled to a surprisingly high accuracy, and graded-index fibers having very low mode dispersion have been manufactured (see Section 6.3).

The presence of layers arising from the deposition process is sometimes clearly evident in the preforms and fibers. This appears to be particularly true of CVD fiber and less so of outside deposition soot fiber. Interferometric analysis of the index profiles of preforms shows the presence of an "oscillating" index profile following a mean curve of the form desired, as illustrated schematically in Figure 9.13. The layering[22] obtained is shown in the scanning electron microscope (SEM) photograph (Figure 9.14) of an etched fiber end. The reason why the effect is so marked in the CVD inside deposition process is believed to be related to the fact that the deposition takes place in an extremely nonuniform environment. The reaction of the gases is initiated as they enter the hot zone, increases as the temperature rises, and decreases again as the reactants are exhausted or the hot zone is terminated. The reaction products are likely to be temperature dependent,

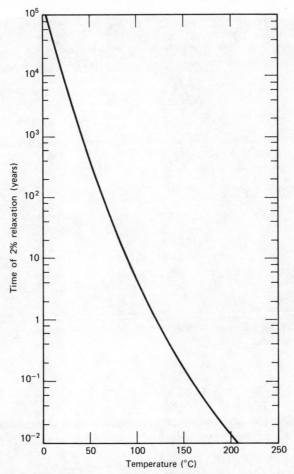

Figure 9.12. Effective lifetime versus temperature for the fiber of Figure 9.11. Reproduced with permission from I. Camlibel, D. A. Pinnow, and F. W. Dabby, *Appl. Phys. Lett.,* **26** 185 (1975).

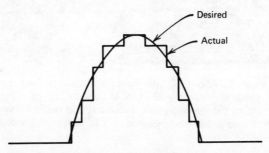

Figure 9.13. A schematic picture of the index distribution of a graded core fiber produced by layer deposition.

Figure 9.14. The end of a silica-CVD fiber viewed in a scanning electron microscope after it has been cleaved, lightly etched, and gold coated to show the imprint of the deposition layers surviving throughout the collapse and pulling processes. Also visible is some distortion of the center of the core. Reproduced with permission of J. E. Midwinter, unpublished work.

and there is always the possibility that previously deposited dopant will be driven off by evaporation as the hot zone approaches a previously deposited region. The resulting layers can thus fluctuate widely in their composition, in this process. In the soot process the new material presumably deposits on the previously deposited material at a lower temperature, minimizing the problem of loss of dopant by evaporation, while the reaction is initiated not by a localized (and spatially varying) hot zone, but within the plasma arc or flame.

9.4.5 Production Factors in Soot and Chemical Vapor Deposition Processes

Both processes offer one great advantage over techniques using bulk glasses melted from powders, namely, the ability to produce easily material which is exceptionally pure, simply because it is formed directly from the vapor phase. The vapors themselves are derived from liquids which can be purified by distillation. The CVD process apparently offers the further advantage that the whole deposition and collapse process can be carried out in a completely closed environment, thus minimizing the problems or recontamination by handling or by atmospheric transport (of OH, for example).

A further advantage appears to be that the starting materials, because they are readily purified, are cheaper than the highly purified powders required for glass melting. However, this may be less of an advantage than it seems, since the efficiency of conversion of material from the liquid state through to the solid state in the fiber can be rather low when compared to a multicomponent glass process, such as the double crucible.

Finally, a factor that presents at the production stage major engineering problems that are absent at the laboratory stage with both the CVD and the soot processes concerns the economical size of preform that can be produced. A 3-mm-diameter preform of 1 m length will yield approximately 1 km of fiber of 100 micron diameter. Scaling the preform to 1 cm diameter will increase the fiber length to 10 km, but at the expense of additional sophistication at the pulling stage and at the collapse and deposition stages. Thus the production of, say, 10^5 to 10^6 km of fiber per annum could rapidly run into a very large number of individual preforms.

By comparison the double-crucible type of process is relatively easy to scale up to a continuous process offering very high output. It remains to be seen whether in production the latter will offer a sufficiently significant cost saving to be attractive. In all probability both techniques will find application in different areas.

9.4.6 Other Preform Techniques

The most immediately obvious preform technique is the rod-in-tube preform, in which a rod of core glass material is shaped to fit within a tube of cladding composition material. Unfortunately, this superficially simple and attractive technique runs into severe problems in practice, associated with the difficulty of getting very clean and smooth surfaces on the rod outer and the tube inner surfaces, the ones that ultimately become the core-cladding interface. If the preform material is silica, good surfaces can apparently be produced, primarily, one suspects, because silica is a single-component material and thus etches uniformly. Multicomponent glasses, by comparison, tend to etch out their separate components at different rates, leaving a surface with a very open network that not only attracts traces of impurity but also is of indeterminate composition.

Impressive results with the rod-in-tube approach have been reported by Rau et al.,[14] who have made fiber having a pure silica core with a fluorine doped silica cladding, the latter dopant reducing the index of pure silica. They have reported loss figures as low as 5 dB/km at 850 nm.

In an alternative preform technique described by the Pilkington[15] Research Laboratories, a two-component preform is formed directly from the melt by pulling vertically upwards from a crucible in which the cladding glass has been allowed to flow over the core glass, as illustrated in Figure 9.15. This has the great attraction that the core-cladding interface is formed directly from the molten glasses, so that problems of contamination during working do not arise. It is also possible in principle to produce very long preforms, so that the limitations of the silica CVD batch processes may be much less evident with this process. Fiber losses have tended to be high, 30 to 100 dB/km.

9.4.7 Assorted Preform Techniques

Three other fibers that are effectively preform fibers should also be noted. The first is the liquid core fiber.[16] Here a silica or glass tube is drawn down to yield a bore diameter of 50 to 100 microns, and a higher index liquid is pumped into it after pulling, to form its guiding core. This fiber has achieved exceptionally low loss and low mode mixing, but it does not appear to be of very much practical interest for systems use.

The second is the so-called single-material fiber, which is usually constructed from pure silica with a very complicated preform assembled from numerous optically polished tubes and plates to form the type of structure shown in Figure 9.16.[17] This fiber has achieved a transmission loss of less than 10 dB/km and exhibits a number of interesting and surprising properties, but again seems of little interest for systems use.

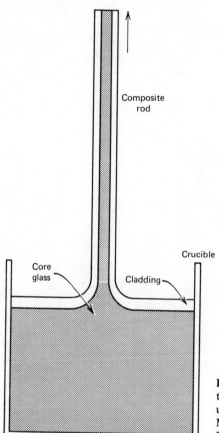

Composite
rod

Crucible

Core
glass

Cladding

Figure 9.15. Technique for preform production described by Pilkington Glass works, using up-draw from a composite melt. After A. M. Reid, D. W. Harper, and A. Forbes, British Patent 50,543 (1969).

The third fiber is of plastic clad silica,[18] in which a pure silica rod is pulled down to a diameter of 50 to 100 microns. This silica filament is then enclosed within a loosely fitting polymer tube, as shown in Figure 9.17a. Once again, very low transmission has been achieved, and the fiber had the attraction that readily available materials are used in its production.

A very similar fiber has been produced in which the plastic is extruded to contact the silica,[19] thus making a solid contacting cladding, as in Figure 9.17b. Other fibers have used silica cores with thin plastic claddings applied by dipping in solutions of plastic or molten plastic baths. All of these are of dubious attraction for telecommunications but offer distinct possibilities for use in less stringent environments, such as many military or commercial applications where either lower data rates or shorter link lengths are used.

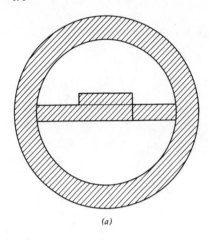

(a)

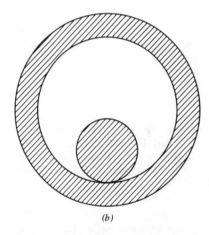

(b)

Figure 9.16. Some typical preforms of silica shown in cross section for pulling into single material fiber.

9.5. PREFORM FIBER PULLERS

9.5.1 General Considerations

Figure 9.18 illustrates a preform fiber puller in schematic form. Casual inspection makes clear that two vital parameters must be controlled if a uniform fiber is to be obtained, namely, the winding speed for the fiber and the feed speed for the preform. Evidently, when averaged over time, material cannot accumulate in the meniscus region of the hot zone of the furnace so that, if R_f and R_p, V_f and V_p are the radii and velocities, respectively, of the fiber and the preform, we must have

$$\langle R_f^2 V_f \rangle = \langle R_p^2 V_p \rangle$$

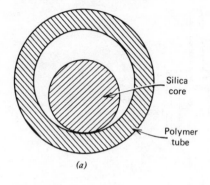

Silica
core

Polymer
tube

(a)

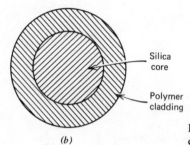

Silica
core

Polymer
cladding

(b)

Figure 9.17. Simple fibers composed of pure silica core within extruded polymer sheaths.

when the average is taken over a period of time of perhaps minutes. However, for short periods the above relation does not need to hold, since the position of the meniscus can fluctuate up and down relative to the furnace winding or heat source. This can lead to slow variations in fiber diameter or, in certain circumstance, to very rapid fluctuations, and must be considered at the design stage. In particular, one can expect to see such fluctuations when a preform is being started, since it is almost inevitable that the meniscus will not form initially with a new preform at the position of long term stability, even if all other parameters are perfectly controlled.

Instability of this nature may be reduced by the use of feedback control, acting upon the fiber winding speed. Equipment to monitor the diameter of the drawn filament immediately below the meniscus region is commercially available, as a development of similar machines used for monitoring the diameters of drawn wires. However, the degree to which such instruments can be used to control transients is not clear, since no results appear to have been published.

Little information appears to have been published on the optimum temperature profile for a preform puller. It is clear (by reference to Figure 9.18) that the preform must pass through several discrete stages. Initially

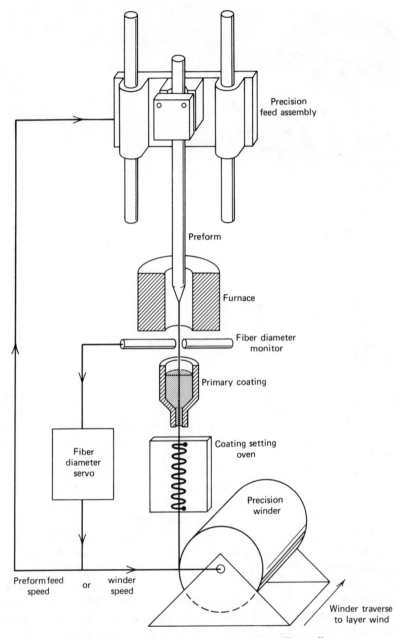

Figure 9.18. A schematic layout of a preform fiber puller.

the preform enters the hot zone, and commences to heat up toward the glass softening point. The meniscus begins to form, typically, when the viscosity reaches a value of 10^4 poise. However, if the hot zone does not rapidly terminate, a fiber filament cannot be formed since a large molten gob will simply detach itself from the preform, or the filament, once formed, will melt off. It is essential, then, in order to neck down and stably form a filament, that the glass cool at the same time, so that there is sufficient strength in the filament to maintain the tension necessary for a stable meniscus.

Analysis of the production of a meniscus is a complex mathematical problem.[20] The meniscus is maintained by a balance of pulling tension, glass surface tension, glass weight, and viscosity. The temperature conditions within the material are controlled by complex heat transfer equations involving conduction, radiation, and transfer through the glass heat capacity since the material itself is flowing. Several analyses have been published for special cases, but for the particular case of interest one usually finds that the data necessary to start the analysis are incomplete. For example, the viscosity-temperature and surface tension-temperature relations are known only for a few selected compositions of glasses. In practice, then, it appears that most workers design a furnace to yield a short hot zone and then experiment with that to obtain stable conditions.

The above discussion has concentrated primarily upon the vertical stability of the meniscus. We should note also that, if the preform is not fed into a circularly symmetrical hot zone, with the preform axis precisely aligned with the center of zone symmetry, there is a real possibility that the meniscus formed will tend to move off center, resulting in a fiber of noncircular cross section.

9.5.2 Furnace Design

The method of production of a suitable hot zone depends critically upon the materials system used. Pulling silica preforms produced by the CVD or other process generally requires a wall temperature in the region of 2200°C. Such a temperature can be produced by the use of electrically heated graphite elements. However, if these elements are not to oxidize rapidly, they must be shielded from oxygen or air, and this is most readily done by using a high melting temperature liner, such as zirconia, leaving the graphite element sealed in an enclosure filled with inert gas. Such furnaces are available commercially and apparently work well, although the penalty must be accepted in operation that the zirconia liners are extremely susceptible to thermal shock and will probably not withstand even gentle thermal cycling. If it is desired to pull preforms of a lower melting temperature glass, at temperatures up to 1200 or 1500°C, a metal wound electric furnace can be used to provide the heat source.

9.5.3   Laser Heating

A particularly ingenious if complex way of providing a very tightly controlled and short hot zone is to use a CO_2 laser as the energy source. Such a puller has been described by Paek,[21] and one version is illustrated in Figure 9.19. The laser beam power must be precisely controlled at the outset. The beam then enters the rotating mirror assembly, so that the emerging beam scans around the surface of a cylinder a few inches in diameter. This impinges upon the tilted mirror, which has a central hole to allow the pulled fiber to pass through it. The scanning beam is then brought to a rotating focus at a single point where the meniscus is formed on the end of the preform by means of the concave mirror. Provided that the preform is accurately centered on the focal point of the optical system,

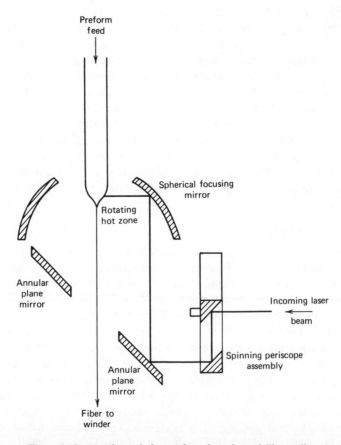

Figure 9.19.   A schematic layout for a laser heated fiber puller.

the meniscus is uniformly heated from all sides and is highly localized because of the very narrow focused beam. Such pullers allow one to overcome the transient fluctuations in fiber diameter seen with long hot zone preform pullers that arise from the mechanical drift of the meniscus position.

9.5.4 Fiber Take-Up

The fiber take-up may be engineered several ways. The simplest is to wind the fiber onto a drum in the manner of wire on a reel. The drum is rotated by a precision drive and is simultaneously scanned sideways to produce a winding at the correct pitch. For a feeling of the scale involved, consider the following. If we take a pitch of 200 microns (50 turns/cm) on a 25-cm-diameter drum, 1 km of fiber can be wound in a single layer on a drum length of about 25 to 30 cm. Similarly, a large winding drum of, say, 1 m diameter and 1 m length will take, in single-layer winding, about 15 km of fiber. Winding speeds can vary greatly but are typically in the range of 1 km/hr at the lower limit to 1 km/min at the upper limit. It is clearly important for the drum to be centered with precision on its axis, for the speed to be well controlled, and for the drum surface to be smooth if the fiber is to be measured on it. It is also important to note that, almost regardless of the drum surface finish, if the fiber is tensioned upon it some microbending loss will be induced in the fiber and the measurements so obtained will be pessimistic. Usually the problem is avoided by cooling the drum so that the fiber becomes slack when measured. Alternatively, the drum may be heated above ambient temperature while winding so that, in storage, the fiber will be under zero tension but will then slip around on the drum surface.

A second way to take fiber off the puller is to use a capstan with the fiber running over a precision capstan wheel under pressure from an idler wheel. In this way precision speed control can be obtained for the fiber as it is pulled, and the fiber flow can be maintained even if short lengths are subsequently wound on individual reels, thus avoiding the starting and stopping introduced by changing reels, which inevitably introduces some dimensional transients in the pulled fiber at each change. Having introduced a capstan take-off, it is possible to interface the fiber puller directly with a plastic extrusion line to package the fiber in dimensions at which it is robust for handling. Very long continuous lengths can then be wound in multiple layers on flanged drums. In the unpackaged state it is desirable to restrict the winding to a single layer per drum since overwinding incurs a very high probability of damage to the fiber surface and subsequent breakage, either from the ensuing weakness of the damaged surface or from entanglement due to turns that have slid over one another.

Measurements on overwound fibers are also, in general, very pessimistic since each overwind point is usually associated with a great deal of mode coupling and loss from the sharp bend in the fiber.

REFERENCES

1. K. J. Beales, C. R. Day, W. J. Duncan, J. E. Midwinter, and G. R. Newns, *Proc. IEE*, **123**, 591 (1976).
2. B. Scott and H. Rawson, *Glass Technol.*, **14**, 115 (1973).
3. T. Inoue, K. Koizumi, and Y. Ikeda, *Proc. IEE*, **123**, 577 (1976).
4. H. S. Carslaw and J. C. Jaeger, *Conduction of Heat in Solids*. Oxford: Clarendon, 1959.
5. K. B. Chan, P. J. B. Clarricoats, R. B. Dyott, G. R. Newns, and M. A. Savva, *Electron Lett.*, **6**, 748 (1970).
6. J. Guttman, H. J. Heyke, and O. Krumholtz, *Opt. Quant. Electron.*, **7**, 305 (1975).
7. K. J. Beales, W. J. Duncan, P. L. Dunn, and G. R. Newns, *Optical Fiber Communication*, IEE Conf. Publ. 132, London, 1975, p. 32.
8. D. B. Keck, P. C. Schultz, and F. Zimar, U.S. Patent 3,737,292.
9. R. D. Maurer, *Proc. IEE*, **123**, 581 (1976).
10. W. A. Gambling, D. N. Payne, C. R. Hammond, and S. R. Norman, *Proc. IEE*, **123**, 570 (1976).
11. W. G. French, A. D. Pearson, and G. W. Tasker, *Appl. Phys. Lett.*, **23**, 338 (1973).
12. S. H. Wemple, D. A. Pinnow, T. C. Rich, R. E. Jaeger, and L. G. Van Uitert, *J. Appl. Phys.*, **44**, 5432 (1973).
13. I. Camlibel, D. A. Pinnow, and F. W. Dabby, *Appl. Phys. Lett.*, **26**, 185 (1975).
14. K. Rau, A. Muhlich, and N. Treber, Paper TuC4, *Optical Fiber Transmission, II*. Williamsburg, Va.: OSA/IEEE, February 1977.
15. A. M. Reid, D. W. Harper, and A. Forbes. British Patent 50,543 (1969).
16. D. N. Payne and W. A. Gambling, *Electron Lett.*, **8**, 374 (1972).
17. P. Kaiser, E. A. J. Marcatilli, and S. E. Miller, *Bell Syst. Tech. J.*, **52**, 265 (1973).
18. P. Kaiser, A. C. Hart, and L. L. Blyler, *Appl. Opt.*, **14**, 156 (1975).
19. K. Yoshimura, T. Nakahara, A. Tsukamoto, and A. Isomura, *Electron Lett.*, **10**, 534 (1974).
20. F. T. Geyling, *Bell Syst. Tech. J.*, **55**, 1011 (1976).
21. U. C. Paek, *Appl. Opt.*, **13**, 1383 (1974).
22. J. E. Midwinter, Unpublished work.

10

Measurements of Losses in Bulk Glasses and Fibers

10.1. INTRODUCTION

The optical losses of materials can be divided into two broad categories, radiative and absorptive. In this chapter we are concerned with the techniques for measuring both mechanisms in glasses that are in the form of either bulk samples or fibers. Measurements of the losses in bulk glasses are generally the more difficult to perform if the loss is at low level because very small signal levels are usually involved, which are difficult to measure accurately. On the other hand, although it is generally easy to measure with precision effects in fibers, the interpretation of these effects may be complex and their repeatability very uncertain. This dilemma arises because in fiber one has an ideal sample shape, namely, a very long, perfect sample, but unfortunately it is not usually valid to assume that, because the sample is uniform with length, the effect under study is equally uniform with length. In practice, in a multimode fiber the effects are usually mode dependent and the integrated effect over all modes is nonlinear with length. This must be borne carefully in mind when designing a measurement procedure.

10.2. MEASUREMENT OF ABSORPTION LOSS IN FIBERS AND BULK GLASSES

In materials having a high absorption loss, say 1000 dB/km or more, measurement is relatively simple, but at the levels experienced in the fibers used for transmission systems, measurement requires both special techniques and extreme care in their use. The ability to make the measurement, particularly in fiber, can be extremely useful to provide backup information for a total attenuation measurement because it is very difficult to measure the fiber's radiative losses in a manner that truly represents the loss rate for a long length of fiber.

197

The measurement of absorption alone, either in a fiber sample or in a bulk sample, must be carried out by a calorimetric technique, so that scattered energy from the transmitted beam can be differentiated from energy directly absorbed. A number of successful techniques have been reported, all having much in common. A general feature is the use of a laser source to obtain enough optical power through a small sample. The reason for this is easily seen. A loss of 1 dB/km by absorption corresponds to a loss of 2.3×10^{-6} cm^{-1}. Thus with a transmitted power of 0.1 W the absorbed power is 230 nW/cm. The temperature rise produced by this power depends upon the sample heat capacity, H, per unit length. Unless this is made as small as possible, the temperature change is so small as to be unmeasurable or measurable only by the most sophisticated techniques. The problems of simultaneously obtaining a very small sample size and a large transmitted power greatly favor the laser source, whose coherence allows the production of a thin pencil beam that is ideal for transmission through a rod shaped sample (or for launching into a fiber). Against this advantage it must be recognized that a penalty is paid through the loss of wavelength tunability, since most dye lasers are insufficiently powerful in the wavelength range of interest to be employed for such a routine measurement. However, this penalty is less of a problem than might be imagined since the total insertion loss of the fiber is measured with ease as a function of wavelength. A few well spaced absorption measurements therefore provide all the additional data necessary. In bulk glasses the limited wavelengths available may be more of an embarrassment.

Some of the early measurements of bulk glass absorption were made using small cube samples of glass with polished surfaces, suspended on thermocouple wires which were used to monitor the temperature rise on irradiation. This technique worked well with samples whose losses were in the range of 300 to 1000 dB/km, but at lower levels repeatable measurements could not be obtained. The major reason appeared to be that the sample surfaces were contributing as much absorption through contamination by water or polishing compound remnants as was the bulk of the sample. However, an advantage of the cubic (5 mm) sample was that a large f number optical system could collect sufficient power from a xenon lamp to allow wavelength tunable measurements to be made over the range of 500 to 1100 nm.

In the author's laboratory the problems of surface contamination were overcome by changing to a long, rod shaped sample so that the surfaces were well removed from the measurement region.[1] At the same time the rod ends were used to mount the sample rigidly by clamping them with O-ring seals into a massive metal enclosure so that the ends were also thermally clamped and any surface heating of the rod end would be thermally short circuited to the enclosure. The sample ends were index

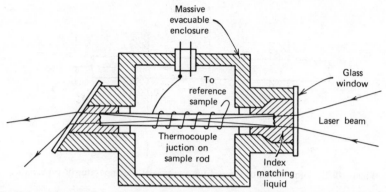

Figure 10.1. Apparatus for the measurement of the absorption loss of bulk glass samples in the form of rods, typically 2 to 3 mm diameter and 50 mm long. Reproduced with permission from K. I. White and J. E. Midwinter, *Opt. Electron.*, **5**, 323 (1973).

matched with liquid so that only minimal end preparation was necessary. (See Figure 10.1.) In this form the temperature rise of the sample was monitored by a platinum based thermocouple glued to the center of the sample. With this method corrections have to be applied for the effects of the thermally clamped sample ends, which distort the temperature measured in the center of the sample by introducing significant longitudinal conduction. This is taken care of in a simple correction factor at the time the results are calculated.

10.2.1   Theory of Measurement

Consider the rod shaped sample shown in Figure 10.2. We will assume that it is infinitely long, for the purposes of analysis, so that longitudinal conduction can be neglected. The heat source takes the form of a uniform energy source along the axis of the rod (the path of the laser beam as it traverses the sample). If the attenuation due to absorption is given by α cm^{-1}, the power dissipated in a unit length of sample (assuming $\alpha \ll 1$) is given by

$$\Delta P = \alpha P_0 \qquad (10.2.1)$$

where P_0 is the power traversing the sample. If the sample is of radius a, density ρ, and specific heat S, the heat capacity per unit length H is given by

$$H = \pi a^2 \rho S \qquad (10.2.2)$$

Initially, when the laser power is switched on, the temperature rise versus

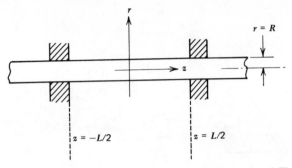

Figure 10.2. Dimensions of the rod for analysis of the apparatus of Figure 10.1

time can be considered linear so that

$$\frac{\partial I}{\partial t} = \frac{\Delta P}{H} \tag{10.2.3}$$

However, as the temperature of the sample starts to differ appreciably from the ambient, heat is lost from the sample to the surroundings. Suppose that the rate of heat loss is given by $\Delta E = B(T - T_0)$, where T_0 is the ambient temperature. Then the net rate of energy input is given by $\Delta P - \Delta E$, and the temperature is described as a function of time by a differential equation of the form

$$\frac{\partial T}{\partial t} = \frac{1}{H}(\Delta P - \Delta E)$$

$$= \frac{1}{H}[\Delta P - B(T - T_0)]$$

$$= \frac{\Delta P + BT_0}{H} - \frac{BT}{H} \tag{10.2.4}$$

The solution of this equation has the form

$$T = \left(\frac{\Delta P}{B} + T_0\right)\left[1 - \exp\left(-\frac{Bt}{H}\right)\right] \tag{10.2.5}$$

After infinite time the temperature reaches a steady state value given by

$$T_\infty = \frac{\Delta P}{B} + T_0 \tag{10.2.6}$$

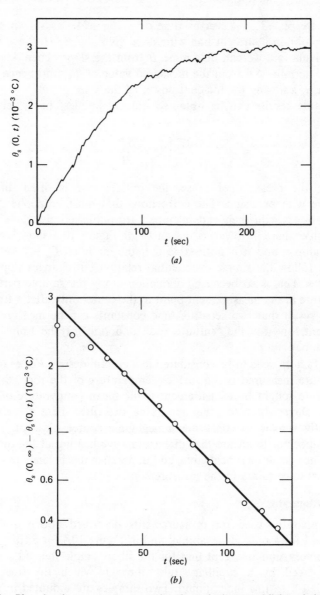

Figure 10.3. Plot of rod temperature versus time from the instant of light switch-on. In (b) the vertical axis is log $[T(\infty) - T(t)]$. This shows the two time constants, of the infinite rod initially and the finite rod when longitudinal heat flow is established. Reproduced with permission from K. I. White and J. E. Midwinter, *Opt. Electron,* **5**, 323 (1973).

Thus, if we plot $T_\infty - T$ versus time on a log scale, as shown in Figure 10.3b, we obtain a straight line with slope given by B/H. Since we know H by calculation, we can determine B from the slope. Then, knowing B, we can determine ΔP from the measured value of T_∞ and hence obtain a value for α, knowing the incident power of the laser.

The same results can be obtained quickly but less accurately by determining

$$\left[\frac{\partial T}{\partial t}\right]_{t=0} = \frac{\Delta P}{H} \qquad (10.2.7)$$

since, in the presence of noise, the initial slope is often difficult to determine with precision. The corrections that must be applied to finite length samples with their ends clamped are published elsewhere and will not be given here.[1] However, the curve of Figure 10.3 shows results for such a sample, and it is noticeable that the curve of $T_\infty - T$ versus time does not follow the simple exponential relation but changes slope after a finite time. This is so because conduction along the sample perturbs the temperature at the measurement point in the center only after a finite time, with the result that two separate time constants can be measured in this experiment, one for the "infinite rod" and one for the finite rod (the second slope).

Other factors need to be considered in a careful design. For example, the temperature measured is the surface temperature of the rod, whereas the temperature sought by measurement is the mean temperature of the rod. Analysis shows that for the apparatus described here these two are sufficiently similar to constitute a negligible source of error. It is also vitally important to ensure that direct radiative heating of the thermocouple does not occur or is negligible, and in practice this is likely to constitute the largest uncertainty in the measurement.

10.2.2 Apparatus

The apparatus used for measurements described above is shown in Figure 10.1 for a bulk glass sample[1] and in Figure 10.4 for a fiber sample.[2] Both are very similar except that for the fiber sample the rod sample has been replaced by a capilliary tube through which the fiber is run, surrounded by a low index liquid. Two samples are mounted in the same enclosure so that one acts as reference for the other, with a junction of the thermocouple attached to each. The electrical signal is amplified by a low noise direct current (DC) amplifier and recorded on a pen recorder.

In operation the bulk apparatus needs considerable care in its assembly and use, particularly in mounting the thermocouples and aligning the optical beam through the sample. The fiber version is somewhat easier to

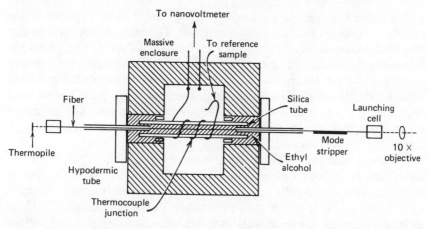

Figure 10.4. Modified version of the apparatus of Figure 10.1 used for the calorimetric measurement of fiber absorption. Reproduced with permission from K. I. White, *Opt. Quant. Electron.*, **8**, 73 (1976).

use since the thermocouple remains permanently mounted and the fiber only has to be led through the measurement region. Also, in the latter case the apparatus has a very low sensitivity to low angle scatter. This can be a severe problem in the bulk rod apparatus since it can lead to direct heating of the thermocouple. Such an effect can, in principle, be distinguished from the true effect sought, but in practice it is often difficult to do so with sufficiently high resolution. In the fiber apparatus low angle scatter is trapped in the sample at the cladding liquid surface since a lower index liquid is used.

In the fiber form of this apparatus, electrical calibration is easily carried out as a cross check on the interpretation method already described. This is done by replacing the optical fiber by a thin resistance wire and passing a known electrical power through it, which can be measured independently by the calorimetric process and by electrical measuring instruments.

In a simpler version of the bulk apparatus described by Pinnow,[11] the sample is suspended on fine threads within a laser cavity, and the temperature rise is monitored by a thermistor attached to the sample. No enclosure of the type previously described is used, although draught shielding is necessary for the sample. Because the sample is mounted within the laser cavity, very high powers traverse the sample so that large temperature rises are obtained. However, two major problems arise. The sample must have precision finished optical surfaces, and it must be of superb optical quality if the laser is not to be extinguished by its insertion within the cavity. These two factors make the technique of limited value to most users.

In another approach to the same problem, described by Zaganiaris,[3] the sample is mounted inside a calorimeter cell. The temperature measured is that of the complete cell rather than the sample within it. This offers the interesting possibility of measuring with the same sample the Rayleigh scattering and the absorption, since the scattered energy hits the cell walls instantaneously but the absorbed energy does not reach the cell walls until later, giving rise to a second, discrete temperature increase. The technique suffers from calibration difficulties, however, and is still under development.

It should also be noted that a number of extremely sophisticated techniques have been described for measuring the total insertion loss of bulk glass samples, usually using alternate long and short samples inserted in a beam of light while the transmitted power is measured.[4] These techniques can be made· to work but depend critically upon the surface finish of the sample and the uniformity of the glass used. They also require very large samples of glass and are thus of little interest as routine tools for material evaluation, but have been used to measure losses below 10 dB/km with reasonable precision—a great tribute to the scientists who operated them.

10.3. LAUNCHING INTO MULTIMODE FIBER

10.3.1 Stable Mode Distributions

We have seen in our discussion of mode coupling in multimode fibers that after a considerable distance ($\gamma_\infty z \gg 1$) a distribution of power among the propagation angles within the fiber develops that is essentially invariate with distance. This is called the stable mode distribution. Such distributions will be generated in multimode graded- or step-index fibers in which there is significant mode coupling. Fibers with values of $1/\gamma_\infty$ greater than the measured length cannot show such a distribution.

In making measurements on such fibers, it is important to recognize that most properties depend upon the actual distribution present. Alternatively, one may say that the losses, both by absorption and more significantly by scattering, together with the group delay, susceptibility to bending loss, and so on, are all mode dependent and thus, when measured for a total power distribution among modes, will depend upon the particular distribution used. Typical distributions that are readily excitable for step-index fibers are shown in Figure 10.5, together with a typical stable distribution for the same fiber. In Figure 10.6 we present the power attenuation that is typically observed for each distribution versus length, showing how errors of interpretation can easily be made in assessing the loss of a low loss fiber. If the fiber is excited by a single-mode laser with low angle launching

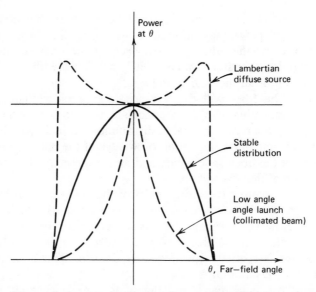

Figure 10.5. Some typical power distributions in step-index fibers for various situations. Reproduced with permission from M. Eve, A. M. Hill, D. J. Malyon, J. E. Midwinter, B. P. Nelson, J. R. Stern, and J. V. Wright, *2nd European Conference on Optical Communications, Paris, 1976.* Paris: SEE.

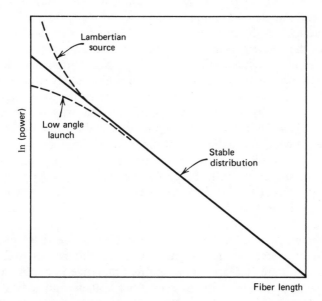

Figure 10.6. Typical curves of transmitted power versus length for the power distributions of Figure 10.5.

205

optics, the radiative loss in the fiber will be less than that of the stable distribution until power has diffused outwards, leading to an artificially low loss figure. At the other extreme, exciting the fiber by means of a Lambertian (diffuse) source that completely overfills the fiber end will lead to the excitation of all the guided and weakly leaky modes so that initially the power in the fiber will decay much more rapidly than the stable distribution. Launching with an intermediate angular and spatial distribution can closely approach the stable distribution, as in Figure 10.5. However, to do this one must have prior knowledge of the natural stable mode distribution for the fiber under study and must experimentally optimize the approximation of that distribution.

In practice, apertured launching optics of the type shown in Figure 10.7 seems to represent the preferred way of generating the approximate stable distribution. However, it retains the disadvantage that precise alignment of the fiber relative to the source is required, and it fundamentally assumes that the source is spatially and angularly uniform. This is often not the case, as with LEDs and GaAs lasers.

An alternative technique is to attempt to rapidly generate the stable distribution immediately after launching. In principle, it might be possible to do this by inducing heavy mode coupling in the fiber immediately after launching. This has been done by sandwiching the fiber between two rough surfaces—sandpaper, in some experiments. The result is to generate a power versus angle distribution in the fiber which is essentially independent of launching alignment and source characteristics and closely approximates the stable distribution. Typical results are presented in Figure 10.8, which shows the distribution after propagation through a 1 km length of mode coupled fiber and the distribution immediately after the launch point.[5] The latter shows somewhat more power in the high angle (lossy) part of the angle distribution, and would thus be expected to give high rather than low figures relative to the true stable distribution, but is nevertheless a good approximation of the required distribution.

Insertion loss studies on multimode fibers having no significant mode coupling are more difficult. Probably the best, although a very time consuming, approach is to measure the propagation loss versus launch angle. From these data the loss for any chosen distribution can then be calculated.

Measurements of radiative losses are evidently the most difficult to make if they are to be meaningful, since the differences in loss rates illustrated in Figure 10.6 are almost entirely radiative in nature, and show that the number obtained for a given fiber will depend almost entirely upon the launched distribution. We will return to these measurements later.

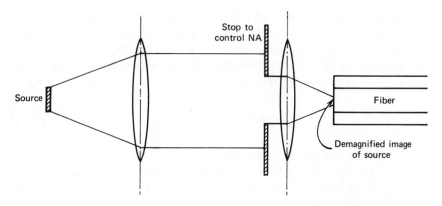

Figure 10.7. Launching optics used for simulating a stable distribution.

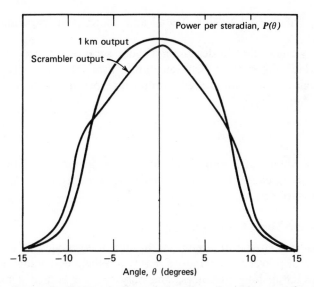

Figure 10.8. Power distributions in a step-index fiber after using a mode-scrambler for launching from a Lambertian source and after travelling 1 km in the fiber. Reproduced with permission from M. Eve, A. M. Hill, D. J. Malyon, J. E. Midwinter, B. P. Nelson, J. R. Stern, and J. V. Wright, *2nd European Conference on Optical Communications, Paris, 1976.* Paris: SEE.

A simple theoretical model has been proposed by Olshansky[6] for modeling the stable mode distribution for the purposes of studying mode filtering and joint loss effects. He proposes a simple relationship for the power in any given mode. It is defined in terms of one adjustable parameter ρ and the single-mode number notation (see Section 7.2.2) and takes the form

$$P(m') = P_0 \left(1 - \frac{m'}{M'}\right)^\rho \qquad (10.3.1)$$

where $P(m')$ is the power in mode m'.

Using the equivalence between the single-mode number notation and the LP_{lm} mode notation leads to a relation for the power in the LP_{lm} mode:

$$P(l,m) = P_0 \left[1 - \frac{(2m+l+1)^2}{M'}\right]^\rho \qquad (10.3.2)$$

Olshansky reports that a value of $\rho \approx 1$ appears to fit the experimental data quite well.

Finally, we note that in designing a mode scrambler for a graded-index fiber in which the spacing between modes available for microbend coupling is constant we should seek to introduce bending with a wavelength given by (Appendix 3)

$$\Lambda = \pi a \sqrt{2/\Delta}$$

For typical values of a and Δ this leads to values of Λ of the order of 1 mm, a wavelength that is conveniently produced by mechanically roughened surfaces.

10.3.2 Single-Mode Launching

At the other extreme to attempting to launch all modes, or a controlled distribution of power among modes that will remain essentially unchanged after propagating long distances, one may seek to launch a pure single mode. The ability to do this is useful if one wishes to study in detail the nature of mode coupling in the fiber, the distribution of $^z k$ values among the modes, and so on. The most successful method for achieving such launching stems directly from the technology of integrated optics, using the prism launcher to excite the fiber modes.

The principle is shown in Figure 10.9. In its simplest form a fiber is laid against a high index prism, usually with an index liquid filling the gaps, and a laser beam is focused into the prism at a carefully controlled angle. Energy from the laser tunnels through the fiber cladding into the guided modes. By changing the angle of incidence of the laser beam, the z

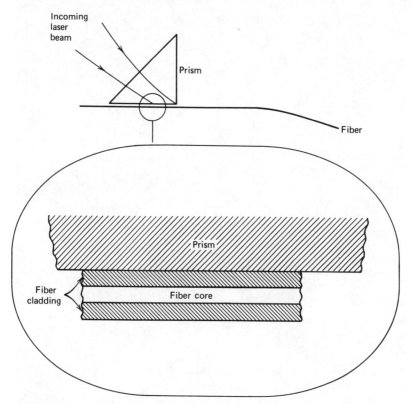

Figure 10.9.   Schematic of prism launching of single modes in a multimode fiber.

component of the incident beam is matched precisely to the $^z k$ value for the desired mode. Careful adjustment of beam widths, skew angles, and fiber prism gap allows one to couple to a single LP mode only.

The technique suffers, however, from one serious disadvantage. Fibers are designed so that their guided modes are well confined to the core, with infinitesimal field at the cladding-air boundary of the fiber. In reverse this means that it is practically impossible to couple significant power to these modes by tunneling. In practice one can excite the modes only very close to cutoff. Such a mode is shown in Figure 10.10, excited on a Corning silica fiber about 85 microns in core diameter.[7] To access the tightly bound modes one must penetrate the cladding in some way.

Two techniques have been proposed. The simplest to implement is to taper the fiber as shown in Figure 10.11 (with the fiber taper length greatly foreshortened to illustrate the geometry).[8] Guided modes traveling along the fiber and entering the taper can be visualized as cutting off in sequence, highest order modes first and the LP_{01}, which continues to be

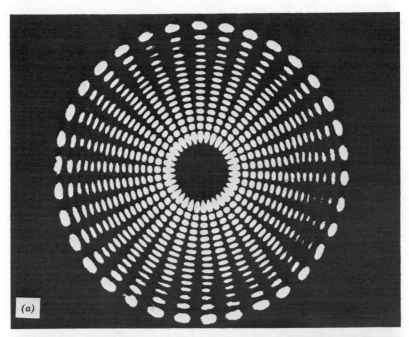

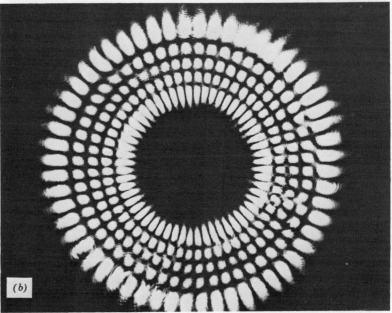

Figure 10.10. Intensity patterns of two LP modes, excited by means of a prism coupler. (*a*) An LP$_{17\ 16}$ mode excited in a step-index fiber. (*b*) An LP$_{28\ 5}$ mode excited in a graded-index fiber. Photos reproduced with permission of W. J. Stewart, Plessey Company Ltd., Caswell, England.

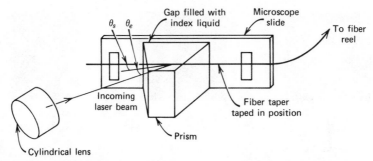

Figure 10.11. The prism-taper coupler for accessing all the guided modes of a fiber. Reproduced with permission from J. E. Midwinter, *Opt. Quant. Electron,* **7,** 297 (1975).

guided, last. As each mode cuts off, its energy tunnels across the cladding into the prism, giving rise to a beam in the prism, one beam per mode. A laser beam passing back down such an emerging beam will couple preferentially to that single mode, and careful adjustment of the incident laser beam will allow single-mode launching to be achieved. In this way the mode set shown in Figure 10.12 was obtained for a step-index fiber having a core diameter of about 15 microns.

Extending this technique to fibers of much larger core diameter is certainly difficult, requiring exceptional control over the tapering process, and to date does not appear to have been done. However, an alternative technique, which involves carefully polishing away of some of the fiber cladding before sandwiching the fiber against the prism, has been described by Stewart.[9] In this way he has obtained a measure of the $^z k$ values for all the guided modes of a Corning fiber, using white light excitation and dispersing optics to collect the power tunneling into the prism, as shown in Figure 10.13. This technique fully characterizes the fiber in principle, although so many data are collected in a single photograph that processing them becomes a major task.

10.4. MEASUREMENT OF RADIATIVE LOSS MECHANISMS IN FIBERS

This section is concerned with techniques for monitoring and quantifying the mechanisms by which energy *leaves* the fiber, excluding absorption, which has already been discussed. In other words, we are concerned with processes which lead to energy being scattered out of the fiber and subsequently lost to the surroundings, be they reel, laboratory, or cable structure.

Since most radiative loss mechanisms are dependent to some extent upon the relative excitation levels of the available guided modes of the

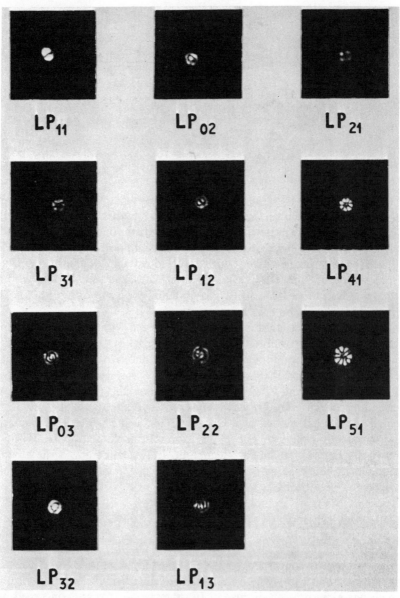

Figure 10.12. Mode set for a small V-value fiber, obtained using the coupler of Figure 10.11. Reproduced with permission from J. E. Midwinter, *Opt. Quant. Electron.*, **7**, 297 (1975).

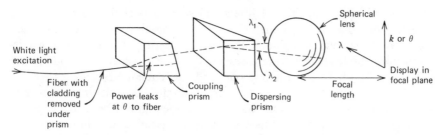

Figure 10.13. Schematic of the apparatus used to display the wavelength dependence of the ^zk values of the guided modes close to cutoff in a large core fiber.

fiber, it is important to exert some control over the mode power spectrum during such measurements if the numbers that are obtained are to mean anything.

Mode-dependent losses by radiative processes generally arise from one of three causes. The most obvious is scattering of energy from large mechanical deformations of the fiber. Another class of loss mechanisms, which in practice may be very hard to distinguish from the first one, concerns applied deformation in the fiber produced after pulling. Such deformations arise from irregularities in the drum surface on which the fiber is wound (either inherent or due to dust, etc.), or they may result from the method of fiber packaging, for instance, irregularities in an extruded plastic coating or deformations arising from the cabling process. These are usually called microbends.

A third class of losses is due to the tunneling of modes very near cutoff through the cladding. Such modes have small exponential decay coefficients for their evanescent fields and thus have appreciable field strengths at the cladding-packaging boundary. Since this boundary will inevitably have a high loss associated with it when compared to the fiber core and cladding materials, these modes will be heavily attenuated. In a perfect fiber such modes will carry no power within a short distance of launching, but if mode coupling is present energy is continually fed into them from the well guided modes and is in turn lost to the packaging. Thus the fiber as a whole exhibits a loss, as discussed in Chapter 7 on mode coupling. Such coupling often arises from microbends.

Experimentally, the separation of these mechanisms can be very difficult, although effects caused by packaging can in principle be separated by measuring the same fiber before and after the packaging is applied, assuming that the coating in question is not applied during the pulling process. Most of the discussion that follows is framed in terms of the study of radiative losses from fibers as pulled, free of coatings and the like, although some of the techniques can obviously be applied in both cases.

We should also note that all fibers will show a background radiative or scatter loss that is fundamental to the material from which they are fabricated, namely, that due to Rayleigh scattering. The loss in transmission from Rayleigh scattering is essentially mode independent, and we will discuss it first before examining our mode-dependent effects.

Measurements of radiative loss processes fall broadly into two categories: those in which all the power lost by a length of fiber is collected by a single detector, and those in which a more selective detection method is used, generally one involving some resolution of the angular distribution of the emerging power. The former measurements will usually use an integrating sphere to minimize the dependence of the detection sensitivity on angle, whereas the latter commonly utilize some form of angle selective, collimating optical system which can be scanned.

10.4.1 Rayleigh Scattering in Fibers

In Chapter 9 we discussed the causes of Rayleigh scattering and noted that it can be composition dependent and will, to some extent, depend upon the rate of quenching of the glass. Thus in practice one expects it to take a numerically different value (in dB/km) when measured in the fiber as compared to the loss in bulk glass of the same composition. The difference in Rayleigh scattering loss between core and cladding glasses is likely to be small (in terms of dB/km), unless the fibers have markedly different compositions in the core and cladding regions. The best known fiber that might show such an effect is the Japanese Selfoc, in which the core is thallium doped. Since thallium is a highly polarizable ion and is present in a matrix of relatively low polarizability, we might expect the core of such a fiber to show a much higher scatter loss than the cladding.

It follows from this preamble that the Rayleigh scattering loss is the loss by scattering from a perfect fiber. It might seem, therefore, that its measurement would in practice be possible only in a perfect fiber, and, strictly speaking, this is true. However, a very good estimate of the Rayleigh component of the total scatter loss can often be made in a less than perfect fiber, by carefully selecting the length of fiber to be studied and by controlling the modes propagating in it, because scattering by the two other most common effects can be largely suppressed or avoided. A sample of fiber, a few meters in length, that is free of discrete scattering centers (inclusions, bubbles, etc.) can be selected by pulling the fiber through an integrating sphere detector to find a uniformly scattering length. Scattering from core or cladding irregularities and tunneling losses from modes very close to cutoff can then be suppressed to a very high degree by deliberately launching low order modes, using a laser source focused onto the fiber and a long (1 to 10 cm) focal length lens. This is followed by a deliberate attenuation of the high order modes, using a series

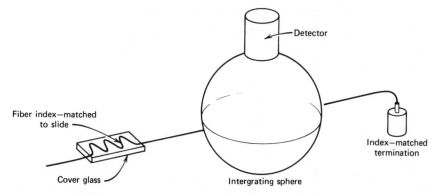

Figure 10.14. Apparatus for the study of radiative loss from fibers showing mode strippers and integrating sphere.

of sharp bends in the fiber (perhaps 1 cm radius), while the fiber is index matched to a glass plate with an index fluid of higher index than the core glass. (See Figure 10.14.) The remaining guided power is now largely restricted to the lowest order modes of the fiber and is thus well confined within the core region, with relatively little power traveling near the core-cladding interface. The integrating sphere following the mode stripper section will now measure the scatter loss from the low order modes of the guide. Use of a multiwavelength source, such as the krypton ion laser, allows the wavelength dependence of the loss to be studied.

Some typical results[10] obtained for a sodium borosilicate fiber by this technique, presented in Figure 10.15, nicely show the expected λ^{-4} dependence of Rayleigh scattering. The loss figures are converted to equivalent loss in units of decibels per kilometer by measuring first the scattered power, as a signal current, and then the total transmitted power through the fiber with the same detection system, by tucking the tail of the fiber back into the sphere. During the first measurement it is important to index-match the fiber tail end into liquid so that reflections are prevented.

In Figure 10.15 two curves are shown, one for the measurements as taken and one lower, because other measurements of the angular dependence of the scattering from the fiber showed that a small, residual, low angle scatter peak was present under the conditions of the measurement, corresponding to about 0.8 dB/km. Subtracting 0.8 dB/km gives the lower curve, so that the true figures for this fiber are assumed to lie between the two curves.

An alternative technique that can be used to measure the Rayleigh scattering in either fiber or bulk samples has been described by a number of authors.[11,12] It makes use of a high resolution Fabry-Perot interferometer to separate the different frequency components in the 90° scattered

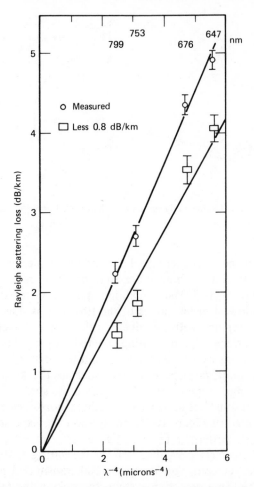

Figure 10.15. Measurements of Rayleigh scattering loss for a sodium borosilicate fiber, using a krypton laser source and integrating sphere. Reproduced with permission from M. H. Reeve, M. C. Brierley, J. E. Midwinter, and K. I. White, *Opt. Quant. Electron.,* **8**, 39 (1976).

light. In this way one can measure directly the ratio of the nonfrequency shifted Rayleigh scattering from the frozen-in fluctuations in the glass to the Brillouin scattering, which arises from the thermally driven density fluctuations present at room temperature and causes frequency shifts of a few gigahertz. The intensity of the Brillouin scattering can be calculated from the photoelastic coefficients for the material, using the relation

$$\alpha_{\text{Brill.}} = \frac{8}{3}\frac{\pi^3}{\lambda^4}kT\left(\frac{n^8 p_{12}^2}{\rho V^2}\right) \tag{10.4.1}$$

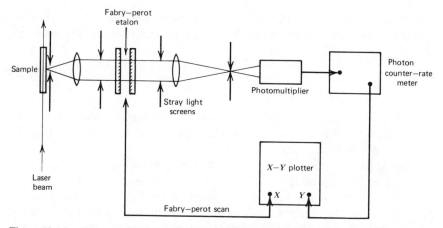

Figure 10.16. Apparatus for measuring Rayleigh scattering by comparison with the level of Brillouin scattering in the same sample, using a scanning Fabry-Perot interferometer to resolve the frequency components in the scattered light.

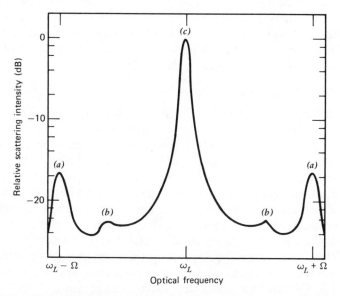

Figure 10.17. Scattered light spectrum from the apparatus of Figure 10.16 obtained when examining a silica sample. The peaks are (a) longitudinal acoustic mode scattering, (b) transverse acoustic mode scattering, and (c) unshifted Rayleigh scattering. Reproduced with permission from T. C. Rich and D. A. Pinnow, *Appl. Phys. Lett.*, **20**, 264 (1972).

217

Here p_{12} is the relevant photoelastic coefficient, which has been carefully measured in silica and can be calculated approximately using the relation[17]

$$p_{12} \simeq \left(\frac{n^2 - 1}{n^4} \right) (1 - \Lambda_0) \qquad (10.4.2)$$

where $\Lambda_0 = -(\rho/\alpha)(\partial\alpha/\partial\rho)$, and α is the polarizability. In equation 10.4.1 the parameter V is the velocity of the longitudinal acoustic mode in the glass, and k is Boltzmann's constant.

The apparatus for carrying out this measurement is shown in Figure 10.16. The 90° scattering is collimated by the first lens and passed through the pressure scanned Fabry-Perot interferometer. The central fringe of the interferometer ring pattern is selected by the pinhole, and the power within it is detected and displayed versus optical frequency (scanned by gas pressure) in an X-Y plotter to yield a trace of the form shown in Figure 10.17. Clearly visible are the Rayleigh line (c) and the scattering from the longitudinal (a) and transverse (b) acoustic modes. Since the intensity of longitudinal mode scattering is calculable, a self-calibrating measurement of Rayleigh scattering is obtained. In practice, this proves to be a useful technique for bulk or fiber samples, since most other scattering mechanisms giving rise to additional scatter losses lead predominantly to low angle loss, which is not seen by this apparatus.

10.4.2 Detection of Scatter Centers

Gross defects in a fiber are readily detected when the fiber is illuminated with a visible laser, such as a HeNe gas laser, and show up as very bright points of light at the point at which the irregularity occurs. Smaller defects in a multimode fiber may also show up as the start of several brighter turns when the fiber so illuminated is wound on a reel. Such defects scatter power into high order modes of the fiber or into cladding modes, which then leak their energy over distances of many centimeters or meters.

Smaller localized defects may be much more frequent and often are not visible to the eye when the laser illuminated reel is visually inspected. These defects can be located by pulling the fiber through an integrating sphere while it is illuminated with a laser source, and monitoring the scattered power as a function of fiber length. The presence of small defects then shows clearly by a local increase in the scattered power. Isolation of the offending length of fiber and its examination under an interference or phase contrast microscope will often reveal the nature of the defect. A number of imperfections can give the same signature in such a scattered power measurement, varying from minute gas bubbles in the core, cladding, or interface regions, through minute inclusions of refractory materials (from a glass melting crucible, for example), to local changes in the dimensions of the fiber, such as a local decrease in the core diameter.

In all such scatter measurements, care must be taken to strip the cladding modes and to attenuate the highest order guided modes of the fiber to ensure that the radiated power measurement is not excessively affected by small local bends in the measurement region. The presence of large amounts of power in the weakly guided high order modes of a multimode fiber will give rise to very large increases in scattered power whenever the fiber is slightly kinked, and could lead one to conclude, mistakenly, that the fiber contained scatter centers. A simple test for such effects is to run the same length of fiber twice through the apparatuus. Bends rarely occur at the same place!

10.4.3 Prolonged Irregularities

It should also be noted that the above measurements serve only to locate an imperfection of the fiber and that a microscopic examination is required to decide its nature. Some classes of defect are not visible under a microscope, for example, minute but prolonged ripples or deformations of the interface such as might be introduced into the fiber by mechanical vibration in the fiber puller. Ripples of this class can be expected to have two effects. First, they can scatter energy directly from the guided modes into the cladding or free space modes, that is to say, straight out of the guided wave system, so that a direct power loss occurs. A secondary effect may be simply to couple guided modes together so that energy continually shuttles between them. Such an effect will in all probability cause energy to be coupled across the guided mode-cladding mode boundary so that a power loss again results. The precise evaluation of such an effect is difficult, although its presence can often be detected fairly readily.

From the theory developed earlier, we know that guided modes of the fiber propagate with $\mathbf{k}$ vector components along the fiber axis (z) direction so that

$$n_1 k_0 > {}^z k > n_2 k_0$$

Waves whose ${}^z k$ is less than $n_2 k_0$ but greater than k_0 will propagate freely in the cladding and will therefore give rise to energy loss. These are cladding modes. To couple a guided mode having a z component of the $\mathbf{k}$ vector of ${}^z k(g)$ to a cladding wave having a component ${}^z k(c)$, an irregularity is needed which has a spatial wavenumber given by $\Delta k = {}^z k(g) - {}^z k(c)$.

Likewise, to couple two guided modes together, the spatial wavenumber of the irregularity must equal the difference in ${}^z k$ values for the two modes. Further symmetry requirements also arise in particular cases of coupling concerning the phi dependence of the irregularity, but we will not discuss these here. The conclusion that we wish to draw is that a guided wave with $^zk = {}^zk(g)$ scattered from an irregularity of wavenumber Δ^zk gives rise to new waves having ^zk components, $^zk(g) \pm \Delta^zk$. In the event that the smaller of these two is less than n_2k_0, a free wave will be generated in the cladding, making an angle θ with the fiber axis, where

$$\theta = \cos^{-1}\left[\frac{{}^zk(g) - \Delta^zk}{n_2k_0} \right] \tag{10.4.3}$$

Here, then, is the basis of a technique for studying such irregularities. The fiber is excited by a single mode, characterized by $^zk(g)$, and the angular distribution of scattering from the fiber, when it is immersed in a known index liquid, is measured. The power distribution of the scattered light outside the fiber·as a function of angle gives a direct measure of the spectrum of ^zk.

It must be remembered that in this experiment the angle actually measured will normally be the angle of light propagating in a liquid surrounding a fiber, rather than an angle within the cladding, since fiber claddings are, in practice, far from infinite in extent! Thus the index of the liquid must be taken into account. If θ_{liq} is the angle measured in the liquid from the fiber axis, and n_{liq} is the liquid index, then

$$\theta_{\text{liq}} = \cos^{-1}\left[\frac{{}^zk(g) - \Delta^zk}{n_{\text{liq}}k_0} \right] \tag{10.4.4}$$

Surrounding the fiber with an index liquid opens two other possibilities. In the first place, if $n_{\text{liq}} > n_2$, the minimum scattering angle within the liquid is given by

$$\theta_{\text{min}} = \cos^{-1}\left(\frac{n_2}{n_{\text{liq}}} \right) \tag{10.4.5}$$

that is to say, grazing incidence scattering in the cladding leaks into the liquid at a finite angle by refraction, so that the scatter power distribution exhibits a hollow cone in the forward direction. This is most useful in practice since it means that $0°$ scattering can be monitored without having to measure radiation emerging from the fiber at $0°$ to the axis, experimentally a near impossibility.

The second possibility again arises when $n_{\text{liq}} > n_2$, in which case "guided modes" can leak by tunneling through the cladding to give rise to a free wave outside, that is, they become leaky. We will return to this situation in the discussion of mode cutoff spectra.

Unfortunately, the angular measurement experiment has to date proved largely impossible to carry out for a variety of reasons. Experimentally, it is extremely difficult to excite a pure single mode in a multimode fiber, although the development of prism couplers used with optical fibers does open some possibilities. A second assumption involved in the above experiment is that, once excited, the mode stays pure.

However, the more serious objection to the experiment is simply that in most fibers direct coupling out of the fiber guided modes appears to be a very small effect compared to leakage by mode coupling through the mode sequence from the launched mode to the highest order guided mode, and thence by coupling of one form or another across the guided wave-leaky wave boundary and out of the guided wave system.

We recall from equation 7.2.16 that the values of $^z k$ for a fiber can be described in terms of the single-mode number, M, in the form

$$^z k = \sqrt{n_1^2 k_0^2 - M^2 \pi^2 / 4a^2} \qquad (10.4.6)$$

At cutoff, $^z k$ is given by $n_2 k_0$, so that M_{max} can be written as

$$M_{max} = \frac{2ak_0}{\pi} \left(n_1^2 - n_2^2 \right)^{1/2} \qquad (10.4.7)$$

In the region of cutoff, it follows that

$$\Delta^z k = -\frac{1}{2} \left(n_1^2 k_0^2 - \frac{M_{max} \pi^2}{4a^2} \right)^{-1/2} \left(-\frac{2M_{max} \pi^2}{4a^2} \right) \Delta M \qquad (10.4.8)$$

Setting $\Delta M = 1$ for microbend mode coupling, we obtain the result that

$$(\Delta^z k)_{co} \simeq \frac{\pi}{2a} \left(\frac{n_1^2 - n_2^2}{n_2^2} \right)^{1/2} \qquad (10.4.9)$$

For $a = 40 \times 10^{-6}$ m and a fiber of 1% index difference this leads to

$$\Delta^z k \simeq 55 \text{ cm}^{-1}$$

$$\Delta \lambda \simeq \frac{2\pi}{\Delta^z k} \simeq 1 \text{ mm}$$

Ripple wavelengths in the range of 1 mm are easily produced in a very small structure like a fiber, so that it is not surprising that effects due to this type of coupling usually predominate in radiative loss measurements of large core multimode fibers. By comparison, to scatter radiation directly from the well guided mode into the radiation field would require very small irregularities, typically of the order of 100 microns (0.1 mm) or less.

In general, we can say that scatter measurements on fibers will show three effects: Rayleigh scattering, scattering from discrete bubbles or inclusions, and energy leakage by tunneling from the highest order modes as the result of a cascade coupling process, the mode coupling process. Energy leaving the fiber by the last process appears as a low angle forward scatter peak at an angle corresponding to grazing incidence propagation in the fiber cladding.

10.4.4 Mode Cutoff Spectra

We have already discussed the fact that, when a fiber is immersed in an index liquid such that $n_{\text{liq}} > n_2$, loss of energy from guided modes can occur by tunneling. A guided mode having a z component, ^zk, gives rise to a free wave in the liquid at angle θ, where

$$\theta = \cos^{-1}\left(\frac{^zk}{n_{\text{liq}}k_0}\right) \tag{10.4.10}$$

When $n_{\text{liq}} \leqslant n_2$, the angle θ remains imaginary for all guided waves, but as n_{liq} becomes greater than n_2 some of the guided modes will leak until the point is reached at which $n_{\text{liq}} > n_1$ and all the guided modes of the fiber leak. In practice, the loss by tunneling, even in the latter situation, will be very small except for the modes very near cutoff. This follows directly by examining the form of the field distribution in the cladding.

The plane wave theory shows that in the cladding, for nominally guided waves, the field has the form

$$E = \frac{E_0}{\sqrt{r}} \exp(\pm i\,^rk\cdot r) \tag{10.4.11}$$

The condition $^zk > n_2 k_0$ makes ^rk imaginary so that the field has an exponential decay of the form

$$E = \frac{E_0}{\sqrt{r}} \exp\left(-\sqrt{l^2/r^2 + {^zk^2} - n_2^2 k_0^2}\ r\right) \tag{10.4.12}$$

Substitution of numbers in equation 10.4.12 shows that for the dimensions chosen for waveguide fibers, in which the cladding is typically 20 microns thick, the attenuation of the field through the cladding is very large except for modes very close to cutoff ($^zk \simeq n_2 k_0$). For example, for a well guided mode in a guide with a 1% index difference we have already shown that the value of $^zk - n_2 k_0 = n_1 k_0 - n_2 k_0 = 10^6$ m^{-1}. Thus we have the factor

TABLE 10.1   VALUES OF V OR ^Tka FOR THE FIRST FEW MODES OF A STEP-INDEX GUIDE AT THE CUTOFF OF EACH

		Mode				
LP	EH	TM	TE	HE	$^Tka = V$	Total Number of modes[a]
01	—	—	—	11	0	2
11	—	01	01	21	2.4048	6
21, 02	11	—	—	31	3.8317	12
31	21	—	—	41	5.1356	16
12		02	02	22	5.5201	20
41	31	—	—	51	6.3801	24
22, 03	12	—	—	32	7.0156	30
51	41	—	—	61	7.5887	34
32	22	—	—	42	8.4172	38
13	03	—	—	23	8.6537	42
61	51	—	—	71	87715	46
42	32	—	—	52	97610	50
71	61	—	—	81	9.9361	54
23, 04	13	—	—	33	10.1735	60

[a]Allowing for orthogonal polarizations.

Note that $^Tka = V_c$ where V_c is the cutoff value of the V parameter. In Figure 10.19 we show the results of such a mode assignment process, in which the measured values of the cutoff peaks have been plotted vertically and the expected values of V_c horizontally (actually V_c^{-1} is plotted to give a straight line plot). With care and some foreknowledge of the fiber dimensions and index difference, a unique assignment of the cutoff peaks to modes can be obtained, and thus the modes of the fibers are identified.

The shape of the peaks can also be used to yield information.[14] If we ensure that the fiber, from launching through to the exit point of the integrating sphere, is maintained straight and immersed in index liquid, we can assume that the loss associated with a particular mode at wavelength λ is constant and is given by $\alpha_s(\lambda)$. We know that α_s will be very small well short of cutoff but will rise rapidly as cutoff is approached. If the length of fiber from the launch point to the integrating sphere is X and the length in the sphere is Y, the power measured in the sphere is given by

$$P_{\text{int sphere}} = P_s \exp(-\alpha_s X)\left[1 - \exp(-\alpha_s Y)\right]$$

where P_s is the power originally launched in the mode s. If we assume that

$'k \cdot r \simeq 10^6 \times 20 \times 10^{-6}$, and the field has decreased by $\exp(-20)$. Howe as cutoff is approached, the factor $'k \cdot r$ starts to decrease rapidly and lo through tunneling increase extremely rapidly.

This effect can be put to good use to enable one to characterize modes of a fiber. The simplest experiment, which is applicable to fib carrying up to perhaps 100 modes, involves illuminating the end of a fi meters of fiber with incoherent light filling the acceptance angle and ar of the core and thus exciting fairly uniformly all the available guide modes and many cladding modes. The cladding modes can then b stripped by index-matching a few centimeters of fiber with high inde liquid ($n_{\text{liq}} > n_1$) to a glass plate. The fiber is then led through an integratin sphere, and the radiated power is measured as the input, exciting radiation from the monochromator/tungsten lamp source, is tuned (a linewidth of a few nanometers is ideal). The scattered power spectrum[13] typically is like that in Figure 10.18. As each mode of the fiber approaches cutoff, its energy tunnels rapidly through the cladding and a peak of radiated power is observed. These peaks are narrow, typically of half-power linewidth, 10 to 20 nm. Furthermore, they do not occur uniformly in wavelength space but form a distinct pattern related to the sequence in which modes cutoff, and this pattern can be used to identify each mode.

We have already noted that the mode cutoff corresponds to the mth condition:

$$J_{l-1}(^T ka) = 0$$

for the LP_{lm} mode. In Table 10.1 we list the values of $^T ka$ and the mode notation for the cutoffs in such step-index fibers for the first few modes.

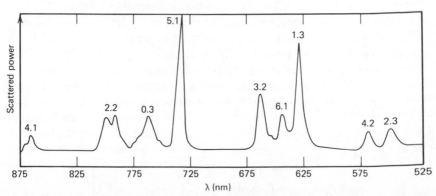

Figure 10.18. The scattered power spectrum versus wavelength for a small core ($V \simeq 10$) fiber when viewed with an integrating sphere. Reproduced with permission from J. E. Midwinter and M. H. Reeve, *Opt. Electron.* **6**, 411 (1974).

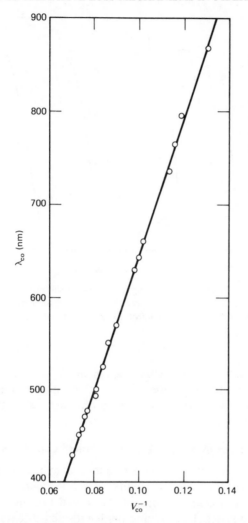

Figure 10.19. The results of a mode assignment for the peaks of Figure 10.18. Reproduced with permission from J. E. Midwinter and M. H. Reeve, *Opt. Electron.* **6**, 411 (1974).

$X \gg Y$, this function has a maximum when $\alpha_s(\lambda) = 1/X$. It is clear also that, knowing X and Y and the form of the observed peak, we can deduce the loss $\alpha_s(\lambda)$ of the mode s over a limited wavelength range.

The results of such an experimental evaluation are shown in Figure 10.20, in which the loss versus wavelength has been deduced using equation 10.4.13 from a measured mode cutoff spectrum. The fiber studied in this work had a core diameter of approximately 13 microns.

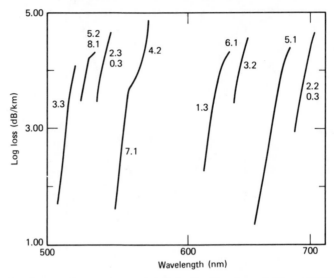

Figure 10.20. Curves of loss versus wavelength for the fiber shown in Figure 10.18 calculated using a simple theoretical model from the scattering data. Reproduced with permission from M. H. Reeve and J. E. Midwinter, *Optical Fibre Communication,* IEE Conf. Publ. 132, London, 1975, p. 16.

The loss versus wavelength curves given in Figure 10.20 can be calculated from a simple model of the tunneling loss mechanisms. A very good approximation to the experiment can be obtained using the plane wave approximation.

10.4.5 Measurement of Scatter Loss as a Function of Fiber Length

We saw earlier in our discussion of mode coupling that it is often of interest to induce controlled amounts of mode coupling in a fiber to modify the modal propagation properties. In general, such mode coupling induces radiative losses. By studying the length dependence of such losses in a multimode fiber, one has a nondestructive method for observing the buildup of a stable mode distribution. An apparatus to do this is shown in Figure 10.21. The fiber under study is wound on a temperature controlled reel. By varying the temperature, the reel can be expanded or contracted to vary the fiber tension and hence the degree of microbending loss. This loss can be monitored by measuring the insertion loss of the whole fiber length.[15]

The radiative component of loss on the reel can be monitored by taking a prism of higher index glass than the core glass of the fiber and by index-matching it across the surface of the drum so that it contacts the

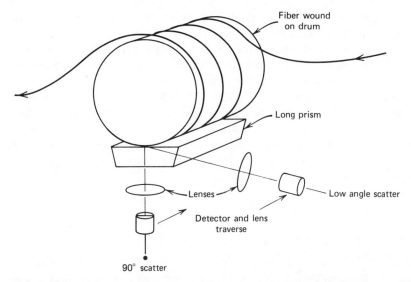

Figure 10.21. A simple apparatus for the study of radiative loss versus length of a fiber wound on a storage reel. Reproduced with permission of K. I. White, unpublished work.

fiber once every turn. The Rayleigh scattered power may then be measured by examining the 90° scattering to obtain a measure of the propagating power at each turn of the reel by looking through the lower face of the prism. The low angle scattering can be monitored by measuring the radiative power coming through the perpendicular face of the prism. Thus, with single-mode launching into a large core fiber, by monitoring both the low angle and the 90° scattered power as a function of power, it is possible to observe the point at which the stable mode distribution is achieved (ratio of low angle to 90° scatter constant), and the length required for it to be built up is also found. The technique is also a very sensitive one for detecting variations in fiber loss or the presence of large mode coupling scatter centers.

10.5. MEASUREMENT OF FAR-FIELD ANGLE FROM A FIBER (NUMERICAL APERTURE)

A simple apparatus to allow one to measure the far-field radiation angle of the energy emerging from a multimode fiber is useful for studying mode coupling. The fiber end is terminated with a smooth surface, using one of the standard end preparation techniques or through index matching. A detector is scanned around at a distance of some 10 to 20 cm from the end to record the angular distribution of the power emerging. As we have seen, this far-field angle is readily related to the modes propagating in the fiber at the point of termination.[18]

Careful measurements of the far-field angle versus length allow the generation and establishment of a stable mode distribution to be detected, and from this an estimate of the coupling length for the guided modes of the fiber to be made. In practice, a simple modification of the fiber mounting in an apparatus designed for studying the angular scatter spectrum from a fiber will allow this measurement to be made, the modification consisting of feeding the fiber into the scattering tank at 90° to the path normally followed by the fiber when under study for the angular scatter spectrum.

10.6. TOTAL FIBER LOSS

10.6.1 Fiber Insertion Loss

Conceptually the easiest of all measurements to make is the fiber insertion loss. All that is necessary is to take a power detector that is linear in its response, launch power from a source into the fiber, measure the power emerging from the far end of the fiber, and then, by cutting off a suitable fiber length, remeasure the power on the shorter length and calculate the attenuation in the section removed.

This technique works well with high loss fibers, say at the 100 dB/km level, provided that the removed length is long enough to involve some 20 dB or more of attenuation. Repeatable measurements are easily obtained, and the apparatus can be readily automated to provide wavelength data for the fiber loss, perhaps with automatic recording of the power data and processing of the data in an on-line computer. Alternatively, a very simple apparatus may be used consisting of little more than an LED source, a photodiode detector, and meter readout for single wavelength measurements.

Meaningful measurements on very low loss fibers are far more difficult to make, for a variety of reasons. If it is assumed that fibers are produced in lengths of about 1 km, modern fibers may have insertion losses as low as 1 dB/km, so that cutting off the full 1 km changes the insertion loss by only 1 dB, thus demanding greater precision of power measurement. At this level the quality of the fiber ends becomes of great significance, and it is usually advisable to prepare several new ends at each of the two fiber lengths (removing a few centimeters for a new end) and to measure the received power for each end. In this way effects due to poorly finished ends or to contamination of the end become clear.

End preparation can follow one of two routes; one attempts to produce a perfect end every time by using a special tool, and the other makes the

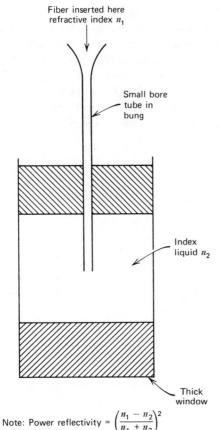

Fiber inserted here
refractive index n_1

Small bore
tube in
bung

Index
liquid n_2

Thick
window

Note: Power reflectivity = $\left(\dfrac{n_1 - n_2}{n_1 + n_2}\right)^2$

Figure 10.22. Index matching cell for terminating an optical fiber when under study.

end termination as insensitive as possible to irregularity by index-matching it. A tool for preparing fiber ends is illustrated in Figure 13.18. It relies on a principle long known by glassblowers for breaking glass rods, namely, to scratch the rod with a diamond and then to tension and simultaneously bend the rod slightly to produce a cleavage that runs smoothly across the end surface. Figure 10.22 shows a simple cell for index-matching a fiber end to make the emergent power insensitive to the end surface of the fiber.

However, given good measurement technique, stable apparatus, and long lengths of fiber to work with, there still remains an uncertainty that is part philosophical when a multimode fiber is to be studied. The reason is simply that there is no single "correct" answer. The loss from the fiber depends upon how one launches power into it, that is, predominantly in low order modes, by exciting all modes, or through some intermediate situation. The number that is most appropriate must depend upon the application. Evidently, if a single-mode source is to be used, the low order mode launch is likely to be appropriate and will be the most favorable. If an LED source is to be used, an all-modes-excited launch is likely to be the most appropriate. Finally, if the fiber is to be jointed to other lengths of fiber to make up a very long transmission line, a loss figure obtained with a good simulation of the stable mode distribution is the most appropriate. Some loss measurements made with different launch distributions are shown in Figure 10.23.[16]

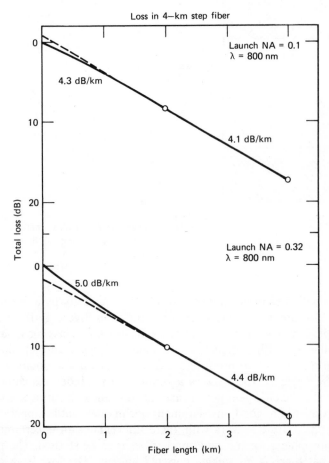

Figure 10.23. Measured loss rates versus length for a step-index fiber, illustrating the effects discussed in connection with Figure 10.6. Reproduced with permission from D. B. Keck, *2nd European Conference on Optical Communications. Paris, 1976.* Paris: SEE.

10.6.2 Remote Measurement of Insertion Loss

In the technique for insertion loss measurement discussed above it is assumed that both fiber ends are available for measurement at the same location. This assumption allows one to avoid the uncertainty associated with using two different detectors for the near and far end measurements, and makes the absolute calibration of both source and detector unnecessary for the measurement, assuming only that they are adequately stable during its duration. Remote end measurements, as encountered in the measurement of cables in ducts, separate the fiber ends by distances of kilometers and thus lead one to consider the use of near and far end detectors. Evidently, with careful use and appropriate design of their associated circuitry, these can be made stable, although as precision requirements increase, their long term stability will almost inevitably become a source of uncertainty. We note here a simple technique that largely avoids the problem.

We postulate both a source and a detector at each end of the fiber, labeled by subscripts 1 and 2. We assume that the launching of power from each source is alignment sensitive so that, of the power from source 1, S_1, only the fraction $a_1 S_1$ actually enters the fiber, and the detected power at detector i is given by $P_{1i} = a_1 S_1 D_i \cdot A_{1i}$, where A_{1i} is the loss in transmission through the fiber. Evidently $A_{11} = A_{22} = 1$, while A_{12} and A_{21} represent the fiber losses that it is desired to measure. It is only a matter of algebra to show that

$$\frac{P_{12} P_{21}}{P_{11} P_{22}} = A_{12} A_{21}$$

The sequence is thus to launch from detector 1, measure the received power at the far end with detector 2, and cut off the fiber cable, without disturbing the launching, a_1, to measure the power at detector 1, P_{11}. The operation is then repeated from end 2 to yield P_{21} and P_{22}, from which the geometric mean loss of the cable is obtained without requiring any long term stability of the optics. The method suffers from the limitation that, if $A_{12} \neq A_{21}$, it does not show this.

REFERENCES

1. K. I. White and J. E. Midwinter *Opt. Electron.*, **5**, 323 (1973).
2. K. I. White *Opt. Quant. Electron.*, **8**, 73 (1976).
3. a. A. Zaganiaris and G. Bouvry, *Ann. Telecommun.*, **5**, 189 (1973).
   b. A. Zaganiaris, *Appl. Phys. Lett.*, **25**, 345 (1974).
4. a. K. C. Kao and T. W. Davies, *J. Sci. Instr.*, Ser. 2, **1**, 1063 (1968).
   b. M. W. Jones and K. C. Kao, *J. Sci. Instr.*, **2**, 331 (1969).
   c. C. R. Wright and K. C. Kao, *J. Sci. Instr.*, **2**, 579 (1969).

5. M. Eve, A. M. Hill, D. J. Malyon, J. E. Midwinter, B. P. Nelson, J. R. Stern, and J. V. Wright, *2nd European Conference on Optical Communications, Paris, 1976*. Paris: SEE.

6. R. Olshansky, International Workshop on Optical Waveguide Theory, Lannion, France, September 1976, Unpublished work.

7. W. J. Stewart, *Optical Fiber Transmission*. Williamsburg, Va.: OSA/IEEE, January 1975.

8. J. E. Midwinter, *Opt. Quant. Electron.*, 7, 297 (1975).

9. W. J. Steward, Private communication.

10. M. H. Reeve, M. C. Brierley, J. E. Midwinter, and K. I. White, *Opt. Quant. Electron.*, 8, 39 (1976).

11. T. C. Rich and D. A. Pinnow, *Appl. Opt.*, 13, 1376 (1974).

12. T. C. Rich and D. A. Pinnow, *Appl. Phys. Lett.*, 20, 264 (1972).

13. J. E. Midwinter and M. H. Reeve, *Opt. Electron.*, 6, 411 (1974).

14. M. H. Reeve and J. E. Midwinter, *Optical Fiber Communication*, IEE Conf. Publ. 132, London, 1975, p. 16.

15 K. I. White, Private communication.

16. D. B. Keck, *2nd European Conference on Optical Communications, Paris, 1976*. Paris: SEE.

17. D. A. Pinnow, *IEEE J. Quant. Electr.*, QE6, 223 (1970).

18. R. Olshansky, S. M. Oaks, and D. B. Keck, *Optical Fiber Transmission*. Williamsburg, Va: OSA/IEEE, February 1977.

11

Measurement of Fiber
Propagation Properties

11.1. INTRODUCTION

In Chapter 10 we concerned ourselves with the study of mechanisms which controlled the power arriving at the detector of the system—in other words, the fiber loss mechanisms. In this chapter we consider the factors which influence the time dependence of the arriving pulse, and measurements of the related parameters. Thus we are concerned with the study of pulse dispersion, the fiber amplitude-frequency response, the fiber phase-frequency response, the index profiles of graded-index fibers, and break detection.

11.2. MEASUREMENT OF PULSE DISPERSION IN FIBERS

It is a simple matter to inject a pulse of optical energy into one end of a fiber, to detect the power emerging from the other end, and to deduce a number for the dispersion (of the half-power points of the pulse, for example). However, with multimode fibers having little mode mixing, such an exercise is generally of rather limited value, since changing the launching conditions, the lay, or the tension of the fiber will change the result.

To characterize a fiber more fully, it is desirable to be able to do one of two things. The first is to obtain a number for the dispersion that will apply under the conditions in which the fiber is to be used, to allow system design to proceed. The second is to obtain a full understanding of the propagation in the fiber so that predictions or modifications can be made for the fiber in some new circumstance. In general, the second goal is only partially possible.

Several factors critically affect the numbers obtained for the fiber dispersion. The source linewidth enters into the results because of the material dispersion, although in multimode fibers this is often a small effect. The dispersion associated with each single mode is also a factor, although, in a multimode fiber the dispersion is usually dominated by the spread in group velocities among the available modes. However, even if it were feasible to measure the delay for each mode individually, this would not necessarily characterize the fiber since in practice the launched power is not uniformly spread among the available modes, the modes may not have the same loss rates, and energy launched into one mode does not necessarily stay in it but can be coupled to other modes through mode coupling. Thus a full characterization should, in principle, involve a study of the presence or absence of mode coupling and the modal distribution of power both as launched and at subsequent points along the fiber. It is immediately apparent that such a comprehensive study is likely to be both difficult and time consuming. Also, in many cases it may not be necessary for the user's requirements, but it is well to remember the limitations of a simple approach.

Of the measurement techniques described in the following sections, the more sophisticated study suggested above can be made with several. We will attempt to indicate the advantages and disadvantages of each approach as it is presented.

11.2.1 Pulse Dispersion Measurements

Several distinct techniques have evolved for studying pulse dispersion in optical fibers. The most direct is to launch a fast pulse into the fiber, to time-resolve the power emitted from the far end of the fiber, and to obtain directly the pulse spreading or distortion. This technique has the attraction of simplicity and directness but also suffers from some severe limitations. The time resolution that can be readily obtained with commercially available optical detectors is typically of the order of a few tenths of a nanosecond; faster than this requires detectors that are currently in very limited production or are made to special order. Thus the pulse spreading, if it is to be measured accurately, must generally be in the region of 0.5 to 1 ns, and with low dispersion fibers this requires a considerable length of fiber for study.

The source for such a measurement may be a GaAs laser or LED, and in either case modulation to produce a pulse width of less than 1 ns is difficult. Faster pulses may be obtained by the use of a mode locked laser source, such as the HeNe, krypton ion, or NdYAG laser. Use of these sources, however, imposes a new difficulty on the measurement, namely, the fact that these sources nearly always operate in single transverse mode and, when coupled to a multimode fiber, will not launch the same mode distribution as a GaAs source. And since in such fibers the delay is mode dependent, a different launched distribution will yield a different dispersion result. However, the best test for the system designer is probably based upon the use of a pulsed source, of the type he or she intends to use, launched in the geometry proposed for the system with a length of fiber close to the design objective. In this way results obtained should be reasonably meaningful for the particular system in question.

The presence of mode coupling in a fiber leads to a pulse dispersion that is not linearly increasing with distance but varies as $\sqrt{L}$ after an initial stabilization length of fiber.

Typical measured values of the stabilization length vary from a few tens of meters for fibers with very irregular interfaces (and high losses) to several kilometers for fibers of excellent mechanical quality. Such mode coupling can be detected by several methods. If the radiation pattern emerging from the far end of the fiber is essentially independent of the launching geometry, it is reasonable to assume that the modes are well mixed in a single transit through the fiber. A more precise study of the mode mixing generally requires further measurements. If the pulse dispersion apparatus has sufficient time resolution and the fiber is expendable, the fiber can be cut back and the pulse width noted at each newly exposed length. A plot of pulse widths versus fiber lengths should then yield a curve of the form given in Figure 11.1, clearly showing the turnover from delay linearly increasing with length to delay varying as the square root, once mode mixing begins to significantly reshuffle the launched power distribution. (But see also Section 5.10.)

If a fiber is not subject to a high degree of mode mixing in its length, some indication of the degree of mode mixing can be obtained by observing the variation of the pulse dispersion with the launching condition. If the pulse is launched through a lens system, only axial modes can be

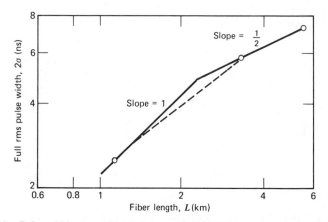

Figure 11.1. Pulse width versus length in a multimode fiber with mode coupling present, showing both linear and square root of length dispersion. Reproduced with permission from L. G. Cohen, P. Kaiser, J. McChesney, P. B. O'Connor, and H. M. Presby, *Appl. Phys. Lett.*, **26**, 472 (1975).

excited by restricting the angular aperture of the fiber that is illuminated, and by launching with a spot that is centrally placed over the fiber end. Alternatively, skew modes can be preferentially excited by aligning the exciting spot close to the core-cladding interface.

A series of measurements of this type is illustrated in Figure 11.2, in which a 1 ns GaAs laser pulse has been launched into a 1 km length of fiber. By varying the launching conditions, the pulse shapes shown were obtained, ranging from predominantly low angle launch to most modes excited.[1] Also shown in the same figure are the results of applying mode scrambling to the launched distribution by sandwiching the fiber between two rough surfaces for a distance of some 10 cm immediately after launching. This has the result of inducing strong mode coupling, which largely removes the dependence of pulse response upon the launching alignment. In this case the mode distribution obtained was close to that expected for the stable mode distribution for this fiber, although the degree of mode coupling was so low that this could not be verified on the length available.

Finally, a note of caution is in order. In all pulse measurements it is most important to note that the avalanche photodetectors are not linear in their response. In general, the avalanche gain is a function of frequency, falling off at high frequencies because of the finite drift times in the device, and the device is not amplitude linear, with its gain saturating at high

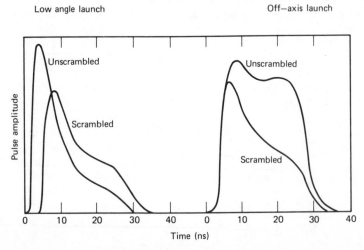

Figure 11.2. Effect of launching on the pulse response of a 1 km step-index fiber with little mode coupling. Reproduced with permission from M. Eve, A. M. Hill, D. J. Malyon, J. E. Midwinter, B. P. Nelson, J. R. Stern, and J. V. Wright, *2nd European Conference on Optical Communications Paris, 1976.* Paris: SEE.

signal levels. As a result a pulse response measurement should normally be used to measure the change in pulse shape only between transmit and receive, at the same signal current and avalanche multiplication voltage settings, with optical attenuators used to match the signal levels. The application of CW techniques in which a single-frequency source is used reduces this problem.

11.2.2   Shuttle Pulse Technique

An elegant method of sampling the pulse shape at regular points during its propagation along a long length of fiber has been described by Cohen and Presby[2] and is illustrated in Figure 11.3. A short length of fiber is terminated with reflectors at each end. A pulse launched through one reflector into the fiber then shuttles back and forth along the fiber and can be sampled at the far end after each of the $2N-1$ transits ($N = 1, 2$ etc.). The technique is very powerful since it allows one to study mode coupling in medium to short lengths of high quality fibers.

The number of reflections that can be accepted depends upon the dynamic range of the detector channel. If we take the range (in dB) to be D, defined as the difference between the power available at the source (into a single fiber) and the minimum detectable power, we have the relation

$$D - 20\log T - \alpha L_0 \geqslant (20\log R + 2\alpha L_0)(N_m - 1)$$

where T and R are the power transmission and reflection coefficients for the mirrors used, α is the loss of the fiber (in dB/km), L_0 is the length (in km) between source and detector, and N_m is the maximum value of N and is the number of pulses detected.

Great care must be taken in assembling an apparatus to use this technique to ensure that the reflectivity of the mirrors is constant over the numerical aperture of the fiber, since otherwise a mode-dependent leakage mechanism will have been introduced into the transmission path that is not related to the fiber itself. This requirement on the mirrors is different from the usual multilayer dielectric mirror specification, which often seeks flat wavelength response at normal incidence.

In the measurements described by Cohen, two sets of mirrors were used, one with $T = 0.3$ and $R = 0.7$, and the other with $T = 0.1$ and $R = 0.9$. The dynamic range of the apparatus, D, was 49 dB. Figure 11.4 shows some results obtained by the use of this technique. In this case the values of the parameters were as follows: $\alpha = 6$ dB/km, $L_0 = 148$ m, and $N = 9$.

Several additional points should be noted about this technique. In operation it is important to align the reflecting mirrors so that they are accurately normal to the fiber end. This can be done by adjustment, but in

238

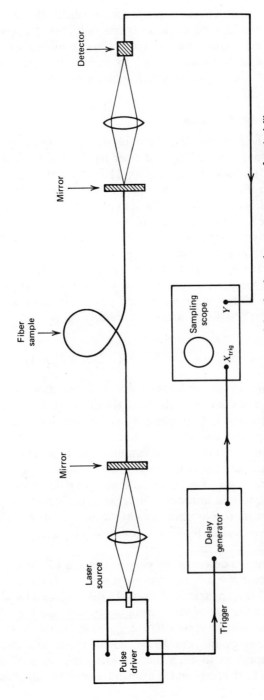

Figure 11.3. Schematic diagram of the apparatus used for shuttle pulse measurements of optical fibers.

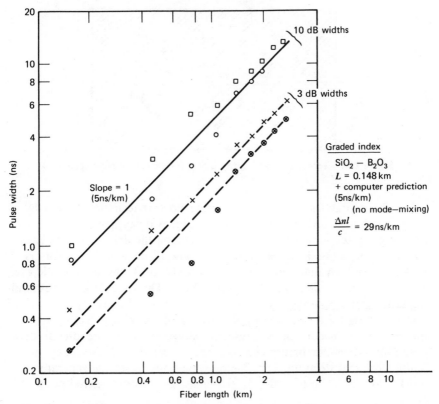

Figure 11.4. Results obtained by the shuttle pulse technique. Reproduced with permission from L. G. Cohen and H. M. Presby, *Appl. Opt.,* **14**, 1361 (1975).

the original experiments it was achieved using a simple fiber termination in which the fiber was held in a fine metal tube contained within a metal V-block with the mirror cemented to its end, as shown in Figure 11.5.

A further point of particular interest in regard to this technique is that it is the only one which allows the pulse shape to be sampled during transmission through a long transmission fiber path, without destroying the fiber in the measurement process. This is particularly important if one wishes to study the level of inherent mode mixing in a fiber, or the variation of mode mixing with some parameter, such as the tension of the fiber on a drum, or impressed bends. On the other hand, the fiber length actually studied is still short even if it is effectively multiplied by N traverses, and the effect of the reflecting terminations on the level of mode coupling is not easy to allow for.

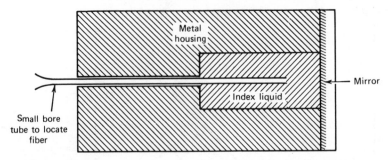

Figure 11.5. Termination and mirror mount for the shuttle pulse apparatus.

Finally, a practical point about setting up this technique is worthy of note. It is highly advantageous to use a digital delay generator to trigger the sampling scope that is necessary to examine the delayed pulses emerging from the multiple passes through a low dispersion fiber. This is so because very high time resolution, of the order of 0.5 ns, is desirable even after a total transmission path of, say, 10 km. The time delay corresponding to the latter path distance is about 45 μs. It will be readily appreciated that searching for a 1 to 10 ns pulse after a 45 μs delay is like looking for the proverbial needle in a haystack. However, with the digital delay generator, after the first pulse has been located and the delay noted, multiplying that delay by 3, 5, 7, etc., allows one to set the generator very quickly to a point in time that is extremely close to the desired point.

11.2.3 Continuous Wave Techniques

11.2.3.1 Modulated LED Source. It is frequently the case that systems designers prefer to work with amplitude-frequency response and phase-frequency response curves rather than with a pulse dispersion measurement in designing a transmission system, particularly if some form of equalization is to be applied to the detected signal. Such curves can be measured directly by means of the apparatus illustrated in Figure 11.6. It utilizes an LED whose drive current is modulated by a sine wave signal from a sweep generator. The intensity modulated light is then launched into the fiber and a signal detected at the far end, by means of a coherent detection channel which extracts the received amplitude (as a ratio of the generated amplitude, by comparison to the reference channel) and also the received phase (by the same process of comparison). The technique suffers the same limitations discussed under single-length pulse measurements in regard to mode distribution, effects of mode mixing, and so on.

The same general measurement can be made using a CW modulated laser source. In this way the material dispersion limitation of the LED

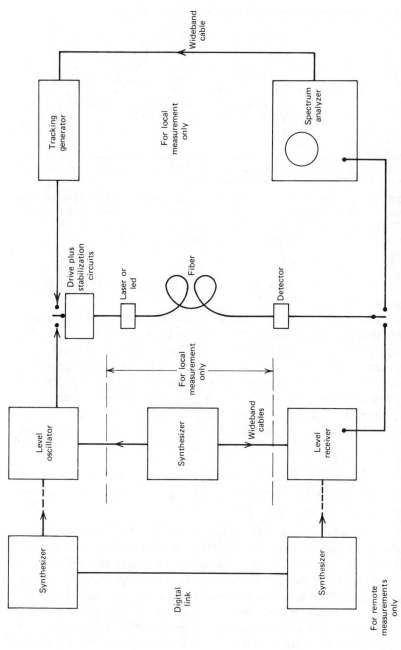

Figure 11.6. Continuous wave modulated LED or laser source with equipment for bandwidth study of fibers. The spectrum analyzer and tracking generator can be used for local measurements where both fiber ends are accessible at one place. For remote measurements, in which phase and amplitude are required, a more complex system using precision frequency synthesizers is needed, with timing provided by atomic clocks. This is not an economic way of studying the phase characteristics!

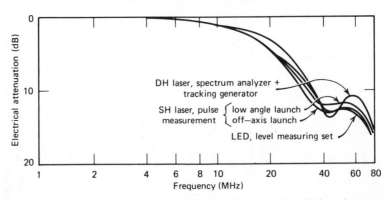

Figure 11.7. Comparison of different bandwidth measurements applied to the same step-index fiber, with mode scrambled launching. Reproduced with permission of M. Eve, A. M. Hill, D. J. Malyon, J. E. Midwinter, B. P. Nelson, J. R. Stern, and J. V. Wright, *2nd European Conference on Optical Communications, Paris, 1976.* Paris: SEE.

source can be avoided, and the higher launch power capability of the laser source exploited.

In Figure 11.7 we show the results of a comparative study of the straight pulse transmission measurement, the LED level measuring set, and a CW modulated GaAs laser measurement made with a spectrum analyzer and tracking generator.[1] The curves of electrical response versus frequency show excellent agreement after the pulse measurements have been Fourier transformed to allow comparison. However, it is important to note that this agreement was obtained only after the launched mode distribution had been thoroughly scrambled immediately after launching by a section of heavily mode coupled fiber. Without such scrambling poor agreement would be obtained.

These measurements were all made on a low loss (~6 dB/km) step-index 80-micron-core-diameter fiber showing a very low level of natural mode coupling over its 1 km length.

11.2.3.2　Free Running Laser CW Bandwidth Study. An elegant technique for studying the amplitude-frequency response of the fiber (but not the phase-frequency response) has been described by Gloge et al.[3] It utilizes the fact that a free running laser usually oscillates simultaneously on a number of lines, spaced by the reciprocal of the cavity round trip time. Each line is essentially a free running oscillator. If this multifrequency source is coupled into a fiber and the output detected by a square law detector, a spectrum analyzer can be used to display directly the amplitude-frequency response. The typical apparatus shown in Figure 11.8 incorporates improvements developed by Jackson.[4]

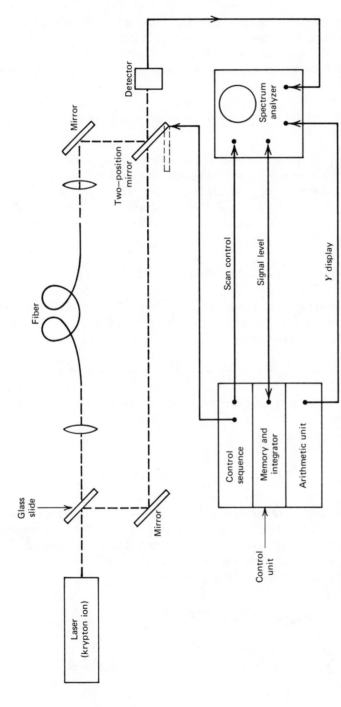

Figure 11.8. Use of a free running krypton laser for fiber bandwidth study. After D. Gloge, E. L. Chinnock, and T. P. Lee, *Appl. Opt.* **13**, 261 (1974) and L. A. Jackson, unpublished work.

The theory of the measurement is easily derived, provided that care is taken in following the various terms involved in the theoretical description. We start by describing the laser output as a summation of free running oscillators with uniformly spaced frequencies assumed to be of the same amplitude:

$$E_L = E_0 \left\{ \sum_{g=1}^{M} \exp\left[i(\omega_0 + g\,\Delta\omega)t + \psi_g \right] + cc \right\}$$

The phases, ψ_g, of the laser modes are considered to be random and slowly varying. We now launch the source power into the fiber, so that the total power is distributed among the guided modes. The power launched into the hth mode is given by $P_h = a_h P_0$, where P_0 is the total power. It follows that $\sum_{h=1}^{N} a_h = 1$. The field now, at the start of the fiber at time t in the hth mode, is given by

$$E_{gh} = \sqrt{a_h}\, E_0 \left\{ \exp\left[i(\omega_0 + g\,\Delta\omega)t + \psi_g \right] + cc \right\}$$

After transmission through the fiber for a distance L the field is attenuated and delayed so that, if the time delay for the hth mode is τ_h, the field detected at time t is given by

$$E_{gh}(\det) = \sqrt{a_h}\, \exp(-\alpha_h L) E_0 \left\{ \exp\left[i(\omega_0 + g\,\Delta\omega)(t - \tau_h) + \psi_g \right] + cc \right\}$$

The current in a square law detector, centered on DC rather than ω_0, is then given by

$$i = AEE^*$$

$$= A \sum_{l=1}^{M} \sum_{m=1}^{N} \sum_{n=1}^{M} \sum_{o=1}^{N} (E_{lm} E_{no}^* + cc)$$

Of the multiplicity of terms in this quadruple summation, only those for which $m = o$ yield finite current components since the others involve beats between dissimilar fiber spatial mode patterns. These are orthogonal to one another and spatially average to zero.

The detector current is therefore given by

$$i = \sum_{m=1}^{N} A a_m \exp(-2\alpha_m L) E_0 E_0^*$$

$$\times \sum_{l=1}^{M} \sum_{n=1}^{M} \left\{ \exp\left[i(l-n)\,\Delta\omega(t - \tau_m) + (\psi_l - \psi_n) \right] + cc \right\}$$

In the spectrum analyzer the electrical power dissipated in the load resistor is displayed as a function of the frequency of the current component. The power is i^2R. This stage of detection, coupled with integration in the display circuits, removes the randomly varying component of i^2 caused by the arbitrary phases and leaves for display only the phase-independent terms arising from the product of each individual term with its own complex conjugate. The result has the form

$$i^2R = A^2R \sum_{n=1}^{N} \sum_{p=1}^{N} a_m a_p \exp\left[-2(\alpha_m + \alpha_p)L\right](E_0 E_0^*)^2$$

$$\sum_{l=1}^{M} \sum_{n=1}^{M} \left\{\exp\left[i(l-n)\Delta\omega(\tau_m - \tau_p)\right] + cc\right\}$$

The electrical power detected at the launching end of the fiber is given by setting $\tau_m = \tau_p = 0$.

At a given frequency, $(l-n)\Delta\omega = K\Delta\omega$, the ratio of the power detected at the receiving end of the fiber to that at the launching end of the fiber is now given by

$$\frac{(i^2R)_r}{(i^2R)_l} = \sum_{m=1}^{N} \sum_{p=1}^{N} a_m a_p \exp\left[-2(\alpha_m + \alpha_p)\right]$$

$$\times \left\{\exp\left[iK\Delta\omega(\tau_m - \tau_p)\right] + cc\right\}$$

which is the response of the fiber to the modulation frequency, $K\Delta\omega$.

This result explicitly contains the factors a_m, which describe the way in which the launched power divided among the available modes, and it also includes modal attenuation coefficients, α_m. These are significant since they affect the measured response in any measurement technique. It is very important, therefore, when making a measurement to ensure, as far as possible, that the conditions of measurement, particularly launching, are those that will apply when the fiber is to be used in a system. Alternatively, one may deliberately launch only a few discrete modes with the clear intention of studying only their propagation.

We should also note that the theoretical model presented above has not allowed for the transfer of power between modes; that is to say, it has been assumed that the mode coupling is zero. The same model can be used to analyze the propagation in a fiber with an amplitude modulated LED, in which similar mode orthogonality and time averaging assumptions are made to extract a result.

11.3. MATERIAL DISPERSION MEASUREMENT

The easiest and most direct method for measuring material dispersion in a given fiber, independent of any other parameters, is to use two separate sources of slightly different wavelength. This is most conveniently done using two different GaAs lasers having different aluminum doping levels and hence different center wavelengths. A pulse from each laser is launched in turn into a known length of fiber, and the total time delay of the pulse in traversing the whole fiber length is measured (by contrast to the pulse spreading considered previously). The two pulses of different wavelength will travel at different mean group velocity because of material dispersion. It follows that, measuring the difference in transit time between the two sources, and knowing the fiber length and the wavelength separation of the sources, one obtains the material component of dispersion directly.

In a typical experiment two GaAs lasers[5] operating at 900 and 860 nm were used. The pulses were transmitted over a distance of ~1.0 km of silica fiber and arrived with a time separation of 3.6 ns, leading to a figure for the material dispersion of that particular fiber of 90 ps/km·nm.

To calculate the total delay in a fiber, making due allowance for the pulse spreading due to mode dispersion and to material dispersion, it is a good approximation to use the relation

$$\sigma_{tot}^2 = \sigma_{mat}^2 + \sigma_{mode}^2$$

where σ is the rms pulse width.

The typical numbers for material dispersion given above are for telecommunications fiber, so that with the GaAs source one is not seriously limited with common lasers at a transmission rate of 140 Mbit/sec out to 10 km of continuous fiber. At higher bit-rates, however, material dispersion becomes a serious consideration, and steps will usually have to be taken to control the linewidth of the source used. Likewise, if an LED source is to be used, typical linewidths are aroung 30 nm, leading to a dispersion due to material alone of 2.7 ns/km.

These problems can be avoided by choosing a longer wavelength, since the material dispersion reduces at longer wavelengths. For example, in the range of 1.2 to 1.3 microns (1200 to 1300 nm), the material dispersion of silica changes sign and goes through zero.[6] This observation has stimulated interest in sources that can opearte at that wavelength, and preliminary work on GaInAsP sources has been reported in the 1.1 to 1.3 micron region. However, system design is further hampered in that wavelength region by the lack of good detectors, although most fibers show lower losses there than at 840 nm, because the Rayleigh scattering has decreased by a factor of about 4.0. (See Sections 5.9 and 8.4.)

11.4. SUMMARY OF BANDWIDTH MEASUREMENTS

We have seen that there are a number of straightforward methods for assessing the pulse dispersion or 3 dB bandwidth of an optical fiber. We have also seen that, given a standardized launching scheme, these techniques can yield repeatable results that compare closely, one with another, for the same fiber. Under these conditions it becomes possible to study the effect on propagation of perturbations to the fiber, arising from such causes as additional tension of the fiber on its drum (caused by heating the drum, and leading to microbending effects), the effects of packaging the fiber, and the effects of cabling the fiber and subjecting the cable to the handling that it must experience in its operating life.

However, we must again caution the reader that extrapolation of the results of such measurements into another situation should be carried out with great care. The shuttle pulse measurement and results (Figures 11.3 and 11.4) show the hazard of linear extrapolation of pulse spreading results from a short fiber length to a longer one, because of the onset of mode coupling and L variation of pulse width. The mode scrambling experiments (Figure 11.2) show the dependence of the short length measurement upon the details of the launching, particularly if one is primarily interested in the response below the 3 dB electrical response region for purposes of equalizer design. The trend in optical fiber systems is for the degree of mode coupling to decrease as the fiber and cabling processes are improved. On the assumption that this continues, the response of the system may become largely launching dependent, although for some time to come it seems likely that the response will be determined by a complex mix of launching, fiber, packaging, cabling, and other effects such as the modal redistributive effects of fiber-fiber joints. (See also Section 13.6.)

11.5. BREAK DETECTION IN OPTICAL FIBER CABLE

It seems inevitable that breaks will occur in optical fiber cables after or during installation, and other imperfections may well develop also, such as high loss joints. A simple technique has been described for locating such imperfections, whose presence would presumably be sensed initially by an unexpected and unacceptable reduction in received signal power in the system, or in measurements after the cable installation.

The technique used[7] involves a simple pulse echo system, as shown in Figure 11.9. A high power, short pulse of light from a GaAs laser is focused onto the end of the fiber, while reflected power from the fiber is monitored via the beam splitter on a avalanche photodiode and sampling oscilloscope. A large pulse is detected from the launching end of the fiber. This pulse can be used to trigger the scope time base and to start the time delay. Pulse echoes from imperfections then show up on the scope trace,

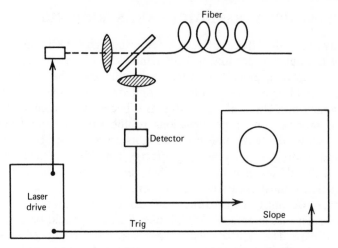

Figure 11.9. Pulse echo equipment for fiber break detection.

delayed by a time proportional to their distance along the fiber. To a first approximation, if the core refractive index is n_1, the light velocity is c/n_1, so that an imperfection distant by an amount L will be seen after a time delay of $2Ln_1/c$ seconds.

With this technique the presence of breaks has been demonstrated in 100 m lengths of optical fiber. An obvious limitation of the approach concerns the dynamic range of the detector, which will most probably be too limited for systems use unless special steps are taken to improve it.

The problem is as follows. The reflection coefficient from a clean glass surface is of the order of 4%, i.e., -14 dB. For a repeater section the insertion loss of the cable is likely to be in the region of 40 to 50 dB, so that the return signal from a perfectly clean break near the far end will be approaching 90 to 110 dB down. In practice, break detection can usually proceed from either end, so that a worst case would seem to be 54 dB down. However, this overlooks the fact that an imperfect end, cleaved at an angle or raggedly, will be a less efficient mirror, while an end immersed in water will have a vastly reduced reflection coefficient. Note that with sufficient sensitivity and dynamic range, back-scatter from the Rayleigh scattering can be detected. This allows an index matched break to be detected and an estimate of the fiber attenuation to be made from the reflected power versus time display as an oscilloscope.

11.6. TECHNIQUES FOR FIBER INDEX PROFILE MEASUREMENT

Interest in graded-index fibers has spurred the development of a number of methods for measuring with great precision the distribution of the refractive index across the core section. The measurement is difficult to make with precision because of the small size of the fiber core itself and the fact that a precision measurement requires a large number of resolved and independent measurement points. A number of techniques have emerged, which can be classified into four groups:

1. End reflection scanning.
2. Interference microscope techniques.
3. Transmitted power profiles.
4. Refracted power scanning.

We will discuss these in turn.

11.6.1   End Reflection Scanning Technique

In this technique[8] a focused laser beam is reflected from the smooth fiber end at normal incidence, and the reflected power as a function of position is measured. The technique therefore measures the Fresnel reflection from the glass-air interface, formed by the freshly prepared fiber end. The normal reflectivity from a fiber end (for any dielectric interface) is given by

$$R = \frac{P_{re}}{P_{inc}} = \left(\frac{n-1}{n+1} \right)^2 \tag{11.6.1}$$

For small changes in the value of the refractive index, we have

$$\delta R = 4 \left[\frac{n-1}{(n+1)^3} \right] \delta n = A \, \delta n \tag{11.6.2}$$

where A may be considered effectively constant, and the reflected power follows the index change. By rearranging equation 11.6.2, we find that the relative accuracy of the refractive index difference, $\delta n / n$, compares to the accuracy of the reflectivity as follows:

$$\frac{\delta n}{n} = \left(\frac{n^2-1}{4n} \right) \frac{\delta R}{R} = 0.208 \frac{\delta R}{R} \qquad \text{(for } n=1.5) \tag{11.6.3}$$

so that the relative error in the index determination is 5 times smaller than that in the reflectivity.

The apparatus for making this measurement is illustrated in Figure 11.10. The $\lambda/4$ plate is necessary to prevent the reflected power from the fiber and optics from modulating the laser output by interference. The laser output power itself is monitored by the solar cell, while the signal reflected from the fiber end is monitored by the pin diode and lock-in amplifier. The power falling on the screen allows visual alignment of the fiber end. To make a measurement, the fiber end is accurately placed at

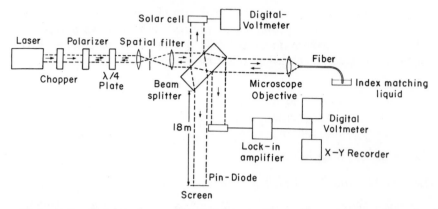

Figure 11.10. End reflection scanning technique for fiber profile measurement. Reproduced with permission from W. Eickhoff, and E. Weidel, *Opt. Quant. Electron,* **7,** 109 (1975).

the focal point of the microscope objective lens. The fiber end is then scanned slowly and with precision across the focal spot, and the reflected power is plotted on a pen recorder as a function of linear fiber movement, to produce a plot of the form shown in Figure 11.11, which gives the index profile of the fiber directly. By use of a 40×0.85 NA microscope objective lens, the focused spot size was found to have a half-power diameter of 0.85 micron, while the estimated uncertainty of 1.5% in the measurement of the reflectivity corresponded to an uncertainty of 0.005 in the estimation of the refractive index.

Experimentally, the technique has the great attraction that it gives

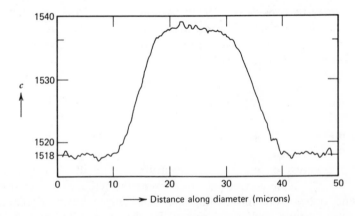

Figure 11.11. The profile of a Selfoc fiber obtained using the technique of Figure 10.10. Reproduced with permission from W. Eickhoff and E. Weidel, *Opt. Quant. Electron.,* **7,** 109 (1975).

directly the required result. However, it requires very accurate alignment and precision scanning of the fiber end if it is to be successful; and since it measures only the optical surface of the fiber, it is also very sensitive to any degradation or film formation of that surface.[9] The latter feature more or less rules out its use with fiber core materials that are attacked rapidly by atmospheric OH and raises uncertainty about its use with any fiber.

11.6.2 Interference Microscope Techniques

A number of reports have been published describing the use of an interference, phase sensitive microscope to measure the refractive index profiles of fibers. In general, all these various techniques require that a thin circular slice of the fiber be prepared with polished, flat, and parallel faces through which the light of the microscope can travel normal to the faces and parallel to the sample and fiber axis. The phase of the emerging light is then compared with that of the incident light, and interference fringes are formed in the field of view so that the sample appears[10] as in Figure 11.12. The photographed fringe pattern shows the phase length of the sample as a function of position, so that, when the sample length is known, the values of $n(r) \cdot L = \phi(r) \cdot \lambda / 2\pi$ can be calculated. Such a calculation is a slow and tedious procedure and would limit this technique to the analysis of very few samples, were it not for the fact that simple procedures for reading and computer-processing these fringe patterns to yield precision plots of the index profile have been developed and described. Figure 11.13 shows the profile computed from Figure 11.12.

In this automated form[10] the interference microscope technique has found considerable application. The major limitation to its use is probably the time necessary for sample preparation.

Figure 11.12. The interference patterns observed in the microscope when measuring a graded-index fiber. Reproduced with permission from L. G. Cohen, P. Kaiser, J. McChesney, P. B. O'Connor, and H. M. Presby, *Appl. Phys. Lett.*, **26**, 472 (1975).

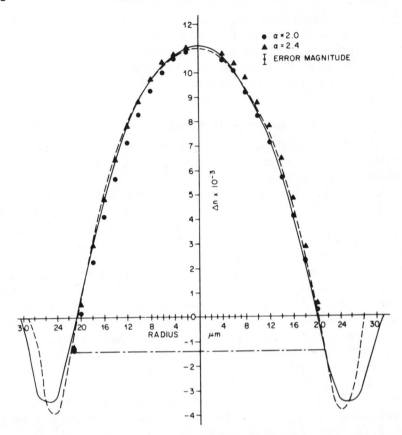

Figure 11.13. Profile deduced from the interference pattern of Figure 11.12. Reproduced with permission from L. G. Cohen, P. Kaiser, J. McChesney, P. B. O'Connor, and H. M. Presby, *Appl. Phys. Lett.*, **26**, 472 (1975).

11.6.3 Transmitted Power Profile Technique

This technique was mentioned in Section 6.5 in our discussion of the excitation of multimode fibers. There we showed that, if a fiber is excited by a diffuse Lambertian source, the power at any point in the fiber cross section in the fully guided modes is proportional to the index difference between that point and the cladding. Thus, if the fiber is excited by a diffuse source, such as a ribbon filament tungsten lamp, a scan of the power distribution with a small pinhole across the fiber end or its image should result in a signal proportional to the index difference and profile.

This is generally true, but unfortunately the results are complicated by two effects. The first involves leaky modes (Section 5.4). When the fiber is excited by a Lambertian source that overfills the end and acceptance aperture, not only will all the guided modes be excited, as required, but all the weakly leaky modes will also be excited. The result will be that far too much energy will appear to be propagating at the periphery of the core, and it will appear that the index is higher near the cladding than it really is, if the power distribution is measured close to the point of excitation.

The alternative would thus appear to be to measure the power distribution well away from the point of excitation, so that the weakly leaky mode power will effectively have decayed away. However, this is likely to lead to a value for the refractive index of the core close to the cladding that is too low, since, in addition to power loss from the weakly leaky modes that

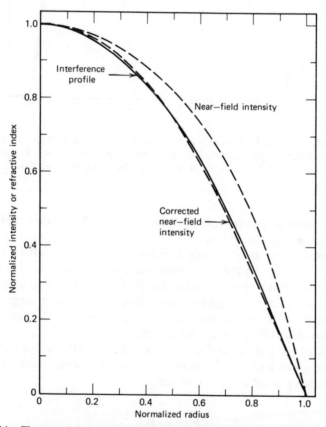

Figure 11.14. The near field intensity pattern as measured, the corrected near field pattern after allowance for leaky modes, and the profile measured with an interference microscope for a graded-index fiber. Reproduced with permission from F. M. E. Sladen, D. N. Payne, and M. J. Adams, *Appl. Phys. Lett.,* **28** 257 (1976).

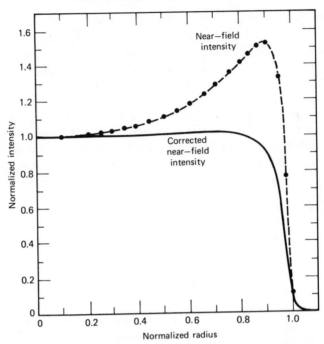

Figure 11.15. The near field intensity and corrected near field patterns for a near step-index fiber, emphasizing the importance of the leaky mode correction factors. Reproduced with permission from F. M. E. Sladen, D. N. Payne, and M. J. Adams, *Appl. Phys. Lett.* **28**, 257 (1976).

were not wanted, power will also have been lost by mode coupling from the guided modes close to cutoff.

In practice, the measurement is made close to the point of excitation since it is possible to apply a correction for the presence of the weakly leaky modes in the "perfect" fiber but it is not possible to allow for the effects of an unknown degree of mode coupling on a long length. Figures 11.14 and 11.15 show the measured and corrected profiles for a graded- and a step-index fiber, using this method.

Details of the correction factors[11] used to interpret these measurements have been given and are shown in Figure 11.16 for some cases of interest. The function plotted is $C(r,z)$, where

$$P(r) = P_0 C(r,z) \left[\frac{n^2(r) - n_2^2}{n^2(0) - n_2^2} \right] \qquad (11.6.4)$$

and $P(r)$ is the power density measured in the near field at radius r. The

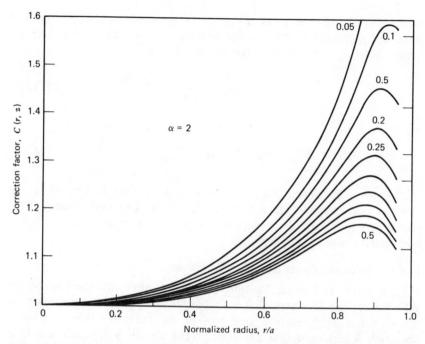

Figure 11.16. Correction factors for the near field scan technique for graded-index fibers. Reproduced with permission from F. M. E. Sladen, D. N. Payne, and M. J. Adams, *Electron. Lett.*, **12**, 282 (1976).

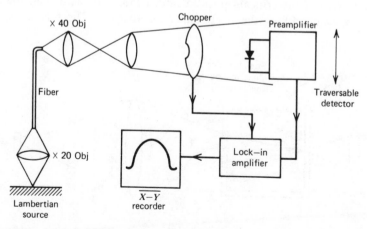

Figure 11.17. Apparatus for carrying out the near field scan measurement. Reproduced with permission from F. M. E. Sladen, D. N. Payne, and M. J. Adams, *Appl. Phys. Lett.*, **28**, 257 (1976).

function is plotted in terms of normalized distances from the launch point; these are specified in terms of the parameter X, which is related to the actual distance, z, from launching by the relation

$$X = \frac{1}{V}\log\left(\frac{z}{a}\right) \tag{11.6.5}$$

where

$$V = \frac{2\pi a}{\lambda}\left[n(0)^2 - n_2^2\right] \tag{11.6.6}$$

In this form the correction factors can thus be directly applied to the measured data to yield values of the index profile. The measurement of the power profile is carried out with the apparatus of Figure 11.17, which is both quick and simple to use.

11.6.4 Refracted Power Scanning Technique

Each of the techniques for fiber profile measurement described so far suffers from the drawback that considerable computation is necessary on the recorded data before a true plot of the refractive index profile is obtained. A technique that gives this profile directly without the need for calculation or correction has been developed by Stewart.[12] Because it does not rely upon guided modes to carry any of the detected power, it avoids the problems of leaky mode corrections.

The apparatus is shown in Figure 11.18, with a schematic diagram in Figure 11.19 of the critical optics. A laser beam is focused onto the fiber end grossly overfilling the fiber numerical aperture. The focused spot is

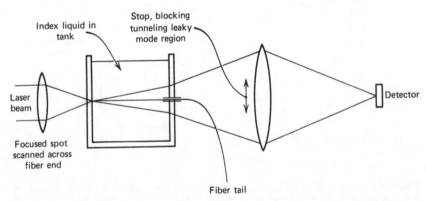

Figure 11.18. Apparatus for the refracted power scanning technique. Reproduced with permission from W. J. Stewart, Plessey Company Ltd., Caswell, England.

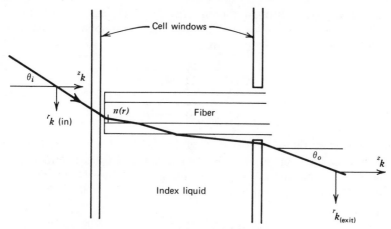

Figure 11.19. Schematic diagram defining the geometry for the analysis of the refracted power scanning technique.

very small, much smaller than the core diameter, and is scanned across the face of the fiber end synchronously with the recording channel (oscilloscope or chart recorder). The fiber is totally immersed within a parallel walled cell in a liquid whose index closely matches that of the cladding or is very slightly lower. Power outside the angle for capture by guided or tunneling leaky modes refracts out of the core into the surrounding liquid and eventually out of the back face of the liquid cell to form a highly divergent cone of light. The diameter of the cone changes as the refractive index of the core changes with the scanning spot.

The detection channel uses a diaphragm placed centrally in the exit cone of light to block the low angle power. The central cone carries no power since that is carried away by the fiber guided modes. The diaphragm is made large enough to more than block the tunneling leaky mode range of angles, so that only purely refracting modes are passed and focused onto the detector. The detected power is then recorded as a function of the focused exciting spot position as it scans over the fiber end. The operation of the device is easily understood simply by considering the following analysis.

We assume that the entrance beam is masked, so that over the outer range of entrance cone angles, up to and including θ_i, the power per unit solid angle in the exciting beam is essentially constant. We then follow the path of the extremum ray through the apparatus.

The radial component of the entrance ray **k** vector is given by $^r k$, where

$$^r k = k_0 \sin \theta_i \qquad (11.6.7)$$

This beam passes through the interface, conserving ^rk, and into fiber material of refractive index $n(r)$, so that the z component of the $\mathbf{k}$ vector for the ray in the fiber is given by

$$
\begin{aligned}
^zk &= \sqrt{n^2(r)k_0^2 - k_0^2 \sin^2\theta_i} \\
&= k_0\sqrt{n^2(r) - \sin^2\theta_i}
\end{aligned}
\tag{11.6.8}
$$

The ray then refracts laterally through the layers of refractive index in the core and cladding materials to enter the liquid. In these refractions ^zk is conserved, so that the ray arrives at the cell entrance face with the same value of ^zk and a value of ^rk given by

$$
\begin{aligned}
^rk(\text{exit}) &= \sqrt{n_0^2 k_0^2 - {}^zk^2} \\
&= k_0\sqrt{n_2^2 - n^2(r) + \sin^2\theta_i}
\end{aligned}
\tag{11.6.9}
$$

Conserving this value of ^rk through the cell exit window, we obtain a relation connecting θ_0 with θ_i:

$$
\sin^2\theta_0 = n_2^2 - n^2(r) + \sin^2\theta_i
\tag{11.6.10}
$$

If we now set $n(r) = n_2 + \Delta$, this reduces to an excellent approximation to the simple relation

$$
\sin^2\theta_0 = \sin^2\theta_i - 2n_2\Delta
\tag{11.6.11}
$$

We note that the solid angles of the entrance and exit cones are proportional to $\sin^2\theta_i$ and $\sin^2\theta_0$. If we decrease θ_i by some small amount, the entrance optical power will decrease by an amount proportional to the change in solid entrance angle. Since all that power refracts out in the hollow cone touching θ_0, and the equation tells us that a change in $\sin^2\theta_i$ is matched by an exactly similar change in $\sin^2\theta_0$, we may conclude that the peripheral illumination of the exit cone is uniform also.

If we now consider the situation in which θ_i is held constant but Δ is varied, we see that $\sin^2\theta_0$ varies linearly with Δ, and hence the optical power present in the region outside the diaphragm and reaching the detector must also vary linearly with Δ. The recorded optical power varies linearly with the refractive index at the entrance point for the optical beam into the fiber.

Thus the technique is capable of plotting the refractive index profile directly without the need for correction factors or interpretation of the recorded data. This makes it a most attractive technique for a routine monitoring of the index profiles of fiber leaving a production line, or for use in experimental work in which the profile is being varied deliberately to study the effects of the production process or the propagation in the fiber.

REFERENCES

1. M. Eve, A. M. Hill, D. J. Malyon, J. E. Midwinter, B. P. Nelson, J. R. Stern, and J. V. Wright, *2nd European Conference on Optical Communications, Paris, 1976*. Paris: SEE.

2. L. G. Cohen and H. M. Presby, *Appl. Opt.*, **14**, 1361 (1975).

3. D. Gloge, E. L. Chinnock, and D. H. Ring, *Appl. Opt.*, **11**, 1534 (1972).

4. L. A. Jackson, Unpublished work.

5. D. Gloge, E. L. Chinnock, and T. P. Lee, *Appl. Opt.*, **13**, 261 (1974).

6. D. N. Payne and W. A. Gambling, *Electron Lett.*, **11**, 176 (1975).

7. a. M. K. Barnoski and S. M. Jensen, *Appl. Opt.*, **15**, 2112 (1976).
    b. J. Guttman and O. Krumholz, *Electron Lett.*, **11**, 216 (1975).

8. W. Eickhoff and E. Weidel, *Opt. Quant. Electron.*, **7**, 109 (1975).

9. J. Stone and H. E. Earl, *Opt. Quant. Electron.*, **8**, 459 (1976).

10. L. G. Cohen, P. Kaiser, J. McChesney, P. B. O'Connor and H. M. Presby, *Appl. Phys. Lett.*, **26**, 472 (1975).

11. F. M. E. Sladen, D. N. Payne, and M. J. Adams, *Appl. Phys. Lett.*, **28**, 257 (1976); *Electron Lett.*, **12**, 282 (1976).

12. W. J. Stewart, Private communication.

12

Fiber Packaging and Cable Design

12.1. INTRODUCTION

One's intuitive reaction to a hair-fine optical fiber is to view it as an excessively fragile component that must inevitably pose serious problems to the cable designer if he or she is to successfully incorporate it into a cable without breakage. We have also seen earlier that small sinusoidal displacements of the fiber from its axis lead to increased losses and modification of the propagation properties of the fiber through the mode coupling so introduced. The term usually used to describe this mechanism is "microbending." Thus, in designing a cable, two factors will usually dominate our thinking: the cable must introduce little or no microbending if the finished product is to be a stable transmission medium, and the packaging must be designed to ensure that the fiber will remain intact for the cable life, which is likely to be 20 or more years.

The discussion which follows is based upon the rather scanty published work concerned with this general area of study and is intended to serve as a basis for thought about the subject rather than as a manual on how to design the optimum cable.

12.2. FIBER STRENGTH: GENERAL POINTS

One's daily experience with glass tends to lead to the conclusion that it is not a strong material, in that glass objects shatter easily in situations where an equivalent metal object would only slightly deform. However, an examination of the numbers listed[1] in Table 12.1 for the breaking stress (in kg/mm^2) for glass fiber samples compared to the yield stress for metals shows that glass can compare favorably, particularly on a strength/weight ratio basis. But there is a vital difference in the behavior of glass and metal, namely, glass does not creep or flow (at room temperature) but rather extends elastically to its yield point, where it fractures.

These points are illustrated in Figure 12.1, which shows the relationship between stress and strain (force per unit area versus change in length per unit length) for a variety of materials of interest in cable design. Several

260

TABLE 12.1 VALUES OF PARAMETERS IMPORTANT FOR CABLE STRENGTH DESIGN

Material	Specific gravity, SG	Young's modulus, E (10^4 MN/m^2)	Yield stress, YS (10^2 NM/m^2)	Elongation at yield (%)	Tensile stress (breaking) (10^2 MN/m^2)	Elongation at break (%)	E/SG	YS/SG ($10 \cdot$ km)
Steel wire	7.86	20	4–15	0.2–1.0	5–30	2–25	2.54	0.5–2
Nylon yarn	1.14	0.6–1.3	78	76	10–15	15–20	0.6–1.0	7
Kevlar 49 fiber	1.44	13	30	2	—	2	9.0	21
S glass fiber	2.48	9	30	3[a]	30	3[a]	3.6	12
Carbon fiber	1.5	10–20	150–300	1.0–1.5[a]	15–20	1.0–1.5[a]	1.0–1.5	100–200

Reproduced with the permission of Standard Telecommunications Laboratories, Ltd. © 1976, from S. G. Foord and J. Lees, *Proc. IEE*, **123**, 597 (1976).
[a]Statistical value only.

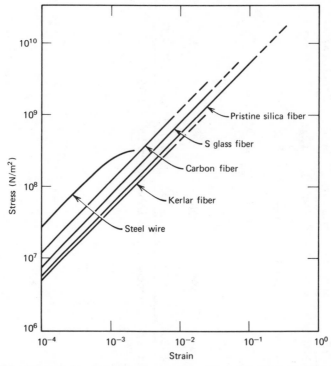

Figure 12.1. Stress-strain relationships for a variety of materials, illustrating the early plastic deformation of steel.

features stand out. Metals have a very small truly elastic range but can suffer very large strains without actual failure. Plastics, particularly in the form of the oriented molecule polymers such as Kevlar,[2] can remain purely elastic for a much greater range, while glasses have the characteristic that they *can* extend elastically to a very large strain (10%) value but more generally fail at rather low strains (1.0 to 0.1%). We will examine the reasons for this behavior more fully in the following section. In the meantime we should bear in mind two facts. The first is that glass can be a very weak material if we handle it incorrectly, but (the second fact), given correct handling and processing, it offers the prospect of very superior mechanical performance, in addition to the superior transmission performance already demonstrated.

12.2.1 Glass Fracture

The theoretical strength of glass is dictated by the bond strengths of the components from which it is constituted, and the breaking strain is related to the number of bonds per square centimeter of material and their mean strength. However, in practice, the measured strengths of glasses fall far short of the theoretical figures. The model used to explain this anomaly is the Griffith microcrack hypothesis.[3] It postulates that on the surface of the glass sample there are exceedingly small cracks. When a uniform strain is applied to the sample, it becomes concentrated around the tip of such a crack and the material at the tip begins to part long before the mean strain has equaled the bond strengths of the material. The effect is particularly marked in glasses because, unlike metals, they have no ductile range. Ductility allows some material to flow, so that strain can be more uniformly spread throughout a sample.

The validity of the foregoing hypothesis is suggested by the success of a number of techniques that are known to increase the strength of glass. The surface of a glass sample may be stuffed full of large ions, using an ion exchange bath between the glass and some molten salt. Typically, one might exchange sodium in the glass with potassium in the salt, with a KNO_3 bath. This puts the surface film into compression relative to the body of glass and results in significant increase in fracture resistance.

Experimental studies of glass samples whose surfaces have been chemically polished using a polishing etch show similar significant increases in strength, presumably because the surface irregularities have been polished away, leaving a pristine surface. And, in the same vein, freshly pulled glass fibers are several orders of magnitude stronger than fibers that have been left unprotected in the laboratory environment for some days.

Glass failure is thus a process that is dependent partly upon the underlying strength of the glass material, but generally to a much greater extent upon the surface state of the sample studied. For this reason it is not surprising to find that studies of large numbers of glass samples show statistical spreads in their failure strains, resulting from the statistical variation of the surfaces.

In designing our cable, we must therefore be aware of the significance of this statistically variable failure process and ensure that we have available the correct data to guide our thinking. This involves the use of Weibull statistics. We must ensure that the data we use are statistically meaningful when applied to the length of cable that we are concerned with; and, of course, we do not just accept what the first valid measurement tells us, but we seek to improve the performance of the fiber so that its strength is not a limiting factor but an asset in our design.

12.2.2 Weibull Probability Distribution

The probability of failure of a given length of fiber is found to be well described by a Weibull distribution.[4] This takes the following form:

$$G(s,f^*,t) = 1 - \exp\left[-\left(\frac{f^*-f_m}{f_0-f_m}\right)^p \left(\frac{t}{t_0}\right)^b \left(\frac{S}{S_0}\right) \right] \quad (12.2.1)$$

where

f^* is failure stress

f_m is stress below which no failure occurs

f_0 is stress above which failure always occurs

t_0 is some time constant for crack growth

S_0 is equivalent surface area scale factor

S is sample surface area

t is duration of applied stress

p,b are experimentally determined powers

and $G(S, f^*, t)$ is the probability that the fiber will fail under stress f^* in time t when tested with length to produce surface area S.

To determine the values of the parameters used in this distribution, one tests to destruction a sufficiently large number of samples to allow one to draw an experimental curve of the form shown in Figure 12.2, which is plotted on Weibull chart paper. Evidently, to evaluate the arbitrary parameters with accuracy in a form that will be statistically meaningful when applied to 1 km samples may be rather difficult. To understand this more fully, we will derive a physical model which gives us some insight into the mechanism behind this statistical distribution.

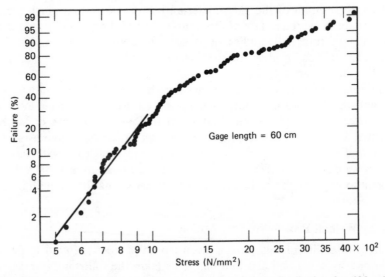

Figure 12.2. A Weibull plot showing a cumulative failure distribution for 125 micron diameter fibers. Reproduced with permission from R. D. Maurer, *Appl. Phys. Lett.*, **27**, 220 (1975).

The physical mechanisms that can lead to a Weibull type distribution are not immediately obvious. However, a number of models have been proposed that produce such results and that point to the underlying mechanism responsible.

One is interested in the failure of the fiber at strengths that fall well short of its ultimate strength. Thus one imagines that the fiber strength has been degraded by surface flaws. Experimentally, one observes that short lengths of fiber break at higher mean values than long lengths. This is perfectly reasonable if it is assumed that there is a statistical distribution of flaw sizes, large flaws leading to low breaking strains and small flaws to higher breaking strains. In any given length the fiber will fail at the breaking strain associated with the largest flaw present in the test sample.

Let us consider a length, l, of fiber and define $g(l,f^*)$ as the probability that it fails at stress f^*. We can immediately write the probability $G(l,f^*)$ that the fiber strength does not exceed f^* as

$$G(l,f^*) = \int_0^{f^*} g(l,f)\,df \qquad (12.2.2)$$

Experimentally, we find that there are rather few large flaws and many more small flaws when many fiber samples are stressed to breaking, so that for purposes of analysis we postulate that, for any particular flaw on the fiber surface, the probability of failing at stress f^* is given by

$$P(f^*) = C(f^*)^m \qquad (12.2.3)$$

This distribution is simply one of many that reflect the physically observed trend. The value of C, the unknown constant, can be estimated by noting that the following boundary limitations apply to the value of $P(f^*)$:

$$P(f^*) = 0 \qquad f^* < 0$$
$$= 0 \qquad f^* > F \text{ (i.e., already broken)}$$
$$\int_0^F P(f^*)\,df^* = 1$$

so that $C = (m+1)/F^{m+1}$, where F is the ultimate stress for the material.

If we now assume a density of flaws per unit length of D, then, in a sample of length l, there are (on average) Dl flaws. Provided that this is a large number, we can say that the probability of the fiber failing when $f = f^*$ must be given by the probability that there will be one flaw of limiting strength f^* and that all other $(Dl - 1)$ flaws will have strength greater than f^*. Mathematically, this is given by the expression

$$g(l,f^*) = lP(f=f^*)[P(f \geqslant f^*)]^{Dl-1}$$

Now

$$P(f \geqslant f^*) = 1 - \int_0^{f^*} P(f)\,df$$

$$= 1 - \left(\frac{f^*}{F}\right)^{m+1} \tag{12.2.4}$$

so that we have the result that the probability of the fiber failing at stress f^* is given (for length l) by

$$g(l,f^*) = \frac{(m+1)Dl}{F}\left(\frac{f^*}{F}\right)^m\left[1 - \left(\frac{f^*}{F}\right)^{m+1}\right]^{Dl-1} \tag{12.2.5}$$

Using equation 12.2.2, we can now calculate $G(l,f^*)$ as

$$G(l,f^*) = (m+1)Dl\int_0^{f^*/F} y^m (1-y^{m+1})^n\,dy \tag{12.2.6}$$

where $y=f/F$ and $n=Dl-1$. Using the binomial expansion for $(1-z)^n$, we can evaluate this in series form as

$$G(l,f^*) = Dl\left[\phi - \frac{n}{2!}\phi^2 + \frac{n(n-1)}{3!}\phi^3 - \frac{n(n-1)(n-2)}{4!}\phi^4\cdots\right]$$

$$\simeq \psi - \frac{\psi^2}{2!} + \frac{\psi^3}{3!} - \frac{\psi^4}{4!}$$

$$= 1 - \exp(-\psi) \tag{12.2.7}$$

Here we have set $\phi = (f^*/F)^{m+1}$ and

$$\psi = (Dl)\left(\frac{f^*}{F}\right)^{m+1}$$

The above result is to be compared to the Weibull result, which takes the form

$$G(l,f^*) = 1 - \exp\left[-\left(\frac{f^*}{f_0}\right)^p\left(\frac{l}{l_0}\right)\right] \tag{12.2.8}$$

assuming that f_m is very small.

Evidently, we should make the following associations between the Weibull parameters and the model parameters. The power is related to the power law describing the flaw probability, $p = m + 1$. The stress at which failure always occurs is the ultimate material strength (unless we have reason to believe otherwise), and the equivalent scale length of area is related to the model through the relation $l_0 = 1/D$ and is thus the mean separation between cracks, measured along the axis of strain. We have left out of the model time dependence, since this depends upon the chemical environment of the fiber, which in turn controls the microcrack growth (see Section 12.2.6).

An important lesson from this analysis is that studies of short fiber lengths will yield data about the small microcracks which occur at high densities. The failures of long sample lengths are not governed by these cracks, however, so that these data are not particularly relevant, except insofar as a power law of the type of equation 12.2.3 can be established to be valid. The failure of a 1 km length of fiber is controlled by the *single* largest flaw present in the test length. The only way to discover its size for certain is to test the whole kilometer sample length to destruction!

12.2.3 Griffith Microcracks

We have already noted that the failure of glass usually occurs prematurely at a stress level well below that associated with the perfect material because of stress concentration around microcracks. The model remains a hypothesis, since the actual microcracks have not been directly observed, although their presence has been deduced from the appearance of etched glass surfaces, in which their size is supposedly enlarged until visible, and is supported by a great deal of circumstantial evidence.

The effect of stress concentration around a microcrack was first postulated by Griffith[3] and subsequently developed by Inglis.[5] They presupposed cracks of certain forms and analyzed their effects. The former author described the effect of a crack of the form of Figure 12.3, having depth x and surface width y, and showed that the stress concentration at the tip was given by

$$f_t = f_a\left(1 - 2\frac{x}{y}\right)$$

Inglis then extended the analysis to cover the case of an elliptical crack having semiaxes a and b and the case in which the ellipse is described by the semidepth, x, and radius of curvature of the tip, ρ. For the latter case he derived the relation

$$f_t = 2f_a\sqrt{x/\rho}$$

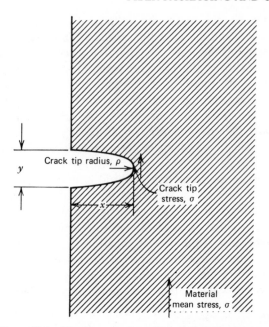

Figure 12.3. The form of a hypothetical Griffith microcrack.

To apply this relation to the situation of glass failure, it is assumed that the stress, f_t, is equal to the ultimate strength of the material and that the tip radius of curvature is given by the interatomic spacing of the lattice constituents. The term f_a is the failure stress for a given crack depth, x.

Evaluating the above relation for silica fiber, Maurer[6] obtained the result

$$f_f = f'(x)^{-1/2}$$

where

$$f' = 10 \times 4 \ \text{N/mm}^{3/2}$$

In these equations x is in millimeters, and f_f in newtons per square millimeter. From a measured Weibull distribution of breaking stresses, Maurer then calculated, using the above relation, the density of flaws of given size in the samples of fiber studied to yield the result of Figure 12.4. It is derived from the Weibull curve of Figure 12.2.

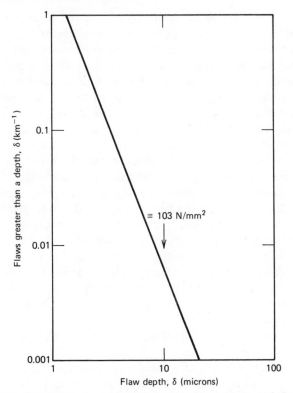

Figure 12.4. The flaw distribution deduced from the data of Figure 12.2. Reproduced with permission from R. D. Maurer, *Appl. Phys. Lett.*, **27**, 220 (1975).

12.2.4 Testing Techniques

The foregoing dicusssion has made clear that, if meaningful statistics are to be obtained, very large lengths of fiber must be tested, lengths equivalent to those that will be used in practice. This immediately raises some difficulties since normal tensile testing machines are hardly designed for use with kilometer sample lengths. However, a number of ways around the problem exist.

The direct way around the difficulty is to scale the sample length to whatever is convenient, perhaps 2 to 5 m, and then to test very many samples, say batches of 200, so that statistics for kilometer lengths are obtained. Great care must be taken in mounting the sample for this type of testing to ensure that the sample is not damaged by the test procedure, mounting, and so forth, and samples that break at the sample mounts should be discarded from the statistics.

The second approach[7] is to simultaneously screen a very large number of short samples. This can be done by taking a kilometer length of fiber on a reel, slitting across the reel to break the samples into lengths of $2\pi r$, and simultaneously mounting the 1000 or so samples thus obtained on a single frame. Tensioning this and straining the multiple samples simultaneously makes it possible to build up the statistics for the whole group during a single test run, since individual fiber breakages can be noted as a function of strain while they occur. To use the test with coated samples, it may be necessary to cycle the tension between the test strain and some negative strain, so that breakages are highlighted by the kink that occurs at the break, since the surface coating may remain intact when the fiber fails.

A third technique which allows the screening of many samples to be carried out involves reeling the fiber from one drum to another at controlled tension.[6] By increasing the tension until breakage occurs, the weakest point on the drum length is found. Rereeling the remainder of the fiber allows the next weakest point to be found, and so on. This approach has one attraction in particular, namely, it would appear to be almost ideal for use as a production line test for rejecting fiber which contains a particularly weak point, thus ensuring that a product reaches or exceeds a specified strength.

A series of results obtained by this technique were shown in Figure 12.2 and include measurements, made both on long lengths (by this technique) and also by testing to failure on a tensile test meachine with shorter sample lengths. The results appear to be fairly typical of present day optical fibers. However, the detailed performance of a fiber depends critically upon the handling, surface damage, degradation, and protection applied, so that we may expect wide variations to be apparent between fibers from different sources and even between fibers from the same source but with different histories.

In summary, then, we find by measuring large numbers of samples to destruction that the statistics of failure of fibers fit a Weibull distribution. This implies, when a straight line is obtained, that the flaw distribution follows some simple power law of flaw size. Some fibers when tested show a double curve of the type shown in Figure 12.5; others, a single curve of the type in Figure 12.6. The former is generally taken as indicating flaws from two sources, such as flaws introduced by the pulling process in addition to fundamental flaws occurring in the glass preform and fiber. Usually, by careful control of the pulling and preparation processes, a single curve of the type in Figure 12.6 can be obtained. It is then a matter of debate as to whether any further improvement can be achieved, whether this curve represents something that is fundamental to the material in fiber form, or whether there is still some physical damage mechanism whose removal or control would lead to a further increase in measured strength.

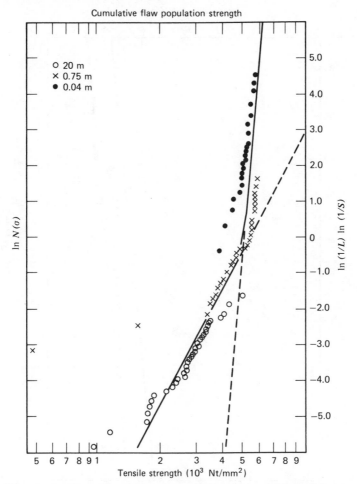

Figure 12.5. A double-mechanism failure process, showing in a Weibull plot two statistical distributions for cracks, one for small and another for large cracks. Note that the vertical axis in this figure labeled $N(\sigma)$ is equivalent in our notation to $\ln\cdot\ln[1/(1\text{-}G)]$. Reproduced with permission from R. Olshansky and R. D. Maurer, *J. Appl. Phys.*, **47**, 4497 (1976).

The importance of the preform preparation and the drawing environment on the ultimate fiber strength has been vividly demonstrated in some recently published work on silica coated fiber. In Figure 12.7 we reproduce the histograms for the failure of fibers drawn from preforms pulled in a graphite resistance furnace. The preforms were of Amersil TO8 silica, and the fiber was coated immediately after pulling with an ultraviolet cured epoxy acrylate coating. The histograms indicate that by fire-polishing the preform (see the code FP on the histograms) substantial improvements in

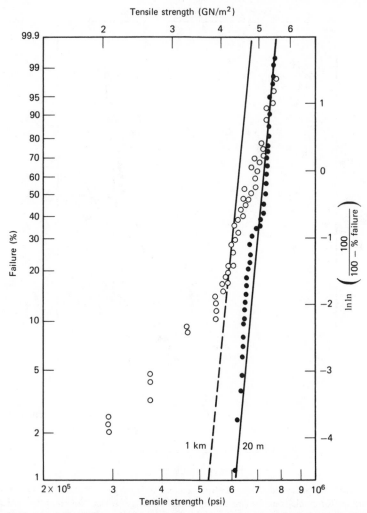

Figure 12.6. A single failure mechanism fiber compared with earlier results showing the presence of two mechanisms, illustrating the great increase in strength obtained on long fiber lengths. Reproduced with permission from H. Schonhorn, C. R. Kurkjian, R. E. Jaeger, H. N. Vazirani, R. V. Albarino, and F. V. DiMarcello, *Appl. Phys. Lett.*, **29**, 712 (1976).

strength are obtained, clearly showing that damage of some form is carried through the pulling process.

Figure 12.8 shows the equivalent histograms for preforms similarly fire polished but pulled using a laser fiber puller. The further increase in strength is again most noticeable, and fiber prepared in this way yielded the single-flaw-mechanism Weibull plot of Figure 12.6. Presumably we

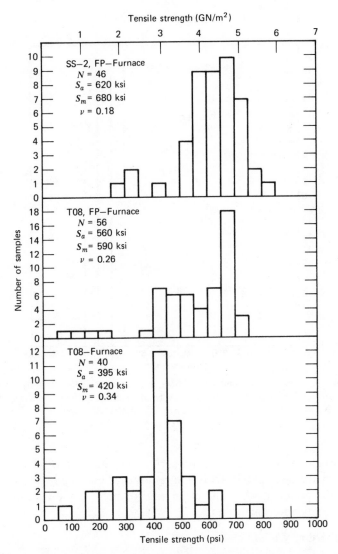

Figure 12.7. Some failure histograms for silica fibers with epoxy-acrylate coatings. FP in the code indicates that the preform was fire-polished before pulling to reduce the density of surface flaws. Reproduced with permission from H. Schonhorn, C. J. Kurkjian, R. E. Jaeger, H. N. Vazirani, R. V. Albarino, and F. V. DiMarcello, *Appl. Phys. Lett.*, **29**, 712 (1976).

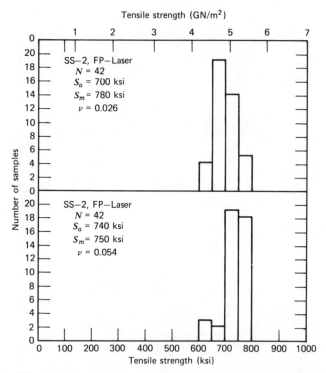

Figure 12.8. Histograms for a similar set of fibers but drawn using laser heating in place of a graphite furnace. Reproduced with permission from H. Schonhorn, C. J. Kurkjian, R. E. Jaeger, H. N. Vazirani, R. V. Albarino, and F. V. DiMarcello, *Appl. Phys. Lett.,* **29,** 712 (1976).

should conclude that the very high surface temperature produced on the preform by the laser puller led to the removal of essentially all preexisting flaws in the preform, while the clean environment of the laser puller did not induce any additional ones.

12.2.5 Plotting Weibull Data

If we wish to examine one set of statistical data for a fixed sample gauge length, it is a simple matter to plot it on Weibull graph paper and to observe the form of the curve obtained. However, if we wish to extract more information from the curves or to superimpose data referring to different gauge length samples, it is useful to know the form of the Weibull scales.

The vertical y axis is a linear scale of $\ln \cdot \ln[1/(1-G)]$, where G is the fraction of samples failed, while the horizontal axis is a simple $\log_{10}$ scale. Using these functions, we can obtain a Weibull plot on linear graph paper.

If we examine the form of the Weibull equation (12.2.8), we have

$$G(l,f^*) = 1 - \exp\left[-\left(\frac{f^*}{f_0}\right)^p \left(\frac{l}{l_0}\right) \right]$$

(12.2.9)

so that

$$\ln \cdot \ln\left(\frac{1}{1-G}\right) = p \ln\left(\frac{f^*}{f_0}\right) + \ln\left(\frac{l}{l_0}\right)$$

(12.2.10)

It follows that to compare data measured at gauge length l_1 to data measured at gauge length l_2, a vertical shift of an amount $\ln(l_2/l_1)$ should be added to the measured values of $\ln \cdot \ln[1/(1-G)]$ before plotting them.

12.2.6 Time of Fiber Stress Crack Growth

We quoted in equation 12.2.1 a result for the probability of failure of a fiber which includes time, indicating that the probability of failure is not only stress or strain dependent but also time dependent. This result greatly increases the complexity of specifying the fiber for a cable that has to last for a long time intact, that is, the normal case.

The physical mechanism underlying this problem is the effect of crack growth in glass under humid conditions and strain. Empirical studies of slow crack growth show that the size of the crack varies with time as

$$\frac{\partial x}{\partial t} = (Af)^b$$

$$f = 2f_a \sqrt{x/\rho}$$

(12.2.11)

Substituting and integrating to find the time for a given initial crack of size x_i to grow to size x_f, the size that will cause failure, we obtain the result

$$t = \left(\frac{2}{b-2}\right)\left(\frac{\sqrt{\rho}}{2Af_a}\right)^b \left[x_i^{(2-b)/2} - x_f^{(2-b)/2} \right]$$

$$\simeq \left(\frac{2}{b-2}\right)\left(\frac{\sqrt{\rho}}{2Af_a}\right)^b \left[x_i^{(2-b)/2} \right]$$

(12.2.12)

The latter approximation is valid because the flaws that cause failure over 20 years are always much smaller than those that will cause failure "instantaneously" at the same stress. We note that the rate of growth of cracks depends sensitively upon the level of applied stress, f_a, on the power,

b, and on the parameter A. Typically we find that b is in the region of 20 to 25 for glasses. Further algebraic juggling leads from the result above to an expression for failure probability as a function of time in the form of equation 12.2.1. The detailed arguments involved have been discussed carefully by Olshansky and Maurer.[8]

Physically, the crack growth described above occurs because of the chemical erosion of material at the crack tip. The reaction is speeded by placing the material under stress. In fiber cable design this leads to the need for accurate measurement of the parameters A and b in the above relations if crack growth is to be estimated and the strength of the cable guaranteed. However, once these parameters have been established over the chemical environment and the strain ranges likely to exist in the operating cable, and for the region of the flaw distribution likely to lead to failure over the life of the cable, it is possible to calculate the maximum flaw size permissible in the fiber at the time of cabling, using equation 12.2.1. To do this the following parameters must be specified:

1. Operating temperature range.
2. Chemical environment and humidity.
3. Residual strain in the fiber after installation.
4. Desired lifetime.
5. Acceptable failure probability per kilometer of fiber.

All the fiber to be used for the cable production is then screened as described in Section 12.2.7 to remove (by breaking) all flaws greater than the allowed maximum size. This may lead to an unacceptable rejection rate for fiber, which in turn will lead to a reexamination of the specifications or the cable design and installation procedures.

12.2.7 Strength Assurance for Production

In the preceding discussion we examined the statistical nature of fiber failure, discussed the physical reasons for it, showed how to analyze and measure data obtained from experimental testing of fiber strength, and, in Section 12.2.6, considered how to extrapolate fiber failure probability into the future under specified conditions of temperature, humidity, and strain (or stress). However, having done this, we are left with the problem of how to ensure that a given fiber in a cable will not fail. An established procedure for achieving this has yet to be agreed upon, but the general guidelines for doing so are already clear.

Careful statistical measurements of the fiber, of the types discussed above, must be made, and the data checked for statistical validity. The operating conditions for the cable design must then be formulated. It must also be established that the protected fiber, tested to produce the above statistics, is not mechanically damaged during handling and cabling. This consideration appears to require primarily that the coatings protecting the fiber be sufficiently durable to be mechanically impenetrable during the manufacturing process.

Given these conditions and data, the temporal analysis (Section 12.2.6) can be used to calculate the smallest initial flaw size that will grow sufficiently to cause failure within the design life of the cable. Once this size is established, it is a straightforward matter to calculate the strain or stress at which a fiber with this size of flaw will break if stressed immediately after drawing but before cabling. The procedure then is to strain all fiber after drawing and coating, but before cabling, to this calculated stress (or strain). If a fiber does not break, it contains no flaws larger than the calculated size, and the life of the cable produced from it can be assured under the chosen operating conditions. Such a procedure is already commonly used during the manufacture of precision glass fiber to ensure operating strength. It will be apparent to the reader, however, that such an approach leads to fiber breakage and hence waste, which ultimately increases the cost of the delivered fiber. It is therefore most important in setting up a manufacturing plant to establish techniques for minimizing the number of large flaws in the as-drawn fiber and for ensuring that the proof stress or strain value is both adequate to ensure long life in service and low enough to allow an acceptable yield of fiber from the plant.

12.3. SURFACE PROTECTION OF FIBERS

In approaching the matter of surface protection techniques, we find that very few published data exist pertaining to optical fibers. However, there is a considerable body of data with regard to improving the strength of structural glass fibers, fibers made for bedding in resins and so forth for use as structural materials. Here the detailed considerations are different, but there is evidently much common ground. From this work we can dintinguish a number of independent factors.

Obviously, we must take great care not to damage the fiber chemically or mechanically before protecting it. The protective layer(s) must provide chemical protection against water or other liquid or gaseous phase materials; it is common knowledge that many optical glasses suffer serious surface attack from the atmosphere. Having chemically protected the surface, we must also mechanically protect it by applying a sufficiently thick and tough barrier to prevent abrasion by grit or other substances. The fiber must then be protected from microbending by enclosing it in a suitable environment. Some or all of these steps may possibly be combined, but each must be considered.

12.3.1 Surface Corrosion

At the atomic level we can consider the surface degradation as largely dominated by the interaction of water with the glass matrix. In an alkali silicate glass the reaction intimately involves the alkali component and proceeds much faster than in pure silica.

The surface corrosion reaction for an alkali glass has been described by Charles[9] and proceeds as follows. The glass structure is considered to consist of an unbroken chain of silicons bonded via oxygens with the alkali ion attached at free bond points. The silica network is very strong, whereas the alkali ions are relatively loosely attached. Water first attacks the alkali ion:

$$\left[-\overset{|}{\underset{|}{Si}}-O-Na\right] + H_2O \rightarrow \left[-\overset{|}{\underset{|}{Si}}-OH\right] + Na^+ + OH^- \qquad (12.3.1)$$

The free hydroxyl ion now formed can attack the strong Si-O-Si bond thus:

$$\left[-\overset{|}{\underset{|}{Si}}-O-\overset{|}{\underset{|}{Si}}-\right] + OH^- \rightarrow \left[-\overset{|}{\underset{|}{Si}}-OH\right] + \left[-\overset{|}{\underset{|}{Si}}-O^-\right] \qquad (12.3.2)$$

The Si-O-Si bond having been broken, the final reaction in the sequence follows to complete the breakage and seal it:

$$\left[-\overset{|}{\underset{|}{Si}}-O^-\right] + H_2O \rightarrow \left[-\overset{|}{\underset{|}{Si}}-OH\right] + OH^- \qquad (12.3.3)$$

This reaction leaves a free hydroxyl radical to initiate another attack on the silica matrix. Overall, reactions 12.3.1 to 12.3.3 might be summarized in a single equation of the form

$$\left[-\overset{|}{\underset{|}{Si}}-O-\overset{|}{\underset{|}{Si}}-\right] + H_2O \rightarrow 2\left[-\overset{|}{\underset{|}{Si}}-OH\right] \qquad (12.3.4)$$

This would be the reaction in the absence of the alkali ion producing the free-OH^- radical, and would thus describe the attack in silica. However, the absence of the-OH^- leads to a much reduced reaction rate, a fact well illustrated by the alkali silicate glasses where the rate of atmospheric attack is reduced by frequent washing, an action which removes the free Na^+ and $-OH^-$ radicals.

12.3.2   Surface Neutralization

Chemical neutralization of the glass surface is frequently achieved by the use of silanol groups. The mechanism of one such coupling compound has been described by Muto et al.[10] in a study of the adhesion of various polymers to a glass surface. Its action is shown in Figure 12.9. The coupling agent, KBM 603, forms OH groups in place of CH_3O groups by hydrolysis, and these in turn bond chemically with Si–OH groups already existing on the free glass surface, liberating water in the process. The coupling agent thus forms a chemical bond to the glass surface; in so doing it attaches itself very firmly and prevents other undesirable materials from doing likewise at a later stage. Many such coupling agents are used in the fiberglass industry, with different groups attached at the X position of Figure 12.9 to form a favorable bond with the matrix material in which the fiber is to be immersed. The effectiveness of this particular coupling agent in bonding to a plastic ethylene tetrafluoroethylene (ETFE) is shown in Figure 12.10, in which the adhesive strength of the ETFE film is shown relative to the soda lime silicate glass substrate after repeated immersion tests in water. Without the coupling agent the film freed itself after a single immersion, whereas with the agent little degradation was observed after repeated immersions.

We note that the critical factor in controlling the choice of the material that interfaces with the glass is the ability of the material to form some sort of bond to the glass matrix. With this in mind, it seems likely that the

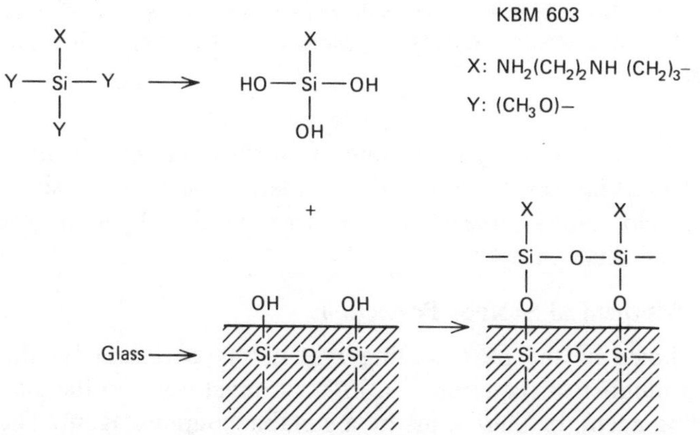

Figure 12.9. The proposed action of one particular agent in neutralizing the surface of a glass to chemical attack. Reprinted with the permission of the North Holland Publishing Company from R. Muto, N. Akiyama, H. Sakata, and S. Iuruichi, *J. Non-Cryst. Solids*, **19**, 369 (1975).

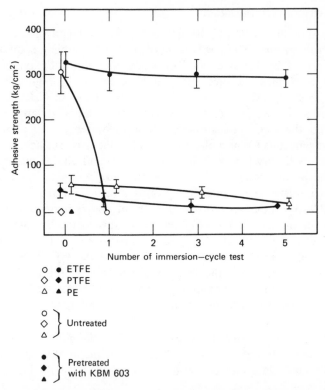

Figure 12.10. Some data to support the claim that the agent has bonded to the glass surface, showing a greatly increased adhesion for an ETFE film when in the presence of water relative to an untreated glass surface. Reprinted with the permission of the North Holland Publishing Company from R. Muto, N. Akiyama, H. Sakata, and S. Furuichi, *J. Non-Cryst. Solids,* **19,** 369 (1975).

functions of mechanically protecting the surface and chemically neutralizing it could be combined in one material, such as a silicone resin. However, no results appear to have been published on the use of such materials in this context.

12.3.3   Mechanical Surface Protection

The choice of the next coating material is governed by the need to satisfy a number of requirements. Either it must bond to the glass without a coupling agent, or there must be a suitable coupling agent. The material must be easily applied, probably from solution, since passing the unprotected fiber through an extrusion die is likely to damage the fiber surface. Once applied, the material must be chemically stable, it must provide an excellent chemical and mechanical barrier, and, in general, it must be possible to remove the material again for the purposes of jointing. The material that meets many of these requirements and appears to be most commonly used is Kynar, which is a vinylidene fluoride polymer widely used for electrical insulation purposes. It is characterized by excellent corrosion resistance, toughness to abrasion, and high dielectric strength and is recommended for a wide variety of applications in the chemical, aerospace, and electronics industries. It is available in two principal forms: the homopolymer, which is a thermoplastic suitable for extrusion, and the copolymer, which in general has slightly lower performance, being less strongly chemically bonded, but as a result is more soluble in acetone and can be applied as a solution.

A simple technique for applying a Kynar[11] coating has been described by France and Dunn,[12] using a crucible with a small nozzle in its base, through which the fiber runs. The Kynar is dissolved in solvent, and the crucible filled with it. By careful design of the nozzle diameter, a coating of constant thickness can be obtained. The nozzle diameter is made small enough with respect to the fiber diameter (see Figure 12.11) so that, under static conditions, the surface tension forces balance the hydrostatic pressure of the crucible liquid head and no material flows. Material flows only when the fiber moves through the nozzle, as a result of viscous drag. Under these conditions the velocity distribution in the liquid in the nozzle region is found by solving Laplace's equation with the appropriate boundary conditions. Thus we have the following relations:

$$\nabla^2 V = 0$$

Since $\partial/\partial\phi = 0$ and $\partial/\partial z = 0$,

$$\frac{1}{r}\frac{\partial}{\partial r}\left(r\frac{\partial V}{\partial r}\right) = 0 \tag{12.3.5}$$

This has a general solution of the form

$$V = B\log Ar + C = \text{velocity} \tag{12.3.6}$$

Now, applying the boundary conditions that at the fiber boundary the fluid velocity is that of the fiber, V_0, and at the nozzle boundary the fluid velocity is zero, we find that

$$V(r) = \frac{V_0}{\log(a/b)}\log\left(\frac{r}{b}\right) \tag{12.3.7}$$

where the fiber radius is a and the nozzle radius is b. The total volume of

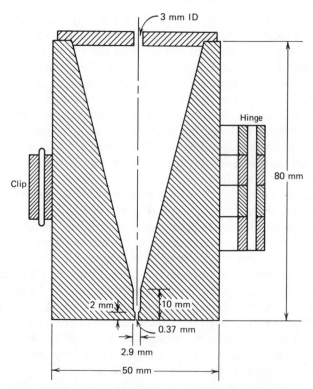

Figure 12.11. A crucible for applying a thin layer of surface protection material to a fiber at the time of drawing. Reproduced with permission from P. W. France and P. L. Dunn, *2nd European Conference on Optical Communications, Paris, 1976.* Paris: SEE.

liquid flow per second is given by

$$Q = 2\pi \int_a^b V(r) r \, dr$$

$$= \frac{\pi V_0}{\log(a/b)} \left[\tfrac{1}{2}(a^2 - b^2) - a^2 \log\left(\frac{a}{b}\right) \right] \qquad (12.3.8)$$

On the assumption that the fiber is central in the nozzles, this volume of material spreads itself uniformly over the fiber surface, the area of which per second is

$$A = 2\pi a V_0 \qquad (12.3.9)$$

A film is thus formed of thickness T, where T is given by the relation

$$T = \frac{1}{2a \log(a/b)} \left[\tfrac{1}{2}(a^2 - b^2) - a^2 \log\left(\frac{a}{b}\right) \right] \qquad (12.3.10)$$

if we assume that T is very much less than a. This film will then dry by losing the solvent to the surrounding atmosphere and in so doing will contract by a factor F, where F is the volume loading factor of the Kynar in solvent, so that the final film thickness is given by

$$T_f = FT \qquad (12.3.11)$$

This result is noteworthy for several features. The coating thickness is independent of pulling speed and varies slowly with fiber diameter. The coating thickness is also independent of the liquid viscosity, depending only upon its dilution.

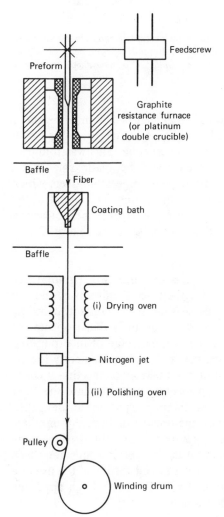

Figure 12.12. Fiber puller shown schematically with coating crucible and drying ovens. Reproduced with permission from P. W. France and P. L. Dunn, *2nd European Conference on Optical Communications, Paris, 1976,* Paris: SEE.

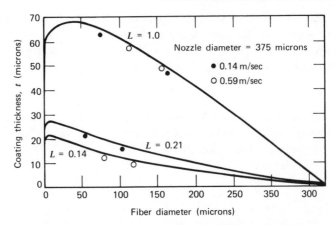

Figure 12.13. The results of some coating studies, showing experimental measurements of coating thickness versus fiber diameter and pulling speed. Reproduced with permission from P. W. France and P. L. Dunn, *2nd European Conference Optical Communications, Paris, 1976.* Paris: SEE.

Results have been described for a nozzle diameter of 300 microns and a variety of fiber diameters with $F = 0.14$. The liquid coating was dried with an oven immediately after leaving the coating crucible. This was found to produce an orange-peel-like surface which could be made smooth by passing the coated fiber through a second, higher temperature oven, which melted the solid Kynar and produced a smooth coating. The layout of the apparatus used is shown in Figure 12.12 and some experimental measurements of coating thickness versus fiber diameter are given in Figure 12.13.

12.3.4 Fiber Cushioning

The fiber packaging or buffering layer has to fulfill two requirements. The first requirement is that it provide for the fiber a uniform cushion to insulate it from impressed microbends arising from contact with the environment surrounding it. The second requirement is that it locate the fiber in a convenient manner both for the original stage of incorporating that particular fiber into a cable making machine and subsequently for the stage of operating with the cable, jointing the fibers or connecting them to repeaters, and so on. In addition, the packaging must insulate the fiber from the effects of longitudinal strain and thermal expansion in the cable, which can give rise to buckling effects on the fiber's nominally straight lay. Experimentally, we find a number of approaches being followed, ranging from no packaging other than a few-micron-thick layer of a substance such as Kynar through the stage of a loosely fitting extruded tube to the use of a solid buffer layer, mechanically in contact with the fiber. Each approach appears to have its attractions and its problems.

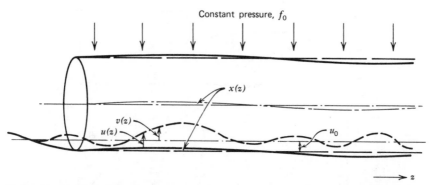

Figure 12.14. The schematic situation used by D. Gloge to analyze the effect of a rough surface causing microbending in a fiber by contact. Reprinted with permission from D. Gloge, *Bell Syst. Tech. J.*, **54**, 243 (1975), copyright 1975, The American Telephone and Telegraph Company.

The effects of a mechanically contacting extruded coating was analyzed by Gloge,[13] who considered the situation illustrated in Figure 12.14. He assumed that the environment can be characterized by a rough surface of some statistically describable form, and that the fiber can be considered as pressed onto that surface with a uniform pressure, P. The fiber then acts as a stiff beam, bent over a curved object by an impressed force, a situation that is readily analyzable. In practice, the lateral force impressing the fiber into the rough surface can arise from longitudinal tension in the fiber, pulling the fiber inwards when it is laid in a curved path, or it can arise from a compressive force from a cable jacket, acting inward onto the fiber and its support structure. Experimentally, the situation can be simulated by tensioning a fiber around a drum surface, a situation well known to produce microbending loss.

Having set up the theoretical model, Gloge proceeded to analyze the effects on the fiber of different composite coatings for a given surface deformation and showed that the best combination is a composite coating consisting of a soft outer shell with a hard and stiff inner layer, or a single cladding of hard, stiff material. The results are shown in Figures 12.15 and 12.16.

The theoretical model analyzed by Gloge considered the fiber within its environment of packaging impressed against a rough drum. Within the package the fiber itself acts as a relatively stiff beam, and with a soft core of inner sheath material surrounded by a hard outer skin, the fiber has considerable resistance to microbending. However, the critical wavelength for graded-index fibers (see Appendix 3) is typically 1 mm, and the fiber can follow this readily unless in a very soft environment. It may therefore be more meaningful to examine the fiber with its extruded coating as a

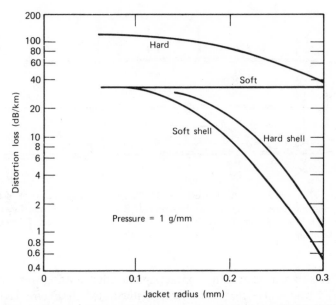

Figure 12.15. Results of the theoretical analysis of the situation in Figure 12.14. A mean lateral pressure of 1 g/mm is assumed as well as an rms jacket thickness variation of 1 micron and a spatial correlation length of 1 mm, with $\triangle = 0.005$. Reprinted with permission from D. Gloge, *Bell Syst. Tech. J.*, **54**, 243 (1975), copyright 1975, The American Telephone and Telegraph Company.

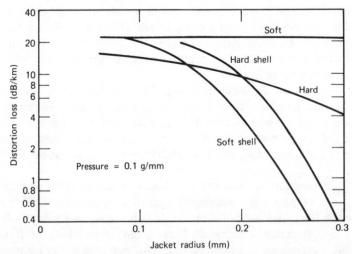

Figure 12.16. Results from the same analysis as in Figure 12.15 but with the lateral pressure reduced to 0.1 g/mm. Reprinted with permission from D. Gloge, *Bell Syst. Tech. J.*, **54**, 243 (1975), copyright 1975, The American Telephone and Telegraph Company.

single stiff beam, interacting with a (isotropic) deformed environment which in reality is the cable structure. Under these conditions the Gloge formulas simplify to yield simple results for the relative propensity to microbending as a function of packaging material.

The argument proceeds as follows. The amplitude of deformation suffered by a stiff beam in an elastic medium is given by

$$X = V\left(1 + \frac{K^4 H}{D}\right)^{-1} \tag{12.3.12}$$

using Gloge's notation, where V is the amplitude of the applied deformation, K is the wavenumber of the critical microbending deformation, $H = EI = \pi E a_{\text{clad}}^4 / 4$, and E and D are the Young's modulus of the "beam" and the effective surface Young's compressibility modulus, respectively.

From Appendix 3 we have

$$K = \frac{\sqrt{2\Delta}}{a_{\text{core}}} \tag{12.3.13}$$

while for D we may use the relation

$$D^{-1} = E_1^{-1} + E_2^{-1}$$

where E_1 and E_2 are Young's moduli for the extruded fiber sheath and the cable surrounding material, respectively. Collecting these results together, we obtain

$$X = V\left[1 + \pi \Delta^2 \left(\frac{a_{\text{clad}}}{a_{\text{core}}}\right)^4 \left(1 + \frac{E_1}{E_2}\right)\right]^{-1} \tag{12.3.14}$$

From this, we may proceed to estimate the material effect alone by defining the relative propensity to microbending as

$$F_{mb} = \left[1 + \left(\frac{A}{1 + A}\right)\frac{E_1}{E_2}\right]^{-1}$$

$$\simeq \left(1 + \frac{E_1}{E_2}\right)^{-1} \qquad A \gg 1 \tag{12.3.15}$$

where

$$A = \pi \Delta^2 \left(\frac{a_{\text{clad}}}{a_{\text{core}}}\right)^4$$

a_{clad} is radius of polymer extrusion, and a_{core} is radius of fiber core.

This argument points strongly to the desirability of using a hard, stiff sheath extruded around the fiber to make a composite structure that is stiff. It should then be bedded into a softer surrounding material. In practice we find that a number of low loss solid sheath materials with good microbending resistance have been used, notably polypropylene and nylon for the extrudant to a diameter of 0.5 to 1 mm. These must generally be encased in a cable structure of similar materials, apparently breaking the above condition, but it seems likely that the effective E_2 operative in such cases is much lower than the sheath value since the contact is likely to be lossy and incomplete.

Experimentally, results have been published of various single-component extrusions of polymers. The most extensive data appear to be those of

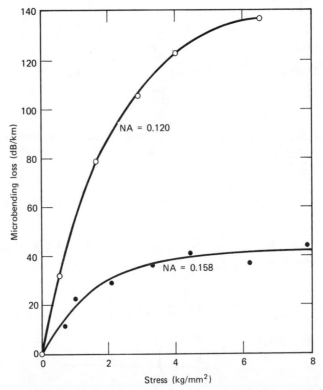

Figure 12.17. Some experimental measurements of microbending losses for fibers of two different numerical apertures but without coatings applied, obtained by tensioning the fibers on a drum. Reprinted with permission from W. B. Gardner, *Bell Syst. Tech. J.*, **54**, 457 (1975), copyright 1975, The American Telephone and Telegraph Company.

Gardner,[14] who has shown the effects on loss of various winding tensions (or contact pressures) for a given fiber (Figure 12.17) and also the effect of a specific buffer coating, a 50-micron-thick layer of Elva, a Du Pont copolymer of ethylene and vinyl acetate (Figure 12.18). Two features stand out from the curves of Figure 12.18. The most immediate is that the buffer coating has reduced the fiber's susceptibility to microbending when subjected to tough operating conditions. However, we should also note that the buffered and unbuffered curves cross at a low level, showing that the buffer coating has had a small degrading effect on the fiber when run free. Typically, a buffer coating appears to increase the fiber loss under favorable conditions (loose on a drum) by up to 5 dB/km; and since systems

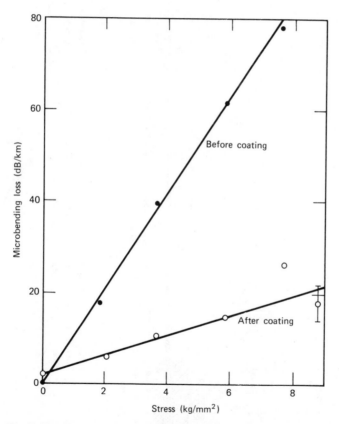

Figure 12.18. Similar curves to those of Figure 12.16 for 0.16 NA fibers but with and without a 50 micron thick buffer layer coating applied. Reprinted with permission from W. B. Gardner, *Bell Syst. Tech. J.*, **54**, 457 (1975), copyright 1975, The American Telephone and Telegraph Company.

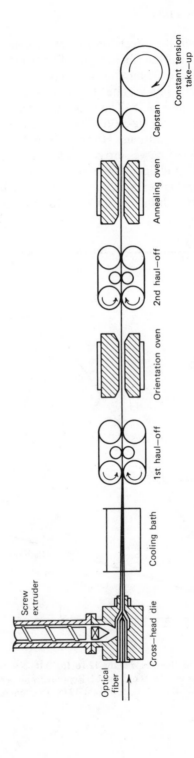

Figure 12.19. An apparatus for applying a loosely fitting oriented-molecule polymer tube around a fiber. Reproduced with permission from L. A. Jackson, M. H. Reeve, and A. G. Dunn, *Opt. Quant. Electron.*, **9**, 493 (1977).

designers are generally seeking cable losses of 5 dB/km or less, we may conclude that many of the current techniques will need improvement before they become acceptable.

The only published composite packaging appears to be that of Jackson[15] et al. who have taken the soft core to its ultimate, using air, and surrounded the hollow core by an oriented molecule polymer tube sheath. The apparatus for this is illustrated in Figure 12.19. A tube of polypropylene is extruded from a screw extruder having dimensions of about 1 mm ID and 2 mm OD. The fiber passes through the die to run loosely inside the tube. The first haul-off runs at about one tenth of the fiber speed. Next, the large tube is reheated in a small oven to soften the polymer, which is then stretched to shrink it and align the polymer molecules. The resulting tube is highly anisotropic in its mechanical parameters and measures about 220 microns ID and 600 microns OD. The longitudinal Young's modulus is increased about tenfold by the molecular alignment, resulting in a very stiff tube for its size, while the fibrillated nature of the material allows simple stripping of it, in the same manner as stripping a banana. This package shows no additional microbending, provided that the polymer shrink-back is correctly controlled by allowing the fiber to slip within the package while the material is annealed. Without such care a shrinkage of some 10% can occur, leading to very serious buckling of the fiber within the tube, associated with very high optical losses.

Other extruded polymer coatings must be expected to show some shrinkage also, by virtue of the cooling and thermal contraction of the material. This can again lead to buckling or microbending of the fiber, and some control must be exerted at the manufacturing stage to reduce such effects to an acceptable level.

12.3.5 Fiber Buckling in Loose Structures

We have already noted that, when a fiber is subjected to tension around a curved surface, a component of that tension normally forces the fiber into the surface. This results in microbending, either from the fiber's following irregularities in the curved surface or from irregularities in the buffer coating, thickness, or compressibility. However, the fiber may also be exposed to local compression forces which will tend to buckle it out of its straight lay. These may arise in one of two general ways. In a helically wound cable that is bent, the fiber on the inside curve side of the neutral axis suffers compression. Alternatively, in a multimaterial cable the fiber is almost certainly the material with the lowest expansion coefficient, so that cooling the cable structure will result in thermal contraction of everything except the fiber.

A fiber that is housed in a loosely fitting tube whose length contracts relative to the fiber will tend to take up a helical path within the tube (see Appendix 7). The results of equation A7.1 can be used to predict the pitch of the helix for a given shrinkage, and some typical results[16] are shown in Figure 12.20 for a fiber of ~120 micron diameter lying in a tube of 350 micron bore. The helical pitch reverses every so often since, in changing from straight to helical lay, the fiber becomes subject to longitudinal

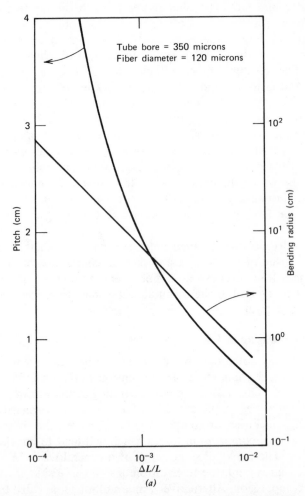

Figure 12.20. Some measurements of fiber losses under conditions of buckling in a tube. (*a*) Calculated helix pitch and maximum bending radius for the situation studied. (*b*) Measured loss (in dB/km) for the same situation. Reproduced with permission from M. H. Reeve, *Electron. Lett.*, **14**, 47 (1978).

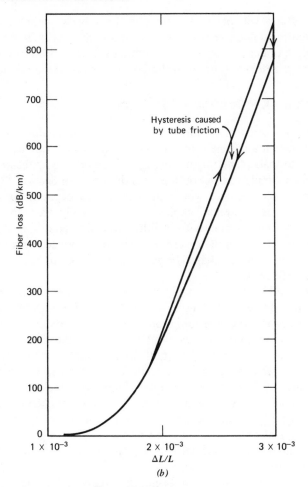

Figure 12.20. (*Continued*)

torque which reaches sufficiently high values to reverse the helix direction. Nevertheless, this type of helical path lay of the fiber leads to additional losses in the fiber by one of a number of possible mechanisms: straight bending loss, mode coupling caused by the helix reversals and their associated curvatures, or microbending due to the forcing of the fiber against the packaging wall by the outward forces of the helix curvature. In designing this type of packaging, the dimensions and expansion coefficients of the materials used must be so chosen as to ensure that these problems are avoided. The use of a material with a low expansion coefficient for the tube, to match that of the fiber, is a good first step.

Similar effects will occur if a fiber is run loosely within a composite cable structure, and a compromise must be struck to provide enough slack space so that, under all operational conditions of expansion or contraction, severe buckling does not occur and a compact cable design is obtained.

12.4. OPTICAL FIBER CABLE DESIGN

A fiber cable is a structure for enclosing a fiber to enable it to maintain its performance for sufficient time in the operating environment in which it is required to work. One might envisage undemanding applications where a fiber with a simple coating on the lines discussed in Section 12.3.4 might be perfectly adequate and would provide a high capacity cable of very small dimensions. However, cables more commonly have to operate in environments where a fiber with so little protection could not be expected to survive for long, even if it were possible to introduce the fiber into the environment without breakage. In general, then, we are concerned with a much sturdier structure.

The cable will be required to withstand the forces upon it during storage, installation, and operation. During storage or transit it may be subjected to considerable heat from solar radiation, to extreme cold, and to rain or very high humidity. During transport and installation the cable will be subjected to vibration, compression (both longitudinal and radial), bending, shearing, and tensioning; and, once installed, it must retain its optical characteristics while stored strains relax, humidity attacks it, the temperature cycles, and the environment vibrates, probably for extended periods of time—20 to 50 years in telecommunications use. Meeting these requirements requires a subtle blend of conventional cable design knowledge and an understanding of the special requirements imposed by the characteristics of glass fibers.

Before a serious design can begin, the requirements for it must be specified, if not totally, then at least in general terms. Tables 12.2 and 12.3 list some of the basic points that must be considered and whose relative importance may vary with different applications. For example, telecommunications cables that are to be pulled in long (500 m) ducts lined with earthenware pipes require different properties from cables that are to be plowed directly into the ground.

In general, the design of the cable proceeds by considering each performance aspect and attempting to meet it by selecting the appropriate material and inserting it in a suitable place in the overall cable structure. Two such hypothetical structures are shown in Figure 12.21. Outer sheaths of extruded polymeric material provide a tough outer skin. The material used must balance the needs to have a low friction coefficient, to allow

TABLE 12.2 FACTORS TO BE CONSIDERED
IN DESIGNING A FIBER CABLE: CABLE
SPECIFICATION

Fiber cable choices

Number of optical fibers
Optical attenuation
Pulse dispersion
Numerical aperture
Loss stability
Bandwidth stability
Ease of jointing
Need for demountable couplings
Need for T or star couplers

Environmental factors

Tension behavior
Radial compression loads
Flex resistance
Cold bend resistance
Abrasion resistance
Vibration resistance
Operating temperature range

Constructional factors

Fiber coatings and location of fibers
Strength member material and disposition
Sheathing and matrix materials
Moisture barriers
Electrical conductors (power feeds?)

easy sliding on installation, and to provide good abrasion resistance, low cost (since this material is used in greatest quantity), and good crush resistance. Armoring may be included in this sheath if the cable is to be operated in particularly stringent conditions, and for some applications a water barrier may also be included between the inner surface of this sheath and the inner structure, using aluminum tape, for example. A popular material for the outer extruded sheath is polyethylene.

Within the sheath the tensile and optical members will be placed. Special members are used to carry the longitudinal load imposed when the cable is

TABLE 12.3   FACTORS TO BE CONSIDERED IN DESIGNING A FIBER CABLE: CABLE DESIGN OPTIONS

Type	Location	Lay
Fiber		
Surface protection	Loosely bundled	Along cable neutral axis
Individually packaged	Rigidly enclosed	Helical around central member
Array packaged		Straight
Strength member		
Steel	Single strand	Neutral axis
Polymer	Bundle	Helical (single)
Glass fiber	Roving	Double helical
Carbon fiber	Plait	Straight, ordered
	Matrix	Straight, random with fibers
Matrix materials		
Cushion layers		
Water or chemical barriers		
Compressive load members (radial loading)		
Fillers and spacers		
Tapes, to package subassemblies		
Jellies, for lubrication and water sealing		
Armor sheathing		

pulled into its operating position. The disposition of these members relative to the optical members may vary widely. Electrical conductors may also be included within the central cable core to carry power to remote electronics or to serve for signaling purposes.

12.4.1   Choice of Fiber

In the choice of the fiber to be used in the cable, a number of trade-offs have to be made, involving core diameter, cladding diameter, index difference, index profile, and coatings applied. These trade-offs are summarized in Table 12.4, which largely refers to the discussion already presented. The major choice to be made concerns the optical performance of the fiber, which will be clear from the application. Knowing the loss and

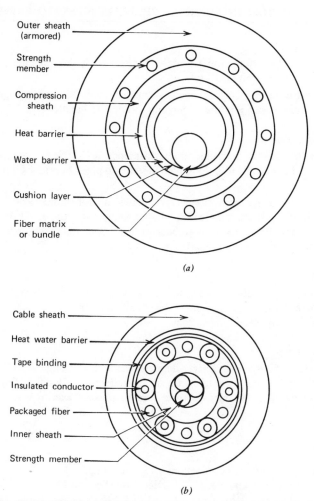

Figure 12.21. Two hypothetical cable designs for use with fibers showing the elements that might be included.

bandwidth required, one can select the fiber type—monomode, graded-index, or step-index—and the design can proceed.

The number and type of coatings to be applied will depend critically upon the fiber disposition and the packing density sought. Fibers that are individually mechanically buffered with thick polymer coatings make for very inefficient utilization of the space in the cable, but if only small numbers of fibers are required in the cable, this may be no problem. Cables employing large numbers (greater than 50) of fibers will probably

TABLE 12.4   FACTORS INFLUENCING CHOICE OF FIBER FOR CABLE

Single-mode fiber
Very small core (3—10 microns). Suitable for long haul, very high bandwidth systems up to 10+ Gbit/sec·km.
Trade-offs. Increasing core diameter—smaller index difference—greater bending losses—easier jointing. Generally difficult to handle.

Graded-index fiber
Core diameter 20—70 microns, index difference 0.7–3.0% at core center. Used for medium bandwidth, long haul systems, typically up to 1 Gbit/sec·km.
Trade-offs. Increasing core diameter—greater microbending loss—easier jointing—possibly lower bandwidth—better launching from LED or laser. Increased index difference—lower bandwidth—lower microbending loss—easier jointing—better launching from LED or laser.

Step-index fiber
Core diameter 50–120 microns, index difference 1–10%. Used for low bandwidth, short link systems, under 100 Mbit/sec·km.
Trade-offs. As with graded-index fiber

Outside diameter
Typically 100–150 microns. Larger diameter leads to a stronger, less flexible fiber and requires more material for its manufacture. Cladding thickness usually greater than 20 microns over core to prevent excessive loss from tunneling.

Fiber coatings

1. Surface neutralization—see Sections 12.3.1 and 12.3.2.
2. Abrasion resistance and chemical barriers—see Section 12.3.3.
3. Mechanical buffer coating for microbend protection—see Section 12.3.4. Combinations of all three types may be used, or in some cases only one. Special metallic coatings may also be applied to perform several tasks. A single coating is usually of type 2.

hold them in a matrix of some sort, so that jointing can be carried out with fiber groups rather than individual fibers.

Cables in which the fibers are carried in a helical lay about the central strength member usually use heavily buffered fibers since they are readily subjected to lateral shear forces between the inner member and the outer sheath.

Fibers placed in cables with only thin coatings applied, where the overall fiber plus the coating diameter is less than 200 microns, are usually run very loosely within the structure, since such fibers are very susceptible to induced microbending from the structure if subjected to lateral pressures.

12.4.2 Strength Member Materials

The fiber cannot usually be regarded as having sufficient strength with or without its coatings to withstand the forces involved in installation. Other members have to be introduced to take the forces imposed on the structure when it is subjected to longitudinal pulling, shear, or bending forces. A minimum specification might be that the cable must be strong enough to support its own weight, a trivial requirement if its length is a few meters but less trivial if the cable is 1 km long. (A pulling force equal to 1 to 5 times cable weight is not unreasonable for a duct installation.)

The choice of the material to be used must therefore take account of the mechanical characteristics of the material governing its operation in the conditions of installation. A high Young's modulus leads to a small strain for a given force. However, if it is coupled with high density, a large pulling force is likely to be needed, so that a good performance indicator is the Young's modulus divided by the specific gravity.

Another factor of importance is the mode of failure of the material. We have already seen that glass fibers extend elastically to their yield points and then fail catastrophically, whereas metals deform elastically at first, then deform plastically, and finally break. In Table 12.1 we list the values of the various parameters of interest for a number of possible contenders for use as strength members. Examination of the column headed E/SG, giving values of Young's modulus divided by specific gravity, shows wide variations and the potential attraction of some of the modern materials.

Steel is an attractive material by virtue of its high modulus. It is available in many grades and alloys, and the figures quoted are for high tensile material. Such materials have the highest yield stresses and strains. However, it is interesting to note that even then they cannot be elastically deformed beyond about 0.1% absolute maximum. It is also noteworthy that, because of the high density of the material and its low elongation at yield, the yield stress/specific gravity ratio, the force required to break a given mass of material is lowest for steel.

The polymer materials, nylon and Kevlar, look most attractive. Nylon monfilament has not been developed for this type of application, although it is used for rope construction, while Kevlar[2] stands out as a prime material. These materials are characterized by the fact that their long polymer molecules are oriented along the filament axis during manufacture, leading to a material that is mechanically anisotropic with a relatively very high longitudinal Young's modulus. They can be obtained as single-filament materials or as rovings in which many strands are loosely held together, often by a size applied to the material.

Glass fibers are well known as strength materials through their use in fiberglass mats or rovings. In the latter, continuous glass filaments of very small diameter (10 microns) are loosely bundled in 1 km lengths. Both forms are widely used as strength materials bonded to a suitable matrix, often formed by an epoxy resin. The resulting material is stiff and very strong. The use of such materials in an optical fiber cable presents difficulties, however, since to prevent the sharp ends of broken filaments from damaging their neighbors, thus leading to premature failure of the roving, a reasonably rigid matrix appears to be necessary, and this can rapidly lead to a cable that is insufficiently flexible for use.

Carbon fibers stand out as the best material in our list, but their marginal availability in very long continuous lengths and their high cost appear to prevent their use at present.

Having examined the mechanical properties of the materials available, we must now seek to make a choice among them. We must know what the failure statistics of our fiber are, from careful measurements on long lengths interpreted by Weibull statistics. From these data we can decide the maximum strain that will be acceptable in our cable *for the fiber*. We must then look at our application and decide the maximum cable length to be handled and the forces to which it is to be subjected. We can then make some preliminary calculations of cable strain. We might postulate that $x\%$ of the cable weight will be a strength member, assume that the remaining material will contribute little strength, and examine the strain induced by our conditions. If this exceeds the fiber strain acceptable but is within the yield strain, we might seek a higher performance material, we might try to increase x, or we might examine a design that reduces the strain on the fiber to less than the cable strain. By a process of successive approximation, we can obtain a feel for the possibilities.

Some useful numbers to bear in mind are given in Table 12.5, in which we have converted yield stress values for our listed materials into maximum vertical lay lengths, the lengths which, hanging vertically, will pro-

TABLE 12.5 NOTIONAL LENGTHS OF STRENGTH MEMBERS THAT COULD HANG VERTICALLY[a] UNDER THEIR OWN WEIGHT

Material	Length (km)
Steel wire	5–20
Nylon yarn	70
Kevlar fiber	210
S glass fiber	120
Carbon fiber	1000–2000

[a]Assuming constant g.

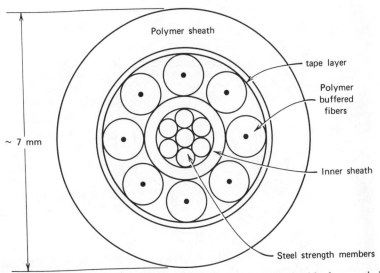

Figure 12.24. A circular cable design due to STC. Reproduced with the permission of Standard Telecommunications Laboratories Ltd. © 1973, from S. G. Foord and J. Lees, *Proc. IEE*, **123**, 597 (1976).

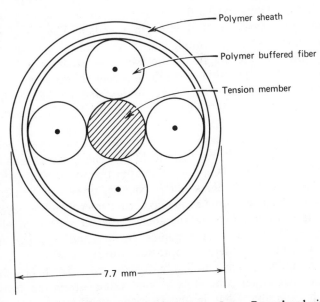

Figure 12.25. A four fiber cable design from Furukawa, Japan. Reproduced with permission from H. Murata, S. Inao, and Y. Matsuda, *Optical Fibre Communication*, IEE Conf. Publ. 132, London, 1975, p.70.

duce the yield stress at the hanging point. These lengths indicate the maximum cable lengths that could be pulled into straight duct if the whole cable was a strength member ($x = 100\%$).

12.4.3 Cable Structure

Having chosen our material, we must next decide in what form to use it and how to place it within the structure. It will usually be sheathed in a polymer. If it consists of multiple strands, these can be twisted or straight, and the material may be run down the neutral axis of the cable, or laid helically or straight along the surface of a sheath further out from the neutral axis.

Helically winding or laying the strength member leads to additional flexibility if the individual strands are free to slide locally over one another. However, in fiber cables which are likely to be of small overall diameter (5 to 10 mm), this is probably not a major consideration. A twisted strand strength member can extend initially with very little load, by allowing the strands to close in on one another, if they are not initially tightly packed and held. Such an extension is highly undesirable. A similar effect may occur if the strength member is helically wound around a deformable central member. Load on the cable tensions the strength members and causes it to compress the central member, allowing the cable to extend far more than a simple calculation of the stress-strain relationship of the strength member would have led one to expect. At the same time, tensioning a helically wound cable will cause it to attempt to unwind itself.

We must now place our fiber in the structure and examine its performance. Two options have emerged: to run the fiber along the neutral axis of the cable, or to wind the fiber helically around a sheath structure containing a strength member. In the former case the fiber must, on average, experience the same strain as the cable itself when pulled into a straight lay, while in the latter case the fiber may suffer substantially less strain than the cable by being wound on a deformable cushion member.

The fiber laid loosely along the neutral axis of the cable is less likely to be perturbed in its optical performance by its surroundings. The helically wound fiber can easily become squashed between inner and outer sheaths, with loss resulting in the transmission, or it can become distorted by bends in the cable which alternately compress and extend it as it winds from the inner to the outer radius of the bend.

The remaining components of the cable must now be considered. Tapes are frequently used to hold subassemblies together during manufacture and to act as heat, water, or chemical barriers. The choice of materials must take into account the sequence of manufacture and ensure that extrusion of an outer layer does not melt an inner layer. Electrical conductors may be needed in the same cable, and these must be accommodated. Also, an outer sheath, possibly coupled with inner sheaths, is required to ensure that loads are transferred acceptably throughout the structure, that it does not crush when subjected to side forces, and that the outer skin remains firmly attached to the inner structure when the whole structure is pulled or handled. Two general approaches to cable structures are shown in Figure 12.21.

We will now illustrate this discussion by examining some designs that have already been published. We should note that almost no data have been given on the expected performance of these designs or on the reasons behind the selections that were made. Nevertheless, they at least demonstrate the current variety of designs being made.

12.4.4 Published Designs

A number of designs have been published in which the fiber is run loosely in a hollow cavity, with little more surface protection than the few-micron-thick Kynar type layer. Two of these are illustrated in Figures 12.22 and 12.23. In the first one, due to British Insulated Callendar Cables (BICC),[17] the fiber runs in two holes down the neutral axis of the cable, which consists of a polymer extrusion containing two steel strength members. The fibers are free to slide in the structure longitudinally, and this limits the strain to which they can be subjected by a local deformation. By controlling the pulling force applied to the cable, the maximum strain on the cable and hence on the fiber (in long lengths) is limited to a safe value. Because the fibers are free to choose their own lay, the cable has proved to have very small microbending losses, with incremental losses typically of less then 0.2 dB/km.

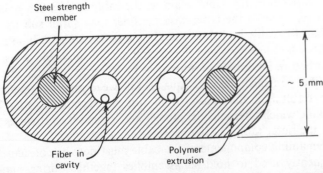

Figure 12.22. The BICC PSP cable design. The PSP optical cable is the subject of British Patent Specification No. 1479426 of BICC Ltd., England, and the diagram of the PSP cable is reproduced with the permission of that Company.

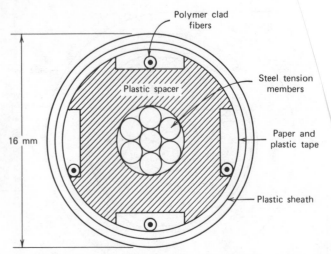

Figure 12.23. An Hitachi cable design. Reproduced with permission from T. Mizukami, T. Hatta, S. Fukuda, K. Mikoshiba, and Y. Shimokori, *Optical Fibre Communication*, IEE Conf. Publ. 132, London, 1975, p.191.

The second design, due to Hitachi,[18] is shown in Figure 12.23. Here fibers having a 100 micron buffer coating are run loosely in large cavities spaced around a large, central strength member. The freedom of the individual fibers to slide, coupled with the helical lay of the cable (assumed), means that the fibers are well insulated from local strains imposed by bends in the cable, while the looseness of the fiber in the cavity ensures that additional fiber can be accommodated, so that the cable may be extended a small amount without imposing any significant strain on the fiber.

We now examine three designs in which the fiber is firmly held within the cable structure (see Figures 12.24 to 12.26). The first is due to Standard Telephones and Cables (STC),[19] and the latter two to Furukawa.[20] In each, the fiber is sheathed in a thick buffer coating, presumably to bolster its strength and to provide some protection from microbending, and is then wound helically around a central member. The cable's mechanical strength arises primarily from the strength member for longitudinal forces and from the sheath structure for lateral compression forces, although in both cases the fibers and their buffering are likely to play a part. The helical lay of the fibers means that under conditions of longitudinal strain the strain in the fiber will be less than that in the strength members which run along the cable axis, since the radius of the fiber helix can contact and compress the buffer coating, giving strain relief.

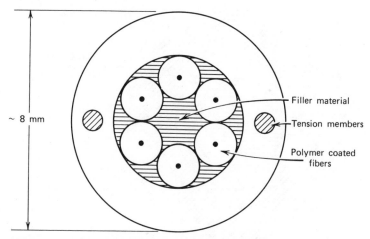

Figure 12.26. A six fiber cable design from Furukawa, Japan. Reproduced with permission from H. Murata, S. Inao, and Y. Matsuda, *Optical Fibre Communication,* IEE Conf. Publ. 132, London, 1975, p.70.

All these designs lead to cables in which the fibers are fixed fairly firmly in the structure, avoiding problems from the "walking" of a loose component under vibration or nonuniform strain. However, the same feature means that the buffered fibers are lightly compressed between the inner and outer sheaths, under conditions that are highly reminiscent of a fiber tensioned on a drum, so that the performances of the individual materials involved in the inner and outer sheaths and the buffer layers are likely to be critically important if microbending is to be avoided. A further feature of this type of design is that, under bending, the helically wound fiber undergoes successive compression and extension as it winds from the inside diameter to the outside. It must withstand both without change of optical properties.

Two cable designs which extend the concept of strain relief in the fiber members are shown in Figures 12.27 and 12.28. The first is due to the British Post Office,[21] and the second to Sumitomo.[22] In both cases the sheathed fiber is wound helically around a central member that is designed to have a very low bulk compression modulus. Thus, as the fiber cable is extended, the fiber in its helical lay moves to progressively smaller diameter helices, compressing the core material but taking up very little additional strain as a result. The central cushion material can be chosen from a range of foam type materials which are strong enough to ensure that the fiber conforms to smoothly varying lay and does not become entangled, but which present so little resistance to compression that strain relief factors of tens or hundreds can be achieved. The Sumitomo design features

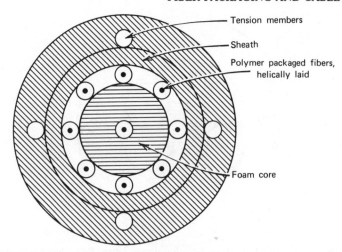

Figure 12.27. A Post Office cable design incorporating a strain relief member. Reproduced with permission from L. A. Jackson, unpublished work.

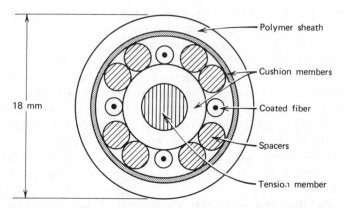

Figure 12.28. A cable design from Sumitomo incorporating a strain relief member for the fiber. Reproduced with permission from T. Nakahara, M. Hoshikawa, S. Suzuki, S Shiraishi, S. Kurosaki, and G. Tanaka, *Optical Fibre Communication,* IEE Conf. Publ. 132, London, 1975. p.81

a central strength member; the British Post Office design, an outer, helically wound member to avoid problems of load transference through the deformable cushion material, but with the result that the cable will be generally less flexible.

Finally, in Figure 12.29 we show a diagram of the 144 fiber cable developed by the Bell Telephone Laboratories.[23] The fibers are placed in 12 flat tapes, each carrying 12 fibers to form a 12×12 array. This structure is wrapped in paper tape with a polyethylene jacket covering. A polyolefin

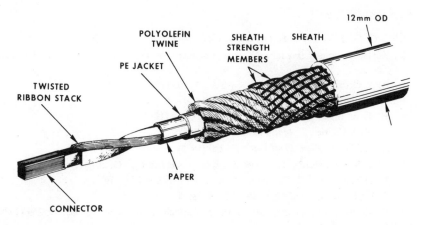

Figure 12.29. Bell Telephone Laboratories 144 fiber cable design containing the fibers in 12 flat tapes each of 12 fibers, formed into a square matrix. Reproduced with permission from M. I. Schwartz, *2nd European Conference on Optical Communication, Paris, 1976.* Paris: SEE.

twine heat shield is laid over this before the strength members are applied in a double helical winding, with an outer sheath extruded in place for final protection. This cable stands out as by far the most efficient design in terms of transmission capacity per square centimeter of cross section. Only careful systems analysis can show what advantage this feature offers the user in practice.

12.5. SUMMARY

We have discussed the aspects of strength and optical stability for fibers in cable, and have outlined the critical decisions that must be made before design can proceed. The maximum strain that a fiber can withstand will govern the choice of strength member material, its disposition in the cable, the maximum continuous length that can be installed at one pull, and the fiber packaging within the cable. Consideration of freedom from micro-bending and buckling, then, influence one's choice of materials to surround the fiber and the way in which they are incorporated into the structure. Evidently, such a multiple choice is unlikely to ever lead to a single logical design utilizing one combination of materials; it is far more likely that a range of designs having particular attributes will emerge. Nevertheless, uncertainties concerning the strength of fiber in long lengths will decrease with more knowledge of advances in technology, and the precise nature of the interactions of materials in causing or avoiding microbending will no doubt also become clearer, so that preferred options will emerge.

REFERENCES

1. S. G. Foord and J. Lees, *Proc. IEE*, **123**, 597 (1976).
2. Kevlar is a trade name for an oriented polymer material marketed by Du Pont Chemical Co.
3. A. A. Griffith, *Phil. Trans. Roy. Soc.*, **A221**, 163 (1920).
4. W. Weibull, *J. Appl. Mech.*, **18**, 293 (1951).
5. a. C. E. Inglis, *Proc. Inst. Nav. Archit.*, **55**, 219 (1913).
   b. R. J. Charles, *J. Appl. Phys.*, **29**,1549 (1958).
6. R. D. Maurer, *Appl. Phys. Lett.*, **27**, 220 (1975).
7. L. A. Jackson and M. H. Reeve, *J. Phys. E: Sci. Instrum.*, **11**, 161 (1978).
8. R. Olshansky and R. D. Maurer, *J. Appl. Phys.*, **47**, 4497 (1976).
9. R. J. Charles, *J. Appl. Phys.*, **29**, 1549 (1958).
10. R. Muto, N. Akiyama, H. Sakata and S. Furuichi, *J. Non-Cryst. Solids*, **19**, 269 (1975).
11. Kynar is a trade name used by Pennwalt Chemical Co.
12. P. W. France and P. L. Dunn, *2nd European Conference on Optical Communications, Paris*, 1976. Paris, SEE.
13. D. Gloge, *Bell Syst. Tech. J.*, **54**, 243 (1975).
14. W. B. Gardner, *Bell Syst. Tech. J.*, **54**, 457 (1975).
15. L. A. Jackson, M. H. Reeve, and A. G. Dunn, *Opt. Quant. Electron.*, **9**, 493 (1977).
16. M. H. Reeve, *Electron. Lett.*, **14**, 47 (1978).
17. The PSP optical cable is the subject of British Patent application 1,479,426 by BICC, Ltd. (United Kingdom). Figure reproduced by permission of BICC.
18. T. Mizukami, T. Hatta, S. Fukuda, K. Mikoshiba, and Y. Shimokori, *Optical Fibre Communication* IEE Publ. 132, London, 1975, p. 191.
19. S. G. Foord and J. Lees, *Proc. IEE*, **123**, 597 (1976).
20. H. Murata, S. Inao, and Y. Matsuda, *Optical Fibre Communication*, IEE Conf. Publ. 132, London, 1975, p. 70.
21. L. A. Jackson, Unpublished work.
22. T. Nakahara, M. Hoshikawa, S. Suzuki, S. Shiraishi, S. Kurosaki, and G. Tanaka, *Optical Fibre Communication*, IEE Conf. Publ. 132, London, 1975, p. 81.
23. M. I. Schwartz, *2nd European Conference on Optical Communications, Paris, 1976*. Paris: SEE.
24. H. Schonhorn, C. R. Kurkjian, R. E. Jaeger, H. N. Vazirani, R. V. Albarino, and F. V. DiMarcello, *Appl. Phys. Lett.*, **29**, 712 (1976).

13

Fiber Termination:
Launching and Jointing

13.1. INTRODUCTION

In the preceding chapters we were concerned with the way that optical power travels in a fiber and with ways of studying it. However, any system requires fibers to be joined, implying the connecting of one fiber to another, and the power in the fiber must have been launched into it from a source of some form. In this chapter we consider these aspects of fiber usage very briefly, so that with the discussion of the receiver in Chapter 14 we will be in a position to view the fiber in its system environment and so to complete our discussion.

Here we present a brief review of some of the techniques and principles involved in handling the terminations of fiber cables. Clean ends must be prepared on fibers, the ends must be jointed to the next cable section, or the cable must be terminated in a unit which includes the detector or the source. The performance of these terminations can play a critical role in the overall system performance, since the loss of a few decibels of power at every termination in a multitermination network rapidly produces an unacceptable loss. We also examine the performance of jointed fiber links.

13.2. JOINTING FIBERS

Essentially any optical fiber system will require fibers to be jointed. Such joints may arise at a number of points. Many manufacturers supply the electrooptical interface devices with fiber tails already attached, so that, instead of having to align a fiber to a laser, only a fiber-fiber joint need be made. Fibers also have to be jointed to other fibers at joins between cables. Some systems may require such joints to be demountable, while for others the joints must be semipermanent or permanent. Most telecommunications systems fall into the latter category. Thus we can distinguish between demountable and permanent joints.

When we examine the many joints proposed, we find a further categorization, between adjustable—in the sense of providing an adjustable alignment between the fiber ends—and nonadjustable. Most users are likely to favor a joint which requires no adjustment, since its setting up is much simpler; but we shall see that for some fibers, namely, those with very small core diameters such as single-mode fibers, adjustable joints appear to be nearly obligatory if jointing losses are to be kept within reasonable limits.

13.2.1 Joint Tolerances

The estimation of the allowable dimensional variation and misalignment in a fiber joint can follow a number of routes, depending upon the type of fiber considered. The calculation of losses in monomode fiber joints involves calculating the mode field patterns for the LP_{01} mode of each fiber and then calculating the field overlap between the two fields as a function of the various types of misalignment. Thus one could study the effect of angular misalignment of fiber axes, spatial misalignment of the fiber ends, the effects of noncircular and various rotational misalignments, the effects of varying core diameters, and the effects of variations in the values of Δn for the two fibers. Some of these effects have been studied,[1] and the results are given in Figure 13.1, which shows that spatial alignment of the fiber ends is likely to be difficult to achieve, since very small tolerances have to be held. Angular alignment to $1°$ is readily achieved.

Most systems are concerned with multimode fibers. The above approach could be followed, calculating the coupling coefficient for each mode in

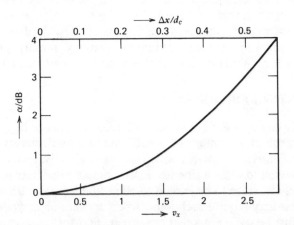

Figure 13.1. Loss of a monomode fiber joint as a function of end displacement in units of core diameter. Reproduced with permission from M. Boerner and S. Maslowski, *Proc. IEE.* **123**, 627 (1976).

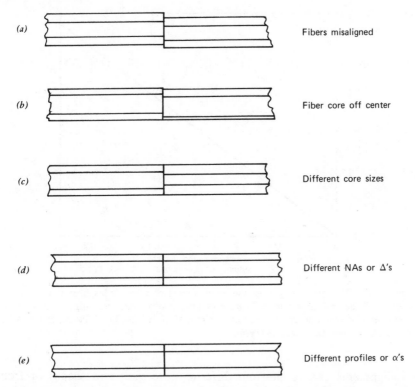

(a)   Fibers misaligned

(b)   Fiber core off center

(c)   Different core sizes

(d)   Different NAs or Δ's

(e)   Different profiles or α's

Figure 13.2. A schematic diagram showing possible cause of loss in multimode graded-index fiber splices.

turn to all other modes of the second fiber and summing the results. In practice, however, such a procedure would be cumbersome in the extreme, and much simpler approaches are used. The effects to be considered in general are shown in Figure 13.2.

The simplest calculation is that for the step-index fiber. From any point on the end of the fiber, guided mode power radiates into the same solid angle. If one assumes that the fiber is uniformly illuminated, the power collected by the second fiber from the first one is given by the intersection of the areas and solid angles of the second fiber with the first. Power that leaves the first fiber end outside the solid acceptance angle of the second or outside the core area of the second is considered lost, and thus the joint loss is found. In practice, the semicone angle for acceptance of light is of the order of 10°. It is generally found to be simple to angularly align the fiber axes to better than 1°, so that losses due to angular misalignment should be very small. The dominant loss thus arises from failure of the core area of the first fiber to overlap fully that of the second. Such

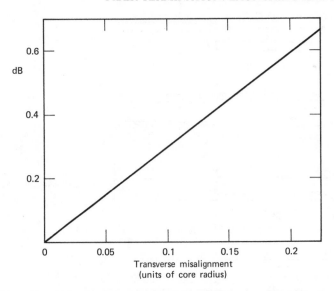

Figure 13.3. The loss arising from the misalignment of a step-index fiber butt joint as a function of the end displacement in units of fiber core radius, assuming all modes to be equally excited. Reproduced with permission from M. J. Adams, D. N. Payne, and F. M. E. Sladen, *Appl. Phys. Lett.*, **28**, 524 (1976).

misalignments can result from differing core sizes and shapes or from imperfect alignment of the ends. The losses due to spatial misalignment of the ends, calculated in this manner, are shown in Figure 13.3.

The calculation of the joint loss for a graded-index fiber proceeds in a similar manner to that for the step-index fiber, but more complex. We must now recognize that the acceptance angle of the fiber is a locally varying parameter because the refractive index is a function of radius (see Section 6.2). Thus we can no longer separate the angular overlap and the spatial overlap calculations. Reference to Figure 13.4 shows that with two spatially misaligned graded profiles power from one side of the first fiber will be readily accepted by the second because it will be entering a material of higher index, while from the other half of the input fiber the power will enter material of lower index and suffer some loss. The calculation of loss thus involves the double integration of the angular and spatial overlap functions over the whole surface of the input fiber.

We should now note that in the above discussion we have neglected two important facts. We have implicitly assumed that for our multimode fiber the WKB approximation holds, so that there is some validity in talking about power per unit area and solid angle, without having to resort to the

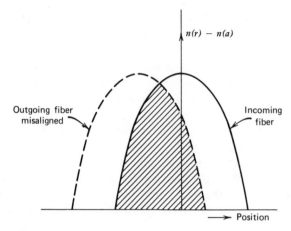

Figure 13.4. Misalignment in a graded-core fiber, indicating the twin effects of nonoverlap of both area and acceptance angle (NA) as a function of core position.

detailed mode field patterns. But we have also assumed the following:

1. The illumination is uniform over the fiber end.
2. The losses in the fiber are mode independent.
3. Weakly leaky modes can be neglected (Section 5.4).

Unfortunately none of these assumptions is likely to be true when we study joints in the laboratory. We know that after power has traveled a significant distance in the fiber a stable (or nearly so) distribution will be generated, and this will have less power per mode in the high order modes than it does in the low order modes (see Section 10.3). This difference will have the effect that at the joint the effects of misalignment will not be felt as sharply, since where the misalignment first shows, in the extreme edges of the spatial or angular overlap, there will be little power. We may conclude that our joint will therefore have a lower loss than is predicted by the calculations indicated above.

In reality, the losses in the fiber are mode dependent, with higher angle modes suffering more radiative loss than low angle modes, so that our misaligned joint, while coupling energy predominantly from guided mode to guided mode, will lead to additional loss in the succeeding fiber length if the net movement of energy is from low order to higher order modes. This additional loss is not detectable at the joint but shows only as additional loss in the succeeding fiber section, which has received a launch distribution that no longer corresponds to the stable distribution (Section 10.3).

Finally, we note that in a misaligned joint some power from guided modes in the first fiber will couple to modes that lie just beyond cutoff in the second fiber. Some of these will be weakly leaky, tunneling guided modes (Section 5.4).

Thus, if we measure the loss of the joint by taking two short lengths of fiber, jointing them, and measuring the power emanating from the first before jointing and from the second after jointing, we will find that a certain loss is registered. But we must recognize that this loss corresponds to the power lost from the particular distribution that we excited in the first fiber to the unguided modes *at the joint*. It takes no account of the additional losses that would occur in the next kilometer of fiber from weakly leaky and high angle modes close to cutoff, nor does it necessarily tell us anything about the losses that would apply for a different power distribution, particularly the stable distribution if that is known.

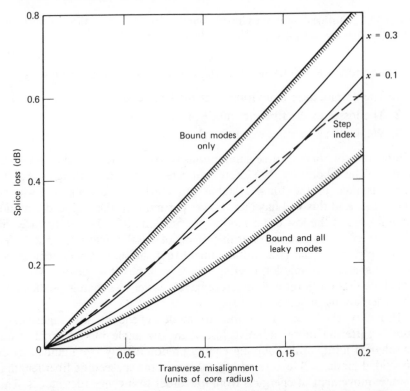

Figure 13.5. Loss of a graded-core fiber joint, calculated for a parabolic profile and assuming that all guided modes are equally excited, showing the loss local to the joint (bound and all leaky modes) and the loss after the leaky modes have decayed (bound modes only). Reproduced with permission from M. J. Adams, D. N. Payne, and F. M. E. Sladen, *Appl. Phys. Lett.*, **28**, 524 (1976).

We present in Figure 13.5 the results of a calculation of the apparent loss from a fiber joint due to spatial misalignment of two perfectly matched ends, for the step-index and parabolic profiles ($\alpha = \infty$ and 2), showing for the latter the effect of including all weakly leaky modes and fully bound modes only, while for the former profile bound modes only are assumed. In both cases uniform excitation of modes is assumed. Also shown are the losses that would be measured at various distances away from the joint, in terms of the normalized parameter X, defined as[2]

$$X = V^{-1} \log_e \left(\frac{z}{a} \right) \tag{13.2.1}$$

where z is the distance along the fiber, and

$$V = \frac{2\pi a}{\lambda} n_1 \sqrt{\Delta} \tag{13.2.2}$$

Since the bulk of the loss in these conditions will occur from high order modes, it is self-evident that, had the calculations been done assuming a stable mode distribution (see Section 10.3.1), lower losses would have resulted. Some recent theoretical and experimental work by Gloge[3] supports this intuitive view. Figure 13.6 shows the joint losses for both

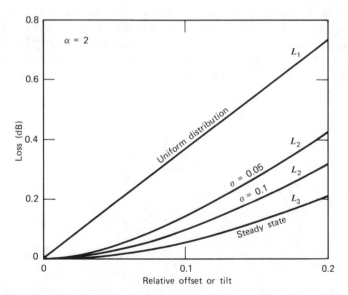

Figure 13.6. The loss arising from a misaligned graded-core fiber joint, calculated for a "stable-mode distribution" (steady state) compared to the predicted loss for a uniform or equal excitation situation. Reprinted with permission from D. Gloge, *Bell Syst. Tech. J.*, **55**, 905 (1976), copyright 1976, The American Telephone and Telegraph Company.

Lambertian and stable mode distribution conditions. Note that, in addition to the losses of the joint being lower when excited by a stable distribution, they no longer vary approximately linearly with fractional misalignment but vary more nearly with the square of the fractional misalignment.

In summary, we may say that simple loss measurements of joints in the laboratory are readily made. Care must be taken to use a standard launch distribution if the results are to be mutually comparable. To transfer such measurements from the laboratory to a working environment, however, and to use them as a basis for estimating the effective mean joint losses in a multiple-fiber link comprising 500 m or 1 km fiber lengths is a hazardous process, and the results should be viewed with a good deal of skepticism unless proper allowance is made for the type of effects discussed above.

13.2.2 Permanent, Nonadjustable Joints

A considerable number of techniques have been proposed for such joints, which are the ones most commonly required in civil telecommunications systems. A large number depend upon a V-groove to provide the fiber alignment. Such a joint is illustrated schematically in Figure 13.7. Two fibers with precision ends preprepared are placed into the V-groove, and their ends are index matched together and sealed in place with some suitable glue, such as an epoxy resin.

A simple apparatus[4] for forming such joints is shown in Figure 13.8. It uses V-grooves that are impressed by a mandrel into copper substrates measuring about 6×2 mm. These are mounted on the central pillar in a jig, while the fibers to be joined are held in the alignment blocks by magnetic clamps. The central pillar is raised so that the fibers are forced

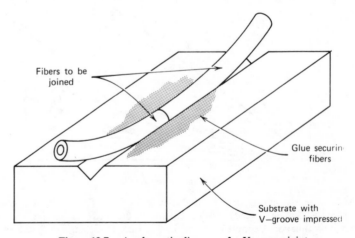

Figure 13.7. A schematic diagram of a V-groove joint.

Figure 13.8. A simple jig for making V-groove joints, showing the small V-groove plate in the center column with guides to feed the fiber on either side. Reproduced with permission of the British Post Office from P. C. Hensel, J. C. North, and J. H. Stewart, *Electron. Power*, February 1977.

down into the V-grooves under their own spring force. A dab of epoxy resin is applied over the fiber ends, which are brought together by moving the right-hand fiber clamp toward the V-groove jig. When the fiber ends touch, a small heater is turned on in the central pillar to set the epoxy resin and fix the fiber in place. The fibers with the V-groove substrates are then ejected and mounted in a joint box. Joints made with this equipment, using identical fibers, show losses of the order of 0.1 dB or less.

A more elaborate application of the V-groove joint has been described by workers at Bell Telephone Laboratories,[5] who were interested in jointing large numbers of fibers quickly. They reported the use of multiple V-groove substrates used in cooperation with fiber tapes, consisting of the same number of fibers set side by side. In this application the fiber tapes must be broken in a single operation to very tight tolerances or cut and polished if the alignment is to be sufficiently precise in the final joint.

The joint uses a precision multiple V-groove chip, shown in Figure 13.9, now made by photolithography on silicon.[6] These chips are used to form a 12×12 array of fibers between 12 fiber tapes, each containing 12 fibers. The ends of the composite assembly are polished flat in a suitable jig to allow two such assemblies to be clamped together to form a single 12×12 array splice,[6] as shown in Figure 13.10.

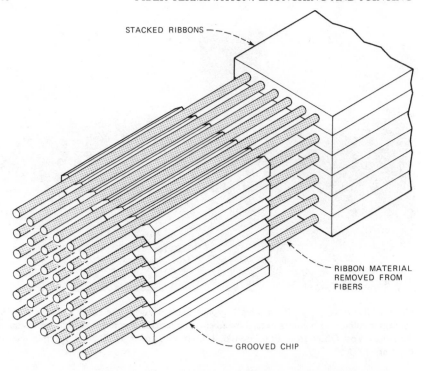

Figure 13.9. A precision multi-V-groove substrate formed of silicon by photolithography for jointing a 12 fiber linear tape array. Reproduced with permission from C. M. Schroeder, *Optical Fiber Transmission, II*. Willamsburg, Va.: OSA/IEEE, February 1977, copyright 1977 by the Optical Society of America, 77ch 12278-GEC

A variation on the V-groove joint uses a square section glass tube[7] which can be pulled down so that its internal cross section is slightly greater than that of the fiber to be jointed. The fiber ends are inserted from either end into the square tube and are curved in the same plane to force their ends simultaneously into the same corner of the tube, as illustrated in Figure 13.11. A similar joint has been described in which the fiber ends are inserted into circular section tubes whose internal bore is only slightly larger than the outside diameter of the fiber. Alignment of the fiber ends occurs by loose contact with the tube walls. In both of these joints the fiber is held in place and index matched with some suitable epoxy resin glue.

An ingenious variation on the circular tube alignment joint has been described by Bell Northern workers.[16] It is designed to operate with fiber having an extruded plastic jacket, so that its OD approaches 1 mm. The joint uses a thin walled metal tube that just slides over the extruded coating of the fiber. Its central section is deformed over a steel mandrel of

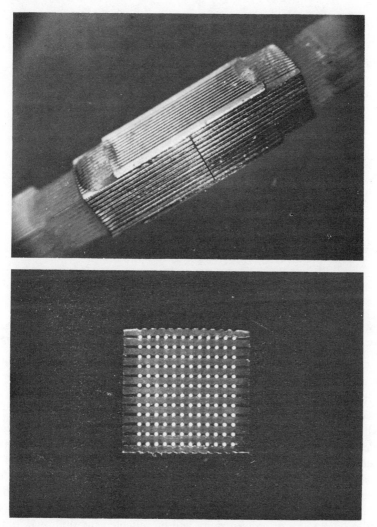

Figure 13.10. A 12×12 fiber array joint using the chips of Figure 13.9 for alignment. Reprinted with permission from C. M. Miller, *Bell Syst. Tech. J.*, **54**, 1547 (1975), copyright 1975, The American Telephone and Telegraph Company.

the same diameter as the fiber to the section shown in Figure 13.12, using a special molding tool. The fiber ends are cleared of extruded packaging, square ends are prepared, and the fibers are inserted into the deformed metal tube, whose mandrel has been removed to leave a central hole that fits the bare fiber closely. The fiber ends are aligned by the contact of the deformed metal wall, they are index matched by some suitable liquid

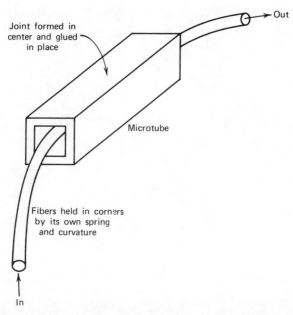

Figure 13.11. The use of a simple square section tube to locate and splice fibers together. Reprinted with permission from C. M. Miller, *Bell Syst. Tech. J.,* **54**, 1215 (1975), copyright 1975, The American Telephone and Telegraph Company.

inserted before jointing, and they are held in place by crimping the outer ends of the metal tube onto the plastic extrusion coating of the fiber.

All of these nonadjustable joints have the attraction that they can be made fairly rapidly and by relatively routine procedures. Equally, they all suffer from the limitation that they can be only as good as the fiber itself and may be much worse, in that the joint loss depends upon the accuracy of alignment of the fiber cores, whereas in these joints the fiber claddings are aligned. Any mismatch between the fibers in core/cladding ratio, cladding diameter, or concentricity will be reflected in additional joint loss because it will result in misalignment of the cores. The accuracy of alignment that is required was discussed in Section 13.2.1 and leads to tight specification of the fiber dimensions if this type of joint is to be used in a system where joint losses have to be held to a low level (less than 0.5 dB).

13.2.3 Demountable Fiber Connectors

The greatest interest in demountable connectors for fiber systems has probably centered on connectors for fiber bundles, in which it is desired to connect a signal that is transmitted on all the fibers of the incoming bundle

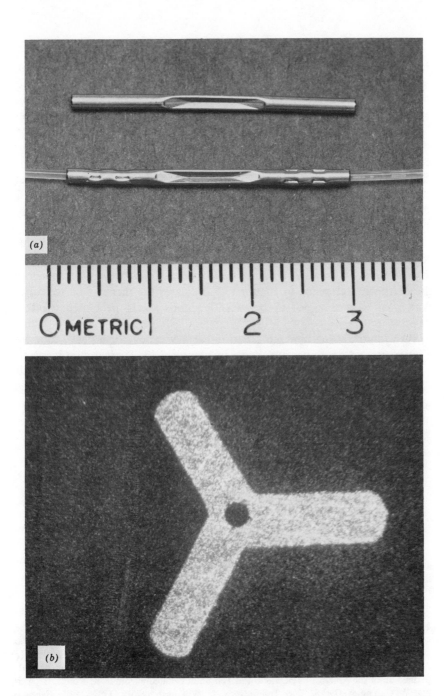

Figure 13.12. A simple crimp joint developed by Bell Northern Research. (*a*) The complete joint. (*b*) The central cross section. Reproduced with permission from J. Dalgleish, *Optical Fibre Communication,* IEE Conf. Publ. 132, London, 1975; photo supplied by Bell Northern Research, Ottawa, Canada.

to all the fibers of the outgoing bundle. With this type of connector no attempt is usually made to achieve registration between individual fibers of one bundle and the other, and it is accepted that some loss occurs from energy that was in the core of an incoming fiber and that, after connection, has coupled to the cladding of the adjoining fiber and is lost. Connectors of this type usually resemble conventional military or civilian multiwire connectors with the fiber bundle replacing the metal conductor in the pins. Even with careful design, required to optimize the alignment tolerances, losses usually remain fairly high, several decibels, and the system is designed to allow for such losses. Such connectors are used in short link systems, for example, in aircraft data buses or computer links.

A need arises in more sophisticated fiber systems for demountable connectors using single fibers. In such joints the need to achieve a very low insertion loss is usually paramount and presents very severe problems to the designer. Reference to Figure 13.5 indicates the mechanical tolerance that must be maintained if the two fiber ends are to be butted together directly.

Such joints[8,9] usually make use of either a V-groove or a sleeve tube for alignment, and a number are shown in Figure 13.13. Typical joint losses lie in the range from 0.2 to 3 dB, although the results obtained evidently depend upon many variables, and may not be representative of what would be achieved if these devices were to be used in a field environment under stringent operating conditions.

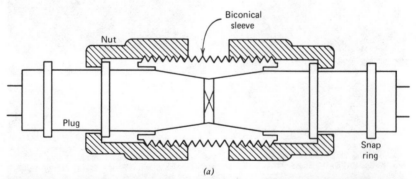

(a)

Figure 13.13. A number of demountable connectors for fibers. (a) and (b) A fiber to fiber connector developed by Bell Telephone Laboratories. (c) A bulkhead mounting version of Bell Telephone Laboratories' design. (d) A multifiber connector design from Telefunken. (e) A photograph of a three fiber version of (d). (a), (b), and (c) are reproduced with the permission of the American Telephone and Telegraph Company from P. K. Runge, L Curtis, and W. C. Young, *Optical Fiber Transmission, II,* Williamsburg, Va.; OSA/IEEE, February 1977, copyright 1977 by the Optical Society of America, 77ch 1227-8QEC, (d) and (e) are reproduced with permission from J. Guttman, O. Krumholz, and E. Pfeiffer, *Electron. Lett.,* **11,** 582 (1975).

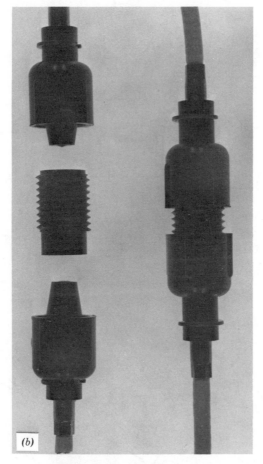

Figure 13.13. (*Continued*)

A second approach to the problem of connecting single fibers in a manner which allows ready separation makes use of some form of lens.[10] The connector is illustrated schematically in Figure 13.14 and is shown in Figure 13.15. The fiber end is placed at the focal point of a molded aspheric plastic lens, so that its emerging beam is collimated into a beam of low angular divergence but larger diameter. The connector splits in the region of the low angular divergence, large area beam and relies for its operation upon the fact that precision angular alignment is more readily obtained in this context than is the precision spatial alignment necessary for a fiber butt joint. By thus trading one tolerance for another, connectors are obtained that can be handled in much the same way as conventional

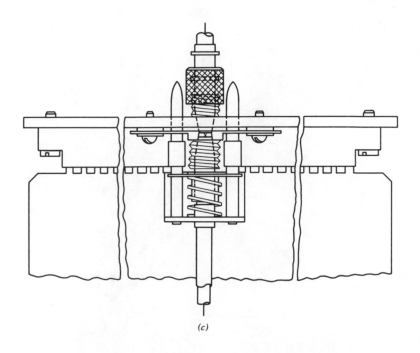

(c)

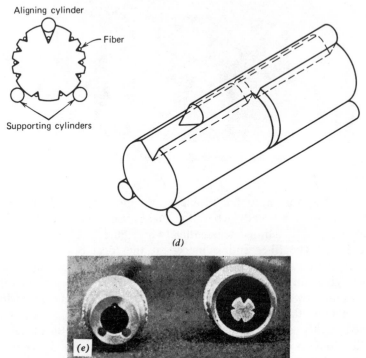

Aligning cylinder

Fiber

Supporting cylinders

(d)

(e)

Figure 13.13. (*Continued*)

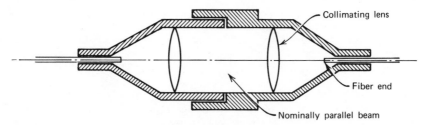

Figure 13.14. The principle of the lens connector. Reproduced with permission from P. C. Hensel, J. C. North, and J. H. Stewart, *Electron. Power,* February 1977.

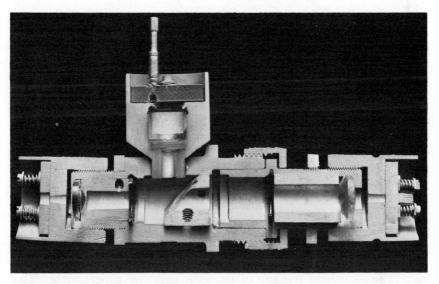

Figure 13.15. Photograph of a lens connector with a monitor side-port. Reproduced with permission of the British Post Office.

connectors and that give repeatable insertion losses. The results of repeated connections and disconnections led to an estimate of joint loss of 0.5 to 1.0 dB.

13.2.4 Adjustable Fiber Joints

The best known adjustable joint design is due to AEG Telefunken[11] in West Germany. It relies upon mounting the two fiber ends to be coupled into eccentric sleeves, which in turn are inserted into a precision housing. The principle of the joint is illustrated in Figure 13.16. To align the fiber ends (3), it is necessary only to rotate each sleeve (2) in turn relative to the

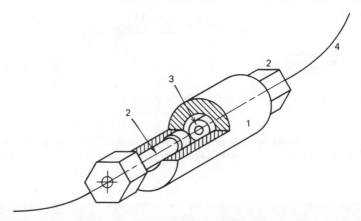

Figure 13.16. The principle of the Telefunken adjustable joint, showing two rotatable eccentric sleeves (2) containing the fibers (3) within the housing (1). Reproduced with permission from M. Boerner and S. Maslowski, *Proc. IEE.* **123**, 627 (1976).

housing (1) so that the fiber end inscribes a circular path. By careful design these circular paths are arranged to intersect so that there is some particular setting of both sleeves that corresponds to perfect alignment. The loss in the joint then depends upon the skill with which the ends are aligned, the perfection with which the ends have been prepared, and the proximity of the two ends, measured along the fiber axis. The presence of an air interface between the fiber ends will lead to reflection losses, but this problem can be overcome by the use of a suitable index matching material probably in liquid form.

The use of these joints has been described for both monomode and multimode fibers, with very low losses being quoted for the monomode joint. The long term stability of the joint does not appear to be documented but presumably can be made adequate by careful choice of materials and mechanical design.

An operational problem associated with any adjustable joint is the fact that, to make the adjustment, some indication must be available to the operator of the result of his or her adjustment. In practice, this requires that a signal be transmitted over the fiber while the adjustment is being made and that the transmitted signal be monitored at the end of the next fiber cable section and be fed back to the operator. Although such operations are clearly possible, they increase the complexity of the installation operation. Alternatively, it is conceivable that the power lost at the joint might be monitored while it is adjusted, through observation of the scatter power. This would avoid the inconvenience of remote monitoring and feedback of information, but does not appear to have been demonstrated as feasible.

The use of adjustable joints appears to be close to a necessity if very wideband, single-mode systems are to be used in the field, but they remain unattractive when compared to the joint that requires no prealignment or adjustment.

13.3. FIBER END PREPARATION

A vital precursor to forming a good fiber joint is to prepare a square and smooth end on the two fibers to be jointed or coupled. This can be done by using the principles long known to every glassblower and used for breaking glass rods. The operation begins by scratching the glass surface, to produce stress concentration and thus initiate cleavage at the desired section. The rod (or fiber) is then subjected to bending and tension, so that the stress is a maximum at the upper scratch point (see Figure 13.17) and the material starts to fail. By carefully controlling the relative and absolute amounts of curvature of the fiber and the tension, perfectly smooth cleaved ends can be produced. Samples of ends are presented in Figure 13.18, taken under a scanning electron microscope, for the correct conditions; also shown are the effects of incorrect cleaving.

The breaking operation is performed using a tool. A number of different versions have been described; two are shown in Figures 13.19a and 13.19b. In the first the fiber is clamped at two points, and the curved anvil is allowed to rotate under a constant tension spring drive, against an oil dashpot. This tensions the fiber correctly around the curved, polished metal anvil surface. The tungsten carbide knife-edge blade then descends and scores the fiber surface, so that failure occurs.

The second tool is simpler, in that the clamping, tensioning, and curvature are all produced in a single operation. The fiber is laid between the

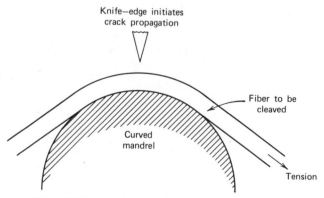

Figure 13.17.   Principle of a fiber end-preparation tool.

(a)

Figure 13.18. Some scanning electron microscope photographs of fiber ends produced by (a) a tool with correct adjustment and (b,c) with incorrect adjustment. Reproduced with permission of the British Post Office from P. C. Hensel, unpublished work.

(b)

Figure 13.18. (*Continued*)

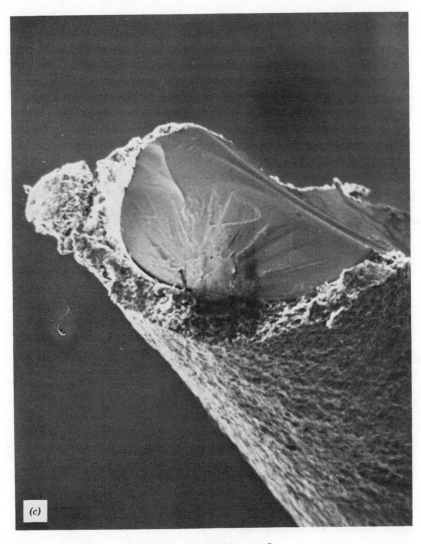

(c)

Figure 13.18. (*Continued*)

(a)

Figure 13.19. (*a*) A bench mounted tool for preparing fiber ends showing the curved anvil and cutting edge. (*b*) A hand held version of the end preparation tool. Reproduced by permission of the British Post Office from P. C. Hensel, J. C. North, and J. H. Stewart, *Electron. Power*, February 1977.

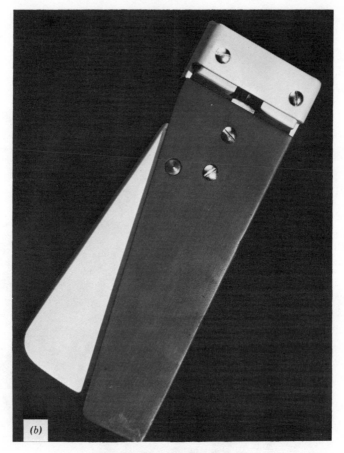

(b)

Figure 13.19. (*Continued*)

heads of the tool. The handles are squeezed so that in sequence the clamps close onto the fiber and the curved anvil tensions the fiber and moves it against the knife-edge. Both tools work well and allow fibers to be produced with nearly perfect ends with a very high success rate.

13.4. POWER COUPLED FROM A LIGHT EMITTING DIODE TO AN OPTICAL FIBER

The light emitting diode (LED) is primarily characterized by its radiance at a given drive current, and the effective emitting area over which that radiance is maintained. Other parameters of vital design interest are, of course, optical linewidth, switch-on and switch-off times, and lifetime versus operating conditions.

The source radiance is usually specified in terms of watts per square centimeter per steradian. To discover the maximum power coupled into the optical fiber, we should first imagine the fiber as butted against the LED emitting surface. If this is not achieved physically, that surface should be imaged with sufficiently large aperture optics onto the fiber end. We can distinguish two cases:

1. The LED emitting area is greater than the fiber core area.
2. The LED emitting area is smaller than or equal to the fiber core area.

In the former case power is accepted into the guided modes of the fiber over the whole core area, so that the effective collection area is πa^2. Power is also collected over the whole solid collection area of the fiber (its numerical aperture), so that the solid collection angle ψ in air is given at radius r by (see Section 6.5)

$$\psi(r) = \pi \left[n(r)^2 - n_2^2 \right] \tag{13.4.1}$$

In the case of a step-index fiber, $n(r) = n_i$ and the acceptance solid angle is given by

$$\psi \simeq 2\pi n_1^2 \Delta = \pi(\theta_1)^2$$

$$\simeq \pi(\text{NA})^2 \tag{13.4.2}$$

where NA is the numerical aperture $= \sin \theta_1$.

Note that we can convert radiance R directly to power per plane polarized mode by multiplying by $\lambda^2/2$ (see Section 7.2.1).[12] We can now proceed to calculate the launched power in a fiber whose end is filled by a radiance R source either by calculating $R\lambda^2/2$ times the number of guided modes or by calculating the product of radiance times local acceptance solid angle and integrating over the fiber core area.

Thus, for an LED of radiance R, the launched power (into guided modes, neglecting weakly leaky modes) is given by

$$P = R(\pi a^2)\psi = 2\pi^2 a^2 n_1^2 \Delta R \qquad \text{(step index)}$$

or

$$R(2\pi^2) \int_0^a r \left[n(r)^2 - n_0^2 \right] dr \qquad \text{(graded index)} \tag{13.4.3}$$

In the situation where the emitting area of the LED is smaller than that of the fiber core, imaging optics can improve the collection efficiency. If a lens is introduced such that the fiber end is demagnified to exactly match the emitting area, the solid angle over which power is collected from the LED is increased by the same magnification factor, assuming a lens of large f number. Thus, if the ratio of emitting areas is $M = a^2/d^2$, where d is the diode radius, the effective solid angle of collection from the diode can be changed from ψ to $M\psi$. In the event that M is greater than unity, this represents a gain in the efficiency of conversion of electrical energy to optical power in the fiber. It has been achieved by confining the same electrical current to a smaller active area of device material and may well be accompanied by a decrease in operating life since higher current densities will have ensued. However, this route is attractive in devices for which life is not a problem. Such a device configuration is shown in Figure 13.20, in which the fiber is cemented onto the device to make a rugged package, but an integral lens has been formed on top of the emitting surface of the active material. All-solid construction has a further attraction over air spaced coupling in that it reduces the problems of Fresnel reflection at the high-low refractive index interfaces that inevitably arise in such a construction.

To summarize our conclusions, then, we may say that, to maximize the power coupled from a given LED into the guided modes of the fiber, we

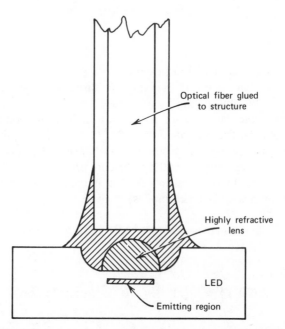

Optical fiber glued
to structure

Highly refractive
lens

LED

Emitting region

Figure 13.20. A neat butt coupling for a fiber interfacing with an LED source, showing the integral spherical lens to couple power from the small emitting area into the fiber core.

should:

1. Use as large an area of fiber core as possible.
2. Use a fiber with as large an index difference, Δ, as possible.
3. Confine the LED drive current to as small an area as possible to increase the radiance.
4. Use imaging optics to demagnify the fiber end onto the device emitting area if the latter is smaller than the fiber core area.

Unfortunately, each of these actions leads to problems in the other aspects of system performance, which may or may not be acceptable, depending on the application. The large core area results in a physically large fiber, with limited bending radius and thus mechanical limitations. The large Δ leads to large pulse dispersion,. and the small LED area is likely to limit lifetime if it is coupled with increased current density.

13.5. POWER COUPLING FROM A GaAs LASER TO AN OPTICAL FIBER

We now discuss the problems of coupling from an LED into an optical fiber. By comparison they are much less with the laser source because the light emission that occurs over 4π steradians below threshold becomes highly collimated above threshold. Thus the light output in stereogram form is typically like that shown in Figure 13.21. The circle is the acceptance angle of the fiber, and the elliptical form is the emitting solid angle of the laser, elliptical because of the planar slit shaped emitting aperture or junction. However, the important point to note is that the emitting angle is contained largely within the fiber acceptance angle, so that butting the fiber onto the end of the laser should result in efficient coupling. Practical problems arise from the difficulties of aligning the fiber with the emitting region of the device, holding it there, and ensuring that its presence does not degrade the device lifetime through damage to the exit (mirror) surface of the laser chip.

The coupling problem is far more severe when single-mode fibers are used. Neither the spatial nor the angular distribution of light from the laser matches that of the LP_{01} mode of the fiber, the situation both angularly and spatially being more as shown in Figure 13.22. Two courses may be taken. One may accept the fact that the overlap of the mode fields is poor and will result in low coupling efficiency, typically -10 to -15 dB, or one may attempt to match the mode fields by the use of optics. One such solution of the latter type has been proposed and is illustrated in Figure 13.23. A cylindrical lens is used[13] to correct the gross astigmatism of the source. The lens is formed by photolithography on the fiber end and

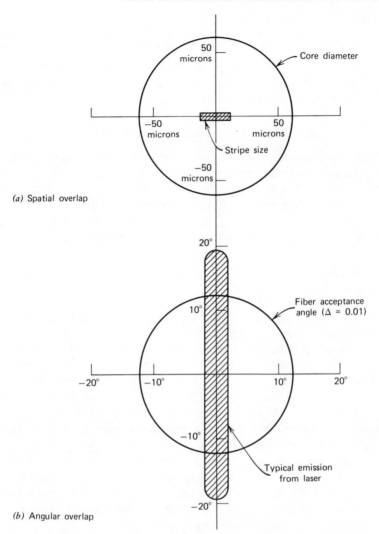

Figure 13.21. A schematic diagram showing (*a*) spatial and (*b*) angular overlap when butt launching a stripe laser into a multimode fiber.

mounted at right angles to the transmission fiber axis. By careful adjustment of the parameters of spacing and alignment, useful gains in launched power into multimode fiber have been made, and with refinement the principle applies to the single mode. However, it seems fair to state that at this time no wholly satisfactory solution to this problem has been found. Perhaps the use of distributed couplers of the type used in integrated optics

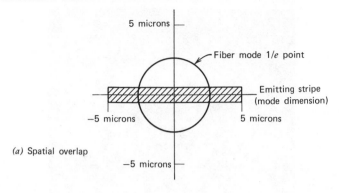

(a) Spatial overlap

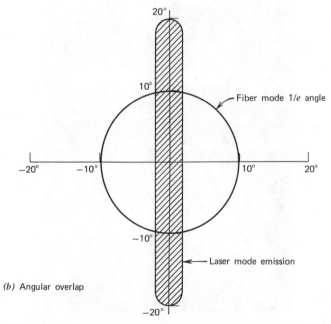

(b) Angular overlap

Figure 13.22. A schematic diagram showing (a) spatial and (b) angular overlap when launching a single mode stripe laser into a single mode fiber.

will provide the answer, although the extreme difficulty of setting them up makes this seem unlikely.

In the quest for higher efficiencies, with either single-mode or multimode fibers, an ingenious astigmatic lens systems has been described which makes use of graded-index diffused lenses rather than polished lenses of uniform material and curved surfaces. The graded-index, diffused glass

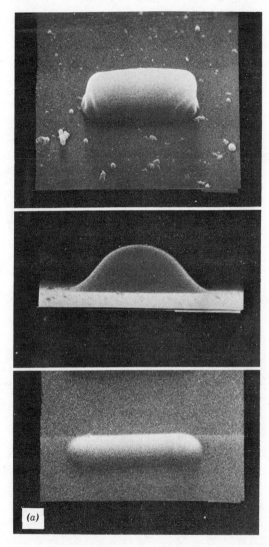

Figure 13.23. (a). Electron micrographs of a cylindrical lens formed lithographically on a fiber end. (b) The far field pattern from fibers with spherical and cylindrical lenses bonded to their ends. Reproduced with permission from L. G. Cohen and M. V. Schneider *Appl. Opt.*, **13**, 89 (1974).

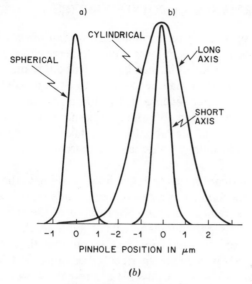

Figure 13.23. (*Continued*)

material allows lens elements to be made that are parallel sided and can therefore be cemented together, since their action does not depend upon the refraction at the glass-air interface. Thus a combination of a circularly symmetrical block lens and a cylindrically symmetrical lens can be cemented together to form a composite element with the source and the fiber end to correct the astigmatism of the laser output and return it to circular symmetry to match the fiber. Such a composite system is shown in Figure 13.24 and has obvious attractions, although the difficulty of constructing and aligning a system of this type is hard to judge.

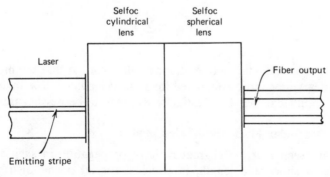

Figure 13.24.   A solid coupler for a laser to fiber using cylindrical and spherical Selfoc lenses to obtain field matching. Selfoc is a trade name for diffused glass lenses made by the Nippon Sheet Glass Co.

13.6. PERFORMANCE OF JOINTED FIBERS

Optical fiber systems for telecommunications purposes will almost exclusively be used for transmission over multikilometer paths. Fiber cables are most unlikely to be installed in such lengths without joints, so that the fiber section between repeaters may be assumed to contain joints. Typically we might envisage a repeater spacing of 8 to 10 km, made up of cable lengths of between 500 m and 1 km. We should also note that nearly all measurements of fibers made in the laboratory are taken on single 1 km lengths, although the shuttle pulse technique does present a way of extending the range.

The systems designer needs to know how to estimate the fiber transmission loss and dispersion from measurements made on cable reels, extrapolating those measurements to the performance of the full jointed section. We have discussed separately the effects of joints on the transmission, but for the purposes of this discussion we assume that they do not play any significant role, that is, they are "perfect." Predicting the fiber performance in the general situation is not possible at present. However, we can make useful predictions in certain cases, and their study will serve to guide thought on the matter. We list the cases individually.

13.6.1   Step-Index Fiber without Mode Coupling

We have seen in the discussion of mode dispersion in the step-index fiber that it can be well explained in terms of a simple ray with the pulse delay proportional to the additional path length acquired as a result of traveling at different angles to the fiber axis. Since in a good joint this angle will be conserved, we may postulate that the dispersion of the ith fiber section will add directly to that of the earlier sections, barring mode filtering effects. Thus, to a first approximation, we anticipate that the mode delays of the sections, σ_i, will add according to the rule

$$\sigma_{\text{total}} \simeq \sum_i \sigma_i \qquad (13.6.1)$$

It is inevitable that some mode filtering will occur, so that the high angle, slow modes will be more attenuated than the well guided, low angle modes, with the result that the above estimate will tend to be pessimistic.

13.6.2   Step-Index Fiber with Mode Coupling

We have seen that the characteristic of propagation in such a fiber is that after a short lead length, in which a special power distribution is established, the pulses spread as $\sqrt{L}$. Evidently, jointing identical lengths would lead to a relationship of the form

$$\sigma_{total} \simeq \sqrt{A \sum_i L_i}$$

$$\simeq \sqrt{\sum_i (\sigma_i)^2}$$

(13.6.2)

A major effect of mode coupling is to destroy the memory property of the mode. Thus energy emerging from mode N does not carry a unique time signature that identifies it as having emanated from that mode, as would be the case in the absence of mode coupling. Rather, the time at which the energy emerges from mode N tells one essentially nothing about the path followed by the packet of energy in question. It follows that, on entering another length of fiber with mode coupling, each packet of energy leaving the length of fiber i will enter the $(i+1)$ length and be operated upon in an essentially similar manner, even if the dispersion characteristics are slightly different. The additional dispersion in the second fiber length will not be correlated with that in the former length, and we may expect that equation 13.6.2 will generally be a good approximation to describe the delays.

13.6.3 Graded-Index Fiber with Mode Coupling

For the same reasons as advanced in Section 13.6.2, we may expect to find the delays adding according to the relationship

$$\sigma_{total} \simeq \sqrt{\sum_i (\sigma_i)^2}$$

(13.6.3)

However, it must be noted that at present there are few published experimental data to either support or refute this statement, and it must therefore be regarded as speculative. In particular, the dispersion in the fiber while a stable mode power distribution is being set up, both initially and after each joint, has not yet been adequately studied.

13.6.4 Graded-Index Fiber without Mode Coupling

The dispersion in the graded-index fiber versus angle to the axis is not a simple linear function, as in the case of the step-index fiber, but can take many forms, depending upon the precise form of the index profile. Consequently, jointing graded-index fibers in which the mode coupling is minimal can lead to unexpected results. One extreme case was discussed in Section 6.3.3, where we pointed out that two low bandwidth fibers can be jointed to yield a high bandwidth fiber if the errors in their profiles are of similar magnitude but opposite sign. This leads one to postulate a model for the jointed bandwidth or pulse dispersion of two fibers of generally similar characteristics but with small errors in profile of the form

$$\sigma_{12} = \sqrt{\sigma_1^2 + 2c_{12}\sigma_1\sigma_2 + \sigma_2^2} \qquad (13.6.4)$$

where c_{12} is a correlation coefficient, taking a value between -1 and $+1$ and closely related to the relative errors in profile of the two fibers. The general validity of such an approach is clearly an expensive matter to prove, requiring extensive data both on long lengths of cable, jointed in a variety of combinations, and on their profiles. However, if proved, such a technique might be of considerable value in designing long fiber links where optimum performance was required.

Equation 13.6.4 can be extended to multiply jointed links having N components by writing it in the form[14,15]

$$\sigma_{\text{total}}^2 = \sum_1^N \sigma_i^2 + \sum_1^N \sum_{\substack{1 \\ i \neq j}}^N c_{ij}\sigma_i\sigma_j \qquad (13.6.5)$$

If the link were totally free of mode coupling ($|c_{ij}| = 1$), we would deduce that this result was order independent. However, small amounts of mode coupling make $|c_{i,i+1}| < 1$, and $|c_{ij}|$ for $|i-j| > 1$ rapidly tends to zero. Thus, to exploit mode equalization effects (Section 6.3.3), we must equalize adjoining fiber pairs.

13.6.5 General Comments

By identifying some extreme cases, we have been able to make some general statements about the nature of jointed fibers and the estimation of mode dispersion in them. To the mode dispersion must be added the material dispersion and the transmit pulse width (Section 5.9), according to the rule for independent mechanisms; thus

$$\sigma_{\text{pulse}} = \sqrt{\sigma_{\text{mode}}^2 + \sigma_{\text{material}}^2 + \sigma_{\text{transmit}}^2} \qquad (13.6.6)$$

Calculation of the value of σ_{material} requires knowledge of the dispersion properties of the material and the linewidth of the source to be used. Extrapolation of the dispersion of fibers from single lengths to jointed lengths, then, requires knowledge of the degree of mode coupling. At present it seems likely that cabled fiber will show significant coupling arising from the fiber deformations impressed by the cabling process, so that the results of Section 13.6.3 are most likely to be applicable for long links ($\Sigma L_i \gg 1$ km).

Note that throughout Section 13.6 we have implicitly assumed Gaussian pulse response in using the rms pulse width, σ. In practice, this assumption is likely to cause smaller errors than are caused by our simple propagation models.

REFERENCES

1. M. Boerner and S. Maslowski, *Proc. IEE*, **123**, 627 (1976).

2. M. J. Adams, D. N. Payne, and F. M. E. Sladen, *Appl. Phys. Lett.*, **28**, 524 (1976).

3. D. Gloge, *Bell Syst. Tech. J.*, **55**, 905 (1976).

4. P. C. Hensel, J. C. North, and J. H. Stewart, *Electron. Power*, February 1977.

5. C. M. Miller *Bell Syst. Tech. J.*, **54**, 1547 (1975).

6. C. M. Schroeder, *Optical Fiber Transmission, II*. Williamsburg, Va.: OSA/IEEE, February 1977.

7. C. M. Miller, *Bell Syst. Tech. J.*, **54**, 1215 (1975).

8. J. Guttman, O. Krumholz, and E. Pfeiffer, *Electron Lett.*, **11**, 582 (1975).

9. P. K. Runge, L. Curtis, and W. C. Young, *Optical Fiber Transmission, II*. Williamsburg, Va.: OSA/IEEE, February 1977.

10. a. J. C. North and J. H. Stewart, Unpublished work.
    b. P. C. Hensel, J. C. North, and J. H. Stewart, *Electron. Power*, February 1977.

11. M. Boerner and S. Maslowski, *Proc. IEE*, **123**, 627 (1976).

12. A. Yariv, *Quantum Electronics*. New York: Wiley-Interscience, 1967, Appendix 5.

13. L. G. Cohen and M. V. Schneider, *Appl. Opt.*, **13**, 89 (1974).

14. M. Eve, *Electron Lett.*, **13**, 315 (1977).

15. M. Eve, *Opt. Quant. Electron.*, **10**, 41 (1978).

16. J. Dalgleish, *Optical Fibre Communication*, IEE Conf. Publ. 132, London, 1975.

14

The Optical Receiver

14.1. INTRODUCTION

The receiver consists of a detector which converts incoming light into an electrical signal, followed by a suitable amplifier to generate a signal of sufficiently large level to be handled by the subsequent processing circuits. In considering the performance of this unit, we must bear in mind the fact that the quantum nature of the radiation cannot be neglected, as it normally can be in a microwave receiver. This is so because at optical frequencies, $h\nu > kT$, the unit quantum of optical energy, the photon, is significantly greater (by a factor of about 10) than the quantum of thermal noise energy associated with room temperature radiation. Physically, this manifests itself in the readily observable fact that our surroundings do not glow in the dark in the visible range, although if we were to "look" at them at longer wavelengths (10 microns or longer) we would find that they were emitting thermally.

The effect of this in designing our receiver is that, in detecting the incoming radiation, the thermally excited noise fluctuations in the detector can well be compared with the fluctuations arising from the radiation itself. The signal we will be detecting will be a finite number of discrete energy packets per bit, typically a few hundred mean, with a large fluctuation from bit to bit about this value. The effect is equivalent to the shot noise in a low level electrical current, which also consists of a small number of discrete units—electrons in that case.

The detector and its associated circuitry therefore play an intimate role in determining the system performance, and their analysis must take account of the properties of both the electrical circuit and the optical signal, together with the electrical signal generated directly by the latter. The analysis of this interface can proceed in a number of ways. In Figure 14.1 we show in schematic form a number of receiver configurations. The one of Figure 14.1a might be realized by using a high gain, low noise photomultiplier detector, feeding directly into a high speed gated counter. The photocounts received in each bit-interval would be collected, and a

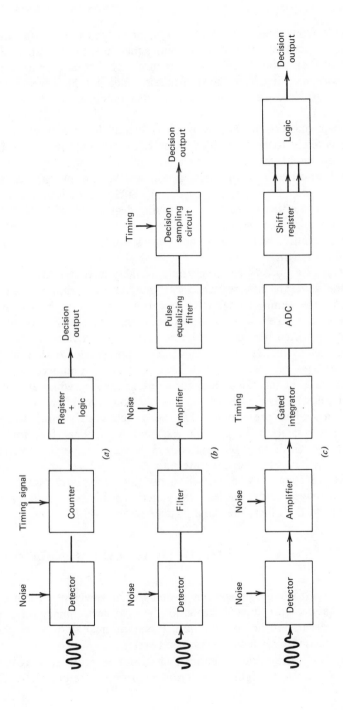

345

decision made from these data. Such a system in which there was some pulse overlap might be improved by forming subcounts, say for a period of $\tau/4$ each, and then using these subcounts in a digital processor to cancel the interference accruing from the transmission medium. However, such a receiver is considered neither economical nor practical for most fiber systems.

The receiver outlined in Figure 14.1b is much closer to the type of circuit that is likely to be assembled in practice. In analyzing its performance, account should be taken of the effects of each of the elements both on the signal power and on the noise power, so that the probability of a correct decision can be optimized by design. Such an analysis has been carried out by a number of authors to varying degrees of refinement, but in all cases complex expressions inevitably arise that make it difficult to achieve much intuitive feel for how the systems would respond to changes in a particular element.

The receiver of Figure 14.1c corresponds roughly to that analyzed here, and is of the general class described as "integrate and dump." Such a design allows some simple modeling to be carried out, although its realization in hardware is likely to be more difficult than that of Figure 14.1b. The analysis presented here of this type of receiver is a hybrid analysis in that it is presented in the terminology most appropriate to that of the Figure 14.1a type of receiver, that is, in terms of photocounts, but allowance is made in the calculation for some of the degrading effects of the receiver shown in Figure 14.1b, arising from amplifier noise and avalanche noise. The presentation has been deliberately simplified to include as few variables as possible, so that a physical feel for the operation of the whole may be obtained. However, despite the apparent wide differences between the integrate and dump receiver and the analogue receiver of Figure 14.1b, we will show that there is in fact a very close analogy between them and that the general predictions made for one will carry over to the other; the same physics covers both!

14.2. NOISE IN A DIGITAL OPTICAL COMMUNICATIONS SYSTEM

Since most optical communications systems being developed for use in telecommunications carry their information in a two-level binary form, we will limit our discussion to this situation. First, we must state some basic principles associated with such transmission systems.

The signal to be transmitted is either 1 or 0, and generally it is signaled by the use of a pulse of light or a period of no light, respectively. The transmission thus consists of a rapid series of light pulses, in the fashion of Morse code with a signaling lamp, although at vastly greater signaling rates. The transmit ON pulse may well *not* fill the full bit-period, τ.

The optical signal is received by a suitable detector, and the light converted to an electrical signal. The signal level in each signaling time slot is then compared to a predetermined "decision threshold level." If the received signal level is greater than the decision level, a 1 is declared to have been received; if below, a 0. Clearly, if the electrical signal from the detector precisely followed the transmit stage drive current, perfect transmission would be achieved. However, in practice, the signal is attenuated in transmission, to become very weak on reception, and thus generates spurious signal (noise), so that occasionally the true signal is overwhelmed by noise and a wrong decision made by the receiver. Thus such a system is usually specified in terms of the acceptable error rate, the number of mistakes or wrong decisions per second. Equipment for monitoring the error rate in digital communications systems is thus widely developed and used to test their performance. We will here examine the nature of the noise mechanisms that can lead to error and, in particular, highlight the way in which they are likely to differ in an optical system from a radio wave or coaxial cable system.

Most communicatons systems are subject to noise whose mathematical distribution is Gaussian. Such noise arises from large numbers of independent small fluctuations in the transmission medium, as thermal noise, and so on. The receiver detects a noise power signal regardless of whether or not the transmitter transmits one. The signal is characterized by the Gaussian or normal probability distribution, so that the probability of detecting a noise level of voltage amplitude x is given by

$$P(x) = \frac{1}{\sigma\sqrt{2\pi}} \exp\left(-\frac{x^2}{2\sigma^2}\right) \qquad (14.2.1)$$

where σ^2 is the variance.

The noise is as likely to be positive as negative, and is thus as likely to increase a transmit 0 to a receive 1 as to decrease a transmit 1 to a receive 0. To calculate the error rate, we must define our decision threshold and our mean received signal level. We set F to be the decision level and x_m as the mean signal level when a 1 is transmitted. Then the error rate when a 0 is transmitted, the probability of signaling 1 when 0 was sent, is given by

$$P(1,0) = \int_F^\infty p(x,0)\,dx \qquad (14.2.2)$$

where

$$p(x, x_m) = \frac{1}{\sigma_m \sqrt{2\pi}} \exp\left[-\frac{(x - x_m)^2}{2\sigma_m^2} \right]$$

$$= \mathrm{Gsn}(x, x_m, \sigma_m) \tag{14.2.3}$$

Similarly, the probability of registering a 0 when 1 was transmitted would be given by

$$P(0, 1) = \int_0^F p(x, x_m)\, dx \tag{14.2.4}$$

Both integrals are well known as the error functions [erf (ψ) and erfc (ψ)] and are widely tabulated. The important point to note is that the width of received signal amplitude probabilities is the same when either a 1 or a 0 is transmitted, the only change being in the mean position of the received signal level. This is illustrated in Figure 14.2.

This situation has to be contrasted with the one in which the received signal is of the very low level optical type. We can envisage the arriving optical signal as a series of discrete energy packets or quanta, *each of the same size regardless of the optical intensity*. The mean arrival rate of these energy packets is proportional to the intensity of the light, or, in statistical terms, the probability of a photon arriving in a given time interval is proportional to the intensity. Thus a strong signal corresponds to a large number of similar sized but randomly timed events, whereas a weak signal corresponds to a small number of equally randomly timed events. The contrast with the normal picture of an electrical signal being detected is obvious.

The action of the photodetector is to accept each packet of optical energy, and with a certain probability (less than unity) to absorb it and produce in its place a photoelectron. Thus an electric current is generated whose individual electrons are randomly spaced in time and whose mean number is proportional to the optical intensity. If the detector has gain, as in an avalanche photodiode or a photomultiplier, each photoelectron is subjected to a random gain mechanism in an avalanche type gain process, a gain that is characterized by some mean value g but varies statistically about this mean.

The analysis which follows is largely presented in terms of counting distributions, assuming that the arrival of individual photoelectrons can be registered as events. However, it also assumes that these events are summed or integrated during a bit-period. If the mean photon arrival rate is N per second and the quantum efficiency of the device is Q, the mean

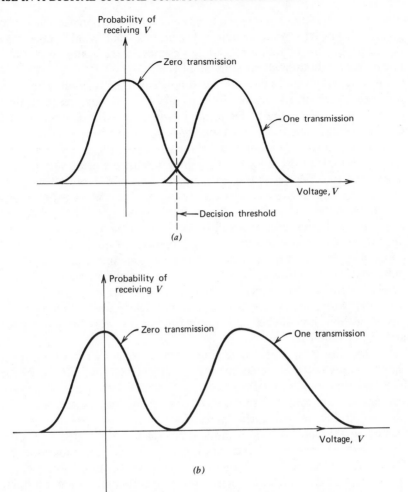

Figure 14.2. Action of the decision circuit in the presence of (*a*) Gaussian nonsignal-dependent noise and (*b*) Gaussian plus signal-dependent Poisson noise, illustrating the asymmetry introduced.

count during the bit-period τ will be $NQ\tau$ and the mean charge accumulated during that period will be $NQ\tau e$. The mean current during a mark-bit will be NQe amperes, so that the relationship between the counting approach and an analogue circuit approach is close. The counting approach allows one to overlook some of the intimate details of the circuitry associated with the device and also emphasizes the importance of the signal-dependent noise in the optical system. Analyses of receiver performance using a more conventional circuit approach have been given.[1,2,4] The electronic circuit configuration that would most closely correspond to the counting procedure described here would be that of the integrate and dump receiver (Figure 14.1c).

Thus we have a picture of the signal now emerging from the photodetector in the form of a series of electron pulses whose occurrence in time is dictated by the statistics of the photon arrivals and whose amplitudes are dictated by the statistics of the detector gain mechanism.

In a more realistic situation we should recognize that the mean arrival rate of photons at the detector is not binary (either a mean rate of m_1 per second or zero per second, according to whether a 1 or a 0 was transmitted), but that the rectangular binary digital signal will emerge from the transmission system rounded, and probably overlapping its neighbors, so that intersymbol interference will occur. Thus a better description of the optical receiver takes account of the fact that the mean arrival rate for photons is a function of time during the pulse time interval, τ, and the Nth pulse may be described in terms of $h(t - N\tau)$, where h now represents the optical intensity. The actual received signal is given by $S = \sum_{N=-\infty}^{\infty} a_N \cdot h(t - N\tau)$, where the a_N are the transmitted digits and the function S now describes the instantaneous optical intensity, or the probability of a photon arriving at time t.

By describing the signal detection in these terms, we have made clear that fundamental noise mechanisms are associated with the signal itself. The instantaneous signal fluctuates significantly, because of the random photon arrivals and the random nature of the gain mechanisms in a photomultiplier or avalanche photodiode. There is also a mean dark current in the detector and a random fluctuation associated with it, and since these currents are all at a very low level, an electronic amplifier must follow the detector. It will generate thermal noise itself, noise which is described by Gaussian statistics and must be added to the photon fluctuation noise. However, it is important to note that the photon fluctuation noise during a transmit 1 period is much higher than during a transmit 0 period, so that the system is not symmetrical.

These effects are illustrated in Figure 14.3. Figure 14.3a shows the output from a perfect optical detector (photomultiplier) when a single photon has been detected and a clean, noise-free pulse is observed. Figure 14.3b shows the output from the same detector when a light pulse was detected consisting of many photons, each giving rise to a single pulse of the type shown in Figure 13.3a but modified by the finite response of the electrical circuit and detector so that a smooth mean envelope pulse is obtained, which follows the mean arrival rate for photons but has superimposed upon it a large signal-dependent noise component.

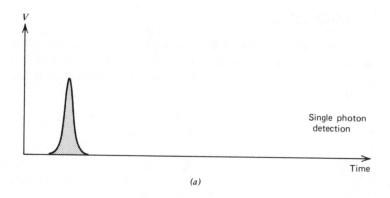

(a)

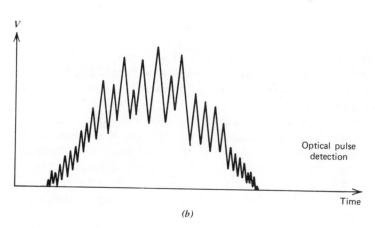

(b)

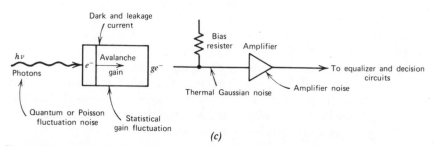

(c)

Figure 14.3. (a) Single photon detection. (b) Single optical pulse detection. (c) Noise sources in the receive circuit.

351

14.3. STATISTICS

If we examine the statistics of laser radiation arriving at a detector, we find that it obeys Poisson statistics.[3] Thus, if in a given time period T we expect to receive on average of x_m quanta, the probability that we will receive x is given by

$$p(x, x_m) = \exp(-x_m) \frac{x_m^x}{x!}$$

$$= \mathrm{Psn}(x, x_m) \qquad (14.3.1)$$

This distribution is characterized by a variance, σ^2, given by

$$\sigma^2 = x_m$$

This is now in marked contrast to the normal distribution, which is characterized by a fixed numerical value of variance. The Poisson distribution also has a finite skewness, $x_m^{-1/2}$, where the normal distribution has zero skewness; it is symmetrical.

We can now proceed to calculate the probability of detecting a signal of given voltage amplitude from our detector-amplifier combination in the presence of Gaussian noise. The conversion of photons to electrons* in the photodetector is a random event, so that a mean number of x_m photons will be converted to Qx_m electrons, and we can expect this distribution to remain Poisson. Thus the probability of generating m photoelectrons is given by $\mathrm{Psn}(m, Qx_m)$.

A signal of m electrons in a load resistor R_L will generate a voltage signal $V_m = meR_L/\tau$, assuming the current pulse to be spread over τ seconds. If a gain factor g (mean) is associated with the detector, the mean signal voltage will be g times greater. For the time being we ignore the random fluctuations in the gain process itself.

We now consider the probability of registering an actual voltage, V, in the detect circuit following the load resistor, which together with the following amplifier will have added some Gaussian noise to the perfect signal, V_m. If the Gaussian noise added is characterized by variance σ_N^2, the probability of detecting signal V is given by

$$p(V, V_m, \sigma) = \mathrm{Gsn}(V, V_m, \sigma_N) \qquad (14.3.2)$$

But we already know that the magnitude of the input signal, V_m, is subject to fluctuation following a Poisson law. We can put these two facts

---

*Strictly photons generate electron-hole pairs in a semiconductor detector.

together to generate an expression for the probability of detecting a signal level V at the detector when a mean photon arrival rate of x_m is expected as

$$p(V, x_m, \sigma_N) = \sum_{m=0}^{\infty} \text{Psn}(m, Qx_m)\text{Gsn}(V, V_m, \sigma_N) \qquad (14.3.3)$$

where

$$V_m = (meR_L/\tau).$$

Electrical circuit noise is taken into account by the term σ_N, while the photon fluctuation noise is accounted for by the Poisson term, which is summed out in the final expression. In practice, account should be taken of the stray capacitance associated with the detector load resistor and the frequency response of the following circuits. For the present we assume these to be negligible effects. Accordingly, we set the thermal noise generated Gaussian variance to

$$\sigma_N^2 = 4kTHB \simeq \frac{2kTH}{\tau} \qquad (14.3.4)$$

where k is Boltzmann's constant, T the absolute temperature, H the circuit noise factor, and B the bandwidth.

If we now reexamine our expression for the Gaussian terms, we see that we can recast them in a form that brings equation 14.3.2 nearer to a counting probability, as follows:

$$\text{Gsn}(V, V_m, \sigma_N) \rightarrow \text{Gsn}(m_c, m, \sqrt{\bar{n}}) \qquad (14.3.5)$$

where

$$\bar{n} = (2kTH\tau)/(g^2e^2R_L) = \bar{q}/g^2.$$

If we now rewrite equation 14.3.3 in terms of the above result, we obtain

$$p(m_c, x_m, \sqrt{\bar{n}}) = \sum_{m=0}^{\infty} \text{Psn}(m, Qx_m)\text{Gsn}(m_c, m, \sqrt{\bar{n}}) \qquad (14.3.6)$$

to be compared to the probability of detecting m photoelectrons in a perfect circuit (thermal-noise-free), which would be given by

$$p(m, x_m) = \text{Psn}(m, Qx_m) \qquad (14.3.7)$$

We have referred the variable m_c back to the primary photoelectron stage to highlight this comparison. However, it is most important to note that, whereas the variable m is an integer, m_c is not and is continuous because of the thermal noise "present in it," in reality added later after various gain stages. The variable m_c is attractive to think in terms of, because it both highlights the difference between itself and the original variable m and because it allows us to concentrate on the "optical" part of the receiver without having to become too concerned with the details of the subsequent electronics.

We could proceed to calculate the receiver performance using equation 14.3.6 for $P(m_c, x_m)$. However, evaluating it is a rather cumbersome procedure because of the necessity of summing over m from zero to infinity. In practice, provided that we are not too concerned with detection in a very low noise detector, such as a photomultiplier, we find that a simple approximation to equation 14.3.6 suffices. The approximation is twofold. We assume that the values of m_1 are sufficiently large for the Poisson skewness to be negligible. We can then approximate the Poisson distribution by a Gaussian distribution and add the variances of the Poisson and Gaussian (thermal) components to form a single Gaussian distribution. Thus we arrive at an expression of the form

$$p\left(m_c, x_m, \sqrt{\bar{n}}\,\right) \approx \mathrm{Gsn}(m_c, Qx_m, \sigma_T) \qquad (14.3.8)$$

where

$$\sigma_T^2 = Qx_m + \bar{n} = \bar{m} + \bar{n} \qquad (14.3.9)$$

giving us the probability of detecting a signal of m_c photoelectrons when a mean number x_m of photons was expected to arrive.

This approximation turns out to be excellent for essentially all cases of interest in present optical fiber systems. The extent of the approximation involved is shown in Figures 14.4 and 14.5, in which we have evaluated equations 14.3.6 and 14.3.8 for the same numerical values. Only in the situation where σ_T is very small does a serious difference emerge. Accordingly, we will use the approximate expression, equation 14.3.8, from now on in the following analysis.

We commented earlier that the effects of avalanche multiplications were twofold, yielding a larger mean signal so that the subsequent Gaussian thermal noise would be relatively decreased, but also adding some noise because the multiplication process itself is a random process. We have neglected this noise to date. However, a simple technique for taking it into account is to replace the gain squared in the variance expression by the gain to the power $2 + X$, where X is a constant characteristic of the device but in the range of 0.3 to 0.5 for silicon avalanche detectors. Thus we make

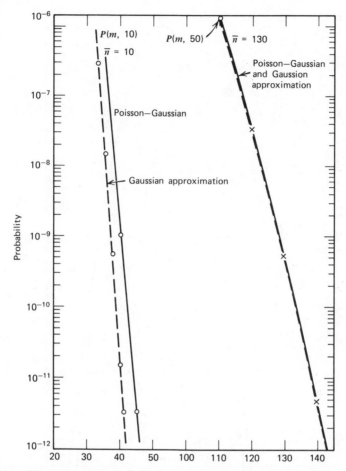

Figure 14.4. Effect of approximating the Poisson noise by a Gaussian distribution of noise when alternately 10 and 50 counts were expected, showing the probability of detecting m counts.

the following substitutions in place of those used earlier:

$$\sigma_T^2 \rightarrow (\bar{m} + \bar{n})g^X$$

$$\bar{n} \rightarrow \frac{\bar{q}}{g^{2+X}} \tag{14.3.10}$$

and the electrical signal/noise (power) ratio at the output takes the form

$$S/N = \frac{(\bar{m}^2)}{(\bar{m} + \bar{n})g^X} \tag{14.3.11}$$

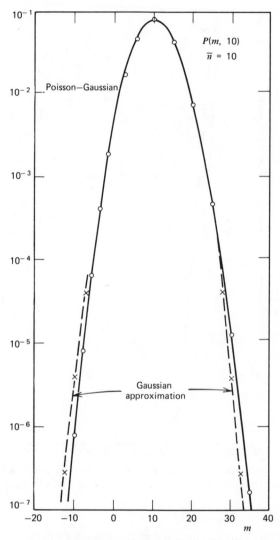

Figure 14.5. Plot of the predicted count probabilities for a mean count of 10 with an equivalent noise "count" of 10, using Gaussian approximation and the Poisson-Gaussian theory.

The S/N ratio and the variance are now dependent in a complex manner on the gain of the photodiode, so that one is naturally led to question whether there is a best setting for the avalanche gain: too little gain and the thermal noise will dominate; too much gain and the avalanche gain will dominate. Thus we seek the condition for the variance to be minimum as a function of g:

$$\frac{\partial(\sigma_T)^2}{\partial g} = 0 \rightarrow g^{2+X} = \frac{2\bar{q}}{Xm} \tag{14.3.12}$$

An optical receiver is usually designed so that, for the lowest signal level expected (smallest m_1), the avalanche diode bias will produce close to this optimum gain.

14.4. RECEIVER ERROR RATE

From the relations already defined, we can now write an expression for the error rate, for a given set of parameter values. Our optical digital receiver will expect to receive either m_1 or m_0 photoelectrons during each bit-period. If we set the decision threshold between these two at F, signals greater than F count as 1 and those less than F as 0. The error probability is then given by the integral of the signal probability outside the decision region. Thus, if we define the probability of detecting state Z when state Y was transmitted as $P(Z, Y)$, then

$$P(0, 1) = \int_{-\infty}^{F} \text{Gsn}(m_c, m_1, \sigma_T)\, dm_c \tag{14.4.1}$$

$$P(1, 0) = \int_{F}^{\infty} \text{Gsn}(m_c, m_1, \sigma_T)\, dm_c \tag{14.4.2}$$

Both these expressions can be written in the form

$$P_{\text{error}} = \tfrac{1}{2}\text{erfc}\left[\frac{|m_Y - F|}{\sqrt{2}\,\sigma_Y}\right] \tag{14.4.3}$$

and can thus be readily evaluated using the standard table for the error function. In practice, if a system is to be designed for a given error rate, the appropriate value for the kernel of error function is established from tables and the parameter values are then adjusted to produce this value. In Figure 14.6 we give the values of ψ for a given error rate where $P_{\text{error}} = \tfrac{1}{2}\text{erfc}(\psi)$. If we are to obtain the best performance (lowest error rate for a given signal power), we must approach as closely as possible quantum-

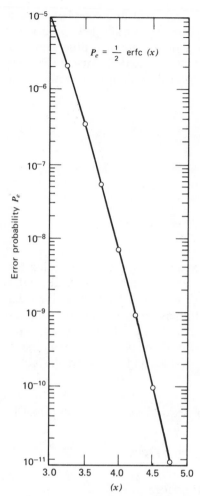

Figure 14.6. Plot of the probability of error $\frac{1}{2}\mathrm{erfc}(k)$ versus the argument of the error function, x.

noise-limited detection. This implies making $\bar{q}$ as small as possible, which in turn implies the use of the best amplifier that is available but also the largest load resistor value, R_L. Increasing the load resistor value ultimately leads to integration of the signal by the stray capacitance of the detector-amplifier input circuit. Goell[5] has shown that it is advantageous to push the circuit into this mode of operation and to correct the frequency response by adding a differentiating circuit after the head amplifier, in order to gain the advantage of the small $\bar{q}$ arising from the large load resistor.

We will now evaluate the above expressions to discover how we should choose the signal levels and threshold (F) values for a given system. Let us assume that we have an error rate target of 10^{-7} and that we find our detector-amplifier combination to be characterized by a Gaussian noise factor $H = 4$. We choose a load resistor value of 1500 Ω and calculate $\bar{q}$ to find a value of 8.6×10^{14} at room temperature (300°K). Assuming that $m_0 = 0$, to give a value of ψ of 3.7 for the desired error rate we must assume a value of g. The optimum value of g depends upon the desired minimum signal level m_1, so that some iteration is required. Assuming a bit rate of 10 Mbit/sec and a quantum efficiency of $Q = 0.5$, we finally obtain a mean (1) receive power level requirement of 1.7 nW. Repeating the calculation for a load resistor value of 1.5 MΩ, we find that the minimum receive power for the same error rate is approximately 0.8 nW.

The evaluations used the parameter values listed in Table 14.1. The same calculations can be done for a whole frequency range. The results of such an evaluation are shown in Figure 14.7, once again with values calculated for 1.5 kΩ and 1.5 MΩ load resistor values. The curves so derived are shown dotted for a part of each frequency range, since for the 1.5 kΩ load situation unrealistic gain values are predicted by the application of the optimum gain condition, equation 14.3.12, while for the high frequency range operating with a 1.5 MΩ load unrealistically small stray capacitance values would be required for this performance to be realized in practice.

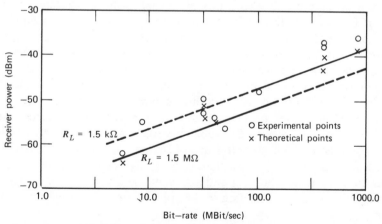

Figure 14.7. Survey of published result for optical receivers together with curves derived from the simple theory presented here. Each receiver was operating at 10^{-9} error rate or probability. Reproduced with permission from J. E. Midwinter, *Opt. Quant. Electron.*, **9**, 299 (1977).

TABLE 14.1 PARAMETER VALUES
USED IN EVALUATING EXPRES-
SIONS

$T = 300°K$
$H = 4$
$\tau = 10^{-7}$ unless specified
$R_L = 1500 \ \Omega$ unless specified
$X = 0.3$
$g = $ defined by equation 14.3.12
$e = $ electronic charge
$k = $ Boltzmann's constant

Accordingly, an intermediate curve (dot-dashed) has been drawn, reflecting more realistic values across the frequency span, and superimposed upon it are a number of experimentally reported results. All these results and the theoretical curves are for 10^{-9} error rates. A remarkably close agreement is seen between experiment and theory when the simplicity of the theoretical model is remembered and the frequency span involved is considered.

14.5. INTERSYMBOL INTERFERENCE

In the analysis presented so far, we have implicitly assumed that during a given bit-period only two forms of "signal" will be registered, namely, power that derives from the transmitted signal *uniquely* associated with that bit-interval, and power that derives from noise sources, typically thermal noise. However, in most real communication systems there is a third source of received power, namely, power that is associated with neighboring transmitted bits, since the transmission medium is most unlikely to be perfect and will introduce some pulse spreading. The power transmitted in a given bit-interval will arrive mainly in the corresponding bit-interval at the receiver, but some will spill over into the nearest-neighbor intervals or even second-nearest-neighbor intervals. This power includes a randomly varying element (the signal-dependent noise associated with the photon statistics[1,2,4]) and also contains a nonrandom component, namely, the mean level relating to the appropriate transmitted bit, be it m_1 or m_0.

A communication system usually contains within the receiver components that are chosen to cancel the mean interfering signal arising from such a situation in the transmission medium. This can be done effectively provided that the response of the transmission medium is known. Two general methods are used, frequency-domain equalization and time-domain equalization. We will discuss these very briefly before considering one in more detail in our own particular context.

In the first case, to correct the frequency response of the transmission medium we need an amplifier with a response that is the inverse of the transmission medium. This will normally boost the high frequency components and correct the overall amplitude-frequency response. However, to produce the desired pulse shape we must also ensure that the phase-frequency response of the system is linear, and this requirement will generally demand that our correction stage apply selective phase shifts to particular frequency components. The equalizer is then a complex amplitude and phase filter chosen to correct the transmission medium response. However, the fact that in the process the high frequency components of the received signal will be amplified means also that the signal/noise ratio of the received signal will be degraded, the penalty for applying this type of correction. It goes without saying that a further penalty is the additional complication and cost involved in the system design.

The second approach to signal interference cancellation is the time domain one. The received signal is fed into a delay line with multiple taps, as shown in Figure 14.8. By accessing the delay line at different points simultaneously, the signals received during different bit-periods become available for inspection. Let us suppose that the mean signal received at the receiver when the jth bit is transmitted is m_j. This signal will be spread among several bit-periods on reception. Suppose that a fraction f of the signal is contained within the jth bit-interval at the receiver and that a fraction h is contained equally in the two neighboring intervals. If we assume that $f \approx 1$, we can set $2h = (1 - f)$ with reasonable confidence. The actual (mean) signal now received during the jth bit-interval is given by

$$\overline{C}_j = f\overline{m}_j + h(\overline{m}_{j-1} + m_{j+1}) \tag{14.5.1}$$

We now feed this signal into the delay line so that at the sampling port some time later we can simultaneously sample $\overline{C}_{j-1}, \overline{C}_{j+1}, \overline{C}_j$. If we now sum these components to form the following result:

$$\overline{B}_j = \overline{C}_j - c(\overline{C}_{j+1} + \overline{C}_{j-1}) \tag{14.5.2}$$

we can choose the constant c to cancel the nearest-neighbor interference. In fact, this is done by setting $c = h/f$. The mean signal B_j is now free of nearest-neighbor interference, but it includes interference from next-nearest neighbors, generated in the processor because the signals C_{j+1} and C_{j-1} included some interference from m_{j+2} and m_{j-2}, although this is now at a much reduced level. Evidently, by adding two further taps to our delay line we could cancel this effect but add third-nearest-neighbor interference,

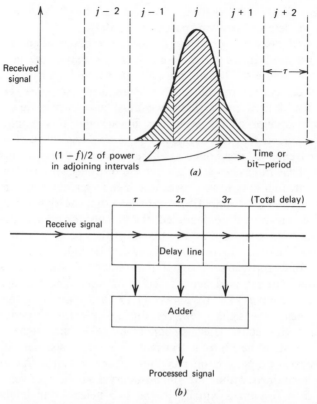

Figure 14.8. (*a*) The receiver signal versus bit-interval in the presence of some intersymbol interference. (*b*) Schematic model of a delay line canceler for correcting the interference.

although at a still lower level. For the purposes of this discussion we will think in terms only of the three-sampling-port system, since the penalty in an optical system of working with heavy interference (small *f*) is very great and seems unlikely to be incurred in any practical system.

In the preceding discussion we have considered the signal per bit as if it were a single number $\overline{C}_j$. In the processor we have described using an analogue delay line and analogue processor; this would not be so, and the detailed pulse shape of the signal would also be included. However, if we consider a receiver which takes the signal from the detector, integrates it for the bit-period, and then dumps the integrated signal into a register, it will much more closely approximate the situation implied by our notation, and we can then consider the effects of a purely digital processing on the signal information in question. This model fits in well with our picture of the incoming light signal as a stream of photons.

For the purposes of analysis, we can treat the C_j values as the signal in place of the m_j used earlier, since the receiver cannot distinguish the source of the individual photons arriving at it. We can consider two signals entering our processor, for the jth bit-interval, the signal C_j, already defined, and the signal A_j, defined as

$$A_j = C_{j-1} + C_{j+1} \qquad (14.5.3)$$

Since these two are statistically independent, we can write the probability of actually receiving A_j when $\overline{A}_j$ was expected and C_j when $\overline{C}_j$ was expected as

$$P(A_j, \overline{A}_j) = \frac{1}{\sqrt{2\pi(\overline{A}_j + 2\bar{n})g^X}} \exp\left[-\frac{(A_j - \overline{A}_j)^2}{2(A_j + 2\bar{n})g^X}\right] \qquad (14.5.4)$$

$$P(C_j, \overline{C}_j) = \frac{1}{\sqrt{2\pi(\overline{C}_j + \bar{n})g^X}} \exp\left[\frac{-(C_j - \overline{C}_j)^2}{2(\overline{C}_j + \bar{n})g^X}\right] \qquad (14.5.5)$$

In the expression for A_j the thermal noise is doubled because two separate bit-interval components have been summed.

We can now calculate the probability of a given corrected signal, B_j, leaving the processor (equalizer). The mean value, $\overline{B}_j$, is formed by the processor as $\overline{B}_j = \overline{C}_j - c\overline{A}_j$. However, since the actual values of both C_j and A_j are free to fluctuate over a wide range, any particular value of B_j can be formed in many different ways—by combining two particularly large values of C_j and A_j, for example, or by combining two particularly small ones, or many other pairs. Thus, to calculate the probability of a given value of B_j being formed, we must sum over all the possible C_j and A_j values that could combine to form it. This is written as

$$P(B_j, \overline{B}_j) = K \int_{-\infty}^{\infty} P(B_{j+i}, \overline{C}_j) P\left(\frac{i}{c}, \overline{A}_j\right) di \qquad (14.5.6)$$

The constant K is adjusted so that

$$\int_{-\infty}^{\infty} P(B_j, \overline{B}_j) dB_j = 1$$

Fortunately, both integrals are analytically soluble so that a simple expression can be obtained. The result is

$$P(B_j, \overline{B}_j) = \text{Gsn}(B_j, \overline{B}_j, \sigma_p) \qquad (14.5.7)$$

where

$$\sigma_p = \sqrt{\sigma_1^2 + c^2 \sigma_2^2}$$
$$\sigma_2^2 = \left(\overline{A}_j + 2\overline{n}\right) g^X \tag{14.5.8}$$
$$\sigma_1^2 = \left(\overline{C}_j + \overline{n}\right) g^X$$

If we collect the terms in σ_p^2 we find that it takes the following form:

$$\sigma_p^2 = g^X \left\{ \overline{m}_j \left[f + c^2(1-f) \right] + (m_{j-1} + m_{j+1}) \left(c^2 f + \frac{1-f}{2} \right) \right.$$

$$\left. + \left(\overline{m}_{j+2} + \overline{m}_{j-2}\right) \frac{(1-f)c^2}{2} + \overline{n}(1+2c^2) \right\} \tag{14.5.9}$$

assuming that $2h = 1 - f$.

We see that the variance (random fluctuation) associated with the signal has been increased by the operation of the processor, both because the signal-dependent noise associated with the interfering signals cannot be canceled, but only the mean level of them, and also because an additional thermal noise term has been added, the factor $(1+2c^2)$. We will comment on that later.

The expression for the probability of detecting a given level, B_j, when $\overline{B}_j$ was expected, equation 14.5.7, is now in a form that allows one to calculate the probability of error in detecting any given symbol group. By analogy with equation 14.4.3 we can write that the probability of error when a threshold decision level F is in operation is given by

$$P_{\text{error}} = \tfrac{1}{2} \text{erfc} \left[\frac{|\overline{B}_j - F|}{\sqrt{2} \, \sigma_p} \right] \tag{14.5.10}$$

We can once again go through the calculations described earlier in assessing the perfect equivalent receive signal level, m_1, necessary for a given set of parameters with the pulse spreading, f, now considered as a variable. This we have done, using the parameter values given in Table 14.1 and holding the value of the parameter $c = h/f$ and of $2h = 1 - f$.

The additional power (m_1) value necessary to hold the error rate constant is plotted versus the parameter f, the fraction of the power received within the bit-interval, in Figure 14.9. We see that the effects of the signal-dependent noise in the system are such that, even with perfect

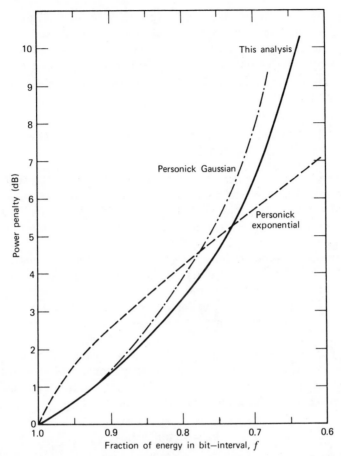

Figure 14.9. Calculated power penalty in dB at the receiver arising from ISI, using this theory and the results obtained by Personick.[1] Reproduced with permission from J. E. Midwinter, *Opt. Quant. Electron.,* **9**, 299 (1977).

canceling of the mean interfering signal levels, the increase in variance requires a substantial increase in signal power with small amounts of pulse spread.

For comparison, we have also plotted in Figure 14.9 results obtained by Personick[1] in a frequency-domain analysis of the same problem. In treating the frequency-domain case, he assumed particular pulse responses, calculated the filter characteristics to correct them for equalization purposes, and then estimated the power penalty associated with each case. Since the calculation involves the convolution of the filter and signal frequency response functions, simple expressions of the type given here do not emerge, although the method is capable of modeling far more accurately the effects of particular electronic components in the detector-amplifier-equalizer chain than is ours. The conclusions, however, are seen to be nearly identical. A simple frequency domain analysis has been given by Smith and Garrett.[2]

14.6. RELATIONSHIP BETWEEN FREQUENCY-DOMAIN AND TIME-DOMAIN APPROACHES

The very close agreement between the results from the two different analytical methods given in Figure 14.9 indicates that both reflect the same underlying physics and that, while the postulated "electronics" are apparently very different, there must be a close analogy between them. That this is so is easily shown.

The frequency-domain filter system postulates that the signal is corrected by use of a filter-amplifier combination, so that the transmission medium response is corrected and the resulting signal is then sampled at the center of the bit-period. The correction is therefore set up to minimize the degree of interference when the analogue signal is at the mid-bit-interval.

In regard to the time-domain system described here, we find that we can model it by a very simple frequency-domain analogue. The frequency response of our integrate and dump filter is found by examining its response to a sine wave input. Thus we consider

$$V_{in} = V_0 \sin(\omega t + \phi) \tag{14.6.1}$$

$$V_{out} = \int_{N\tau}^{(N+1)\tau} V_{in} dt \tag{14.6.2}$$

Carrying through the integration, we obtain the result

$$V_{out} = V_0 \tau \operatorname{sinc}\left(\frac{\omega\tau}{2}\right) \sin\left[\left(N + \tfrac{1}{2}\right)\omega\tau + \phi\right] \tag{14.6.3}$$

which tells us that it is equivalent to the output of a filter having a frequency response given by the sinc function sampled in the center of the bit-period.

If we now consider the effect of the signal processor postulated, we must take the output of the $N-1$, N, and $N+1$ bits and convolve them to form the composite output. This leads to the following expression:

$$V_{equ} = V_0 \tau \, \text{sinc}\left(\frac{\omega\tau}{2}\right)(1 - 2c \cos \omega\tau)$$

$$\times \sin\left[\left(N + \tfrac{1}{2}\right)\omega\tau + \phi\right] \qquad (14.6.4)$$

The new filter response function is now given by

$$h(\omega) = \text{sinc}\left(\frac{\omega\tau}{2}\right)\left[1 - 2c \cos(\omega\tau)\right] \qquad (14.6.5)$$

The effect of this new filter response function was to cancel the interference, as we showed earlier in the time-domain analysis. To show that it corresponds to an increased bandwidth, or amplification of the higher frequency signal components, we work out the integral of $h^2(\omega)\,d\omega$. This leads to the simple expression

$$H = \int_0^\infty h^2(\omega)\,d\omega = G(1 + 2c^2) \qquad (14.6.6)$$

where G is a constant. Since c is a parameter that increases with increasing requirements for intersymbol interference correction, we see that the power bandwidth of the filter also increases as the interference is corrected. We also note, by reference to equation 14.5.9, that the factor $(1 + 2c^2)$ is precisely the factor by which the thermal noise term in the variance for the signal emerging from the signal processor was increased.

14.7. SUMMARY

In this discussion we have stressed the role of the signal-dependent noise that arises in the optical system because one is detecting photons having an energy that is large compared to kT, and we have shown how this leads to heavy penalties being paid for operating a system in which the transmission medium is bandwidth limited, so that pulse spreading is a serious factor. In a digital repeater other factors which will still further reduce the efficiency of operation must also be taken into account. We have implicitly assumed that the receiver "knows" when each bit-interval starts and finishes. In practice, the receiver must discover this itself from the incoming signal power, and since it cannot ever do this with absolute precision, there will be a small timing error or jitter. This will have the effect of introducing an additional "intersymbol interference," which will further degrade the quality of the information received.

We have also implicitly assumed that the transmission medium is stationary in time, that its loss and bandwidth do not change. However, our discussion of cable design and mode coupling effects makes clear that this is most unlikely to be so, although the amount of fluctuation in the medium that will occur in a given cable must be discovered with the cable in its operating environment.

The designer must therefore leave further surplus power margins so that these other effects can occur without causing unacceptable degradation in system performance, and he or she must also design the electronics and source drive circuits to take account of degradation with age and temperature of the components used.

REFERENCES

1. S. D. Personick, *Bell Syst. Tech. J.*, **52**, 843, 874 (1973).
2. I. Garrett and D. R. Smith, *Opt. Quant. Electron.*, **10**, 211–221 (1978).
3. R. M. Gagliardi and S. Karp, *Optical Communications*. New York: Wiley-Interscience, 1976.
4. J. E. Midwinter, *Opt. Quant. Electron.*, **9**, 299 (1977).
5. J. Goell, *Bell Syst. Tech. J.*, **53**, 629 (1974).

15

Optical Fiber Systems

15.1. INTRODUCTION

This book is primarily concerned with examining the properties of the optical fiber, considered as an information transmission medium. While we have not specifically excluded other applications, we have had in mind throughout the transmission of large quantities of data over long lengths of fiber without the use of more amplifiers or repeaters than are necessary.

Thus we have had in mind systems in which the fiber link lengths extend from 1 to 10 km and transmission rates are in the range from 1 Mbit/sec to 1 Gbit/sec. Such systems span the range of interest in civil telecommunications use, and the presentation has been influenced by this application. However, many other systems can be considered, such as data links for aircraft, ships, or offices, business information systems in general, or computer-computer and computer-sensor links in automatic control situations. In each of these applications the same principles will apply, but the emphasis is likely to be different.

A system is composed of a number of discrete components, connected together in such a way as to perform some task in the desired manner. Thus an optical fiber, be it in cable or on a drum, constitutes not a system, but merely a component, albeit a complex one. To produce an information transmission system we must bring together other components as well. We will assume that the input of information is electrical, in analogue or digital form, and that the task is to transmit this information from point A to point B as faithfully as possible (or as specified).

Our system is thus to consist of some drive electronics, driving an optical source, whose light output is coupled into a fiber in a cable. The light emerging from the far end is detected, and the electrical signal so generated is amplified until it becomes comparable to the one sent. The system designer has to select each of the individual components carefully, consider how each interacts with the others, and then evaluate how the assemblage of components will perform. At this stage he or she will usually have to reexamine the whole design and seek ways of optimizing its performance, trading something here to gain something there.

It is the purpose of this chapter to assemble some basic data on the individual components and to initiate some thought on their interactions. To treat system design and optimization exhaustively would require another book.

15.2. PRELIMINARY SYSTEMS DESIGN

We have already seen in our discussion that the fiber transmission medium has a finite loss rate for optical transmitted power and a finite dispersion for optical pulses which will limit the bandwidth that can be transmitted. We have also seen in our discussion of the optical receiver that the limitations imposed by the physics of photon detection require certain minimum received powers for any given error rate and data rate. Most systems are constrained to use a semiconductor laser or LED source, by virtue of its small size, easy modulation, and relatively high efficiency. However, these devices limit the power that can be transmitted to the order of 1 mW for the laser and typically 100 μW for the LED. Evidently, these facts taken together restrict the system designer, since power considerations along will limit the fiber length he or she can interpose between source and detector. We will now formalize these arguments so that we can examine the nature of these limitations more quantitatively.

Figure 15.1 reproduces the information given earlier (Figure 14.7) to illustrate, for the case of the laser source, launching 1 mW power at 840 nm wavelength into the fiber, the total power available for the system operating link. The figures were calculated for an error rate of the order of 10^{-9}, although the resulting numbers for allowed insertion are rather insensitive to the exact values taken (less than 1 dB power for one order of change in the receiver error rate). The total power available to the system under the conditions of Figure 15.1 may be described as $F(T)$, where T is the bit-rate interval and is given by (curve fitting) approximately as

$$F(T) = 128 - 4.35 \ln(T^{-1}) + P_L \text{ dB} \qquad (15.2.1)$$

where P_L is the launched power (dBm). This power is used in various ways in the operating system. An operating margin must be maintained for the electronics so that small fluctuations in the system performance do not lead immediately to an unacceptable decrease in performance. We will assign M decibels to the margin. Typically, we might find that $M = 5$ dB. The fiber itself is characterized by a loss in decibels per kilometer, so that for a link of length L km the total loss is given to first approximation by αL dB. Some correction of this is required to allow for the effects of the establishment of a stable distribution after launching. We might also in a more sophisticated analysis separate out the losses arising from mode coupling and mode filtering effects, both of which might interact back upon the dispersion of the fiber. For simplicity, however, we will use the single parameter α and add to it an equivalent loss per kilometer, α_J, which arises from the effects of joints. We could assign this as an integer variable loss, proportional to the number of joints, but since the power lost from a disturbed mode distribution is not lost only in the immediate vicinity of the joint, it can be more realistic to view α_J as a distributed loss. In addition, this has the advantage that it leads to a simpler description. Thus our fiber and joints contribute a total loss of $(\alpha + \alpha_J)L$ decibels.

We must now recognize that we cannot design our system unless we take into account the temporal response of the medium. We showed in Section 14.5 that pulse spreading in the fiber arising from a finite data bandwidth in the optical system could be accommodated by equalization in the

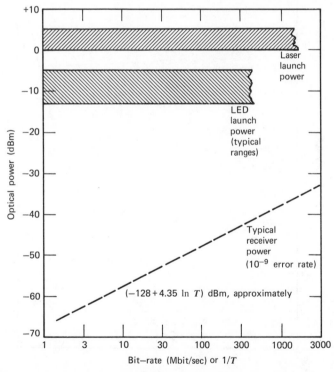

Figure 15.1. A simple picture of the power available to the transmission medium in an optical fiber system, showing the launch power from an LED or laser source and the receive power versus bit-rate for a 10^{-9} error rate.

detector circuits, but only at a penalty. Either a higher error rate must be tolerated, or more optical power must be received to restore the error rate to the design figure. Figure 15.2 repeats the data of Figure 14.11, plotting them in terms of τ_e, the $1/e$ pulse width (see Appendix 5), and considering the Gaussian shaped pulse, as was done by Personick. Also plotted on the same curve is the analytical function, $F_{ISI} = 2(\tau_e/T)^4$. It is seen that the analytical function provides an exceptionally close fit to the desired curve, and we will use it to model the effects of intersymbol interference in our system. Since the penalty, F_{ISI}, is measured in decibels, we can include it in the same formula with our available power, F, and the losses and margin.

The quantity F_{ISI} is a function of a number of variables, through the dependence of τ_e on various effects. Three main factors control τ_e. First, the initial pulse width transmitted must be accounted for. Let us assume

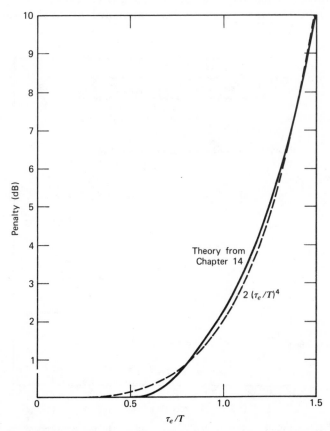

Figure 15.2. Penalty arising from the ISI of Figure 14.9, with the simple analytical expression superimposed for comparison.

that the transmitter sends a pulse of width $T/2$ that is rectangular in shape. The rms pulse width is defined as

$$\sigma^2_{\text{rms}} = \int_{-\infty}^{\infty} t^2 h(t)\, dt - \left[\int_{-\infty}^{\infty} t h(t)\, dt \right]^2 \qquad (15.2.2)$$

where the function $h(t)$ not only defines the pulse shape but also is normalized so that $\int_{-\infty}^{\infty} h(t)\, dt = 1$. Using this relation, we find that the $T/2$ rectangular pulse corresponds to an rms pulse width of $T/\sqrt{48}$ and $\tau_e = 2\sqrt{2}\ \sigma = 0.408\,T$, using the Gaussian approximation (Appendix 5).

The second factor contributing to the final value of τ_e is the material dispersion of the fiber. This may be written in the form

$$\tau_{\text{mat}} = \phi_M \cdot \Delta\lambda \cdot L \qquad (15.2.3)$$

where ϕ_M is a parameter characteristic of the fiber materials, $\Delta\lambda$ is the linewidth of the source used, and L is the link length (see Section 5.8).

The third factor—and often the dominant one—is the mode dispersion of the fiber. We have seen that to describe the mode dispersion of a fiber adequately is a complex task if we attempt it from a purely theoretical point of view. Given that we are almost always concerned with multimode guides, we must know not only the relative values of the group velocities of the different modes but also the distribution of power launched into them, the effects of filtering and interchange among them during transmission, and the redistribution and the redistributive effects of joints. We may say with some confidence that, in practice, we will not be able to acquire all this information. However, we have seen from the discussions in Chapters 6 and 7 that mode dispersion in the extreme cases of no mode coupling and strong mode coupling varies linearly with L and as the square root of L, respectively. Thus in general we can state that our mode dispersion will be of the form

$$\tau_{\text{mode}} = \psi (L)^b \qquad 1 \geqslant b \geqslant \tfrac{1}{2} \qquad (15.2.4)$$

where ψ is an experimentally determined constant, as is b, and assuming no equalization effects (Section 6.3.2).

We can now bring these factors together to derive an expression for τ_e of the form

$$\left(\frac{\tau_e}{T} \right)^2 \simeq (0.408)^2 + \left(\frac{\phi_M \Delta\lambda L}{T} \right)^2 + \left(\frac{\psi L^b}{T} \right)^2 \qquad (15.2.5)$$

We can now combine these results into a simple equation which includes the effects of the source, detector, and fiber and allows us to determine L for a given bit-rate. The design equation is, equating powers,

$$F(T) = 128 - 4.35 \ln(T^{-1}) + P_L$$

$$= (\alpha + \alpha_J)L + M + F_{ISI} \qquad (15.2.6)$$

In general, the solution of this equation must proceed numerically, but in two cases much further simplification is possible. Both cases correspond to the situation, often realistic in practice, that one term in τ_e dominates. Then F_{ISI} takes a particularly simple form which becomes more apparent if we choose to examine the two extreme cases of $b = 1$ and $b = \frac{1}{2}$. Before doing this, we note a further point of interest.

In designing an information transmission system, it is frequently the objective to maximize the amount of information transmitted over a link of

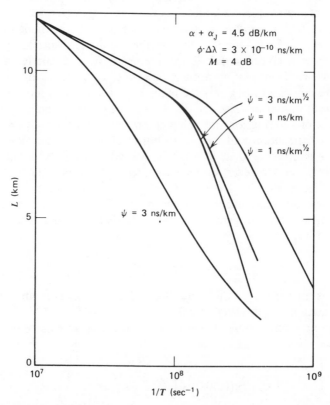

Figure 15.3. Curves of the calculated repeater section for a fiber system versus the bit-rate for a variety of bandwidth limited conditions.

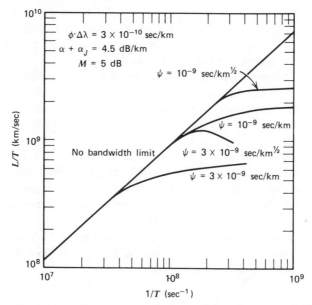

Figure 15.4. The calculated curves for the bit-rate times kilometer product for the same systems as in Figure 15.3 versus the bit-rate.

given overall length. In our terminology this corresponds to maximizing the product of the bit-rate and the repeater spacing, namely the factor L/T. If we concentrate first on the special case that the mode dispersion dominates the total dispersion, and b takes the value $\frac{1}{2}$ (the fiber is well mode mixed), equation 15.2.6 can be written in the form

$$\left(\frac{2\psi^4}{T^3}\right)\left(\frac{L}{T}\right)^2 + (\alpha + \alpha_J)\left(\frac{L}{T}\right) - \frac{F(T) - M}{T} = 0 \qquad (15.2.7)$$

so that

$$\frac{L}{T} = \frac{\left\{-(\alpha + \alpha_J) + \sqrt{(\alpha + \alpha_J)^2 + 8\psi^4[F(T) - M]/T^4}\right\}^{T^3}}{4\psi^4} \qquad (15.2.8)$$

In this form the expression is directly evaluated for L/T, and hence L as a function of T. In the extreme case of no bandwidth limitation, the term $2\psi^4/T^4$ is negligible, and we find that

$$L = \frac{F(T) - M}{\alpha + \alpha_J} \qquad (15.2.9)$$

and the system is purely loss limited. As the bit-rate is increased, the bandwidth limitation of the fiber becomes apparent, L begins to shorten rapidly, and increasing amounts of power are required to maintain the error rate at an acceptable level at the receiver.

Figure 15.3 shows a plot of the variation of L with $1/T$ for a number of different fibers in this limiting case of τ dominated by mode coupled mode dispersion, and the onset of bandwidth limitation is clearly visible. The same data are replotted in Figure 15.4, but in the form of the range bit-rate product, L/T, versus $1/T$ for the same fibers. A maximum in transmission capacity is visible, the exact position of the maximum depending upon the relative interplay of loss and dispersion for the particular fiber.

In the case that the dispersion of the fiber is linear in length and is again dominated by the mode (or material) dispersion, equation 15.2.6 again reduces to a simple form and can be solved for values of L and L/T versus $1/T$ directly. The results of such an evaluation are also shown in Figures 15.3 and 15.4 for ostensibly similar fibers. The beneficial effects of the $\sqrt{L}$ bandwidth dependence are clearly seen.

15.3. SOME TYPICAL SYSTEMS

We will now use the data derived above to examine two systems based upon fiber of readily available performance to discover what is likely to be realistic. We will take fiber of measured mode dispersion 1 ns on a 1 km length and loss 4 dB/km. Let us assume that 0.5 dB/km of effective additional loss is introduced by joints, and that for an additional mode coupling loss of 1 dB/km the fiber dispersion can be considered to extrapolate as the square root of the length. If we first examine the options for systems operation at 30 Mbit/sec, we find the two cases listed in Table 15.1. Clearly, we are better off to choose the fiber in cable that is free of mode coupling, since in both cases a negligible bandwidth limitation occurs and the additional loss associated with the mode coupling merely results in a reduced section length with additional unused bandwidth.

If we now turn to a system to operate at 300 Mbit/sec using the same fibers, the options that are available are as presented in Table 15.2. We see that the system is bandwidth limited in either case; and the more efficient way to use the additional power available is in mode coupling, generating additional bandwidth, rather than in the receiver for overcoming the effects of intersymbol interference. In reaching this conclusion we have assumed a relationship between the additional loss and the effects on the mode dispersion for our fiber that may not be realizable in practice. Nevertheless, with minor modifications to the numbers, we may expect to be able to observe similar effects in practice.

TABLE 15.1   TYPICAL VALUES FOR A SYSTEM OPERATING AT 30 MBIT/SEC

Power budget

Assuming 1 mW launch power, from equation 15.2.1	53.1 dB
Operating margin	5.0 dB
Available to transmission medium	48.1 dB

Assuming dispersion linear with fiber length (from Figure 15.3)

4.5 dB/km × 10.6 km	47.7 dB
Intersymbol interference penalty	0.02 dB
Total	47.72 dB

Assuming dispersion linear with fiber length$^{1/2}$ (from Figure 15.3)

5.5 dB/km × 8.5 km	46.75 dB
Intersymbol interference penalty	0. dB
Total	46.75 dB

Summary

The linear fiber allows 10.6 km while the mode coupled fiber only allows
   8.5 km because of the additional loss in the fiber

Finally, let us examine the trade-offs for a system operating at longer wavelength. We note that the only fundamental loss mechanism in a glass fiber is due to Rayleigh scattering. If we take presently available silica materials, the Rayleigh scattering is given to within a factor of about 1.5 by the expression

$$\alpha = R \cdot \lambda^{-4} \text{ dB/km} \tag{15.3.1}$$

where λ is expressed in microns, and the constant R takes the value 1.5 dB/km·micron^4. For a fiber with negligible absorption losses it will be to our advantage to move to a longer operating wavelength. If we wish to continue to use a silicon avalanche detector, we are limited to a long wave limit of about 1.1 microns. However, this gives us a decrease in Rayleigh scattering loss relative to a 840 nm wavelength of about 3 times, reducing the loss from about 3 to 1 dB/km. It seems reasonable, therefore, to consider cables having losses of the order of 2 dB/km, compared to the values of 4 to 5 dB/km considered at 840 nm. (See Section 8.3.2.)

TABLE 15.2   TYPICAL VALUES FOR A SYSTEM OPERATING AT 300 MBIT/SEC

Power budget

Assuming 1 mW launched power, from equation 15.2.1	43.1 dB
Operating margin	5.0 dB
Available to transmission medium	38.1 dB

Assuming dispersion linear with fiber length (from Figure 15.3)

4.5 dB/km × 5.3 km	23.85 dB
Intersymbol interference penalty	14.25 dB
Total	38.1 dB

Assuming dispersion linear with fiber length$^{1/2}$ (from Figure 15.3)

5.5 dB/km × 6.5 km	35.75 dB
Intersymbol interference penalty	0.685 dB
Total	36.44 dB

Summary

The linear fiber allows 5.3 km while the mode coupled fiber allows
  6.5 km because of its superior bandwidth

Other system parameters are also changed. Material dispersion is greatly reduced, by a factor of about 10, while a new source is clearly required. Sources at about 1.06 micron wavelength fall into two categories, those based upon the semiconductor quaternary system Ga-In-As-P and those based upon the optically pumped transition metal ion neodymium in a variety of crystal host lattices, most commonly YAG (yttrium-aluminum-garnet). Neither has yet been fully developed for use in fiber systems, but it does not seem unreasonable to expect CW output powers of the order of 1 to 10 mW from lasers of the former material system and upwards of 1 W from neodymium based lasers. (See Section 5.9.)

In the detector some new problems appear since at 1.06 micron wavelength the band edge of silicon is being approached. For this reason the absorption coefficient or quantum efficiency for a detector of given thickness is reduced. This can be compensated for by increasing the thickness of the detector's absorbing layer, but the result is to increase the drift time of the photoelectrons, reducing in the same proportion the high frequency response of the device. Generally speaking, it appears to be realistic to expect a penalty of about 3 dB for the same speed of response.

TABLE 15.3   NOTIONAL VALUES FOR A LONG WAVELENGTH
SYSTEM OPERATING AT 1.3 MICRON WAVELENGTH AND EITHER
10 OR 100 MBIT/SEC

| | Transmission rate | |
Parameter	10 Mbit/sec	100 Mbit/sec
Total power (dB)*	58	48
Fiber loss (dB/km)	1	1
Fiber length (km)	50	40
Fiber type	Precision graded	Monomode

*Assuming equation 15.2.1 still applies

A further adjustment that must be made to our theoretical model
concerns the photon noise term for the receiver, since the value of $h\nu$, the
quantum energy, has been reduced by a factor of $1060/840 = 1.26$. Taking
these factors together, we can make some tentative estimates of the system
shape that might emerge for transmission at 10 and 100 Mbit/sec at this
longer wavelength. Most striking is the increase in section length that
becomes possible because of the reduced fiber loss and the shift of
emphasis from loss being a limiting factor at the higher bit-rate to fiber
bandwidth. Moving to 1.3 micron wavelength and assuming that equation
15.2.6 remains valid leads to even more impressive possibilities, as indicated
in Table 15.3.

15.4.   OPTIMIZATION OF THE TRANSMISSION MEDIUM

In the analysis presented so far, we have assumed that a transmission
medium is available with specified performance and that our task is to
transmit over it the maximum amount of information. In the process we
have seen that the problem is complex and that, in general, such a situation
is unlikely to arise; rather, we will be faced with the problem of designing
the transmission medium to carry out a specific job. In this situation we
must examine how its performance will vary with the parameters that are
adjustable at the manufacturing stage and see whether some rational
choice of parameters can be made.

We can discern some variables that can be varied easily. These include
the fiber index difference, Δ, core diameter, and cladding diameter. The
cable design can be varied to increase or decrease the degree of microbending, and the index profile might conceivably be modified to change
the bandwidth. Tolerances on the fiber must be specified if joint losses are
to be controlled, and coatings will certainly have to be specified if the fiber
is to remain intact for any reasonable time. We will examine some of these

variables to see how the fiber performance may vary with them. In general, experimental data are rather sparse, although much theory has been published.

15.4.1 Mass of Material Used in Forming a Fiber

Ultimately, the cost of a fiber cannot be reduced below the cost of the materials used in its construction. Thus, in a very high productivity fiber plant, there may well be some correlation between the materials cost and the fiber cost, and this would lead one to choose a smaller diameter fiber. For CVD fibers the time of production may also be directly related to the volume of material to be deposited, again arguing for reduced diameter. Smaller core diameters may also be advantageous in situations in which the core material is more expensive than the cladding material, as may well be the case for the CVD fiber. In all cases mass of material relates to radius squared.

15.4.2 Dispersion

The waveguide dispersion is controlled by three factors: mode dispersion, mode coupling, and material dispersion. The last is similar for all the presently used fiber materials and thus sets requirements on the source linewidth, but not on the fiber.

The mode dispersion is controlled by the index profile and the index difference. Given that most systems will be using graded-index guides, we can examine the condition for dispersion in the minimum dispersion profile situation. We find that, taking equation 6.3.15,

$$\sigma_{rms} = \frac{LN_0}{c} \frac{\sqrt{3}}{24} \Delta^2 \tag{15.4.1}$$

so that the rms pulse width varies as the square of the index difference. This relation evidently favors small index difference guides for wide bandwidth operation. However, it should also be pointed out that most graded-index guides are probably well away from this limiting value, because of imperfect profile control. Notice also that this result, which typically leads to about $360\Delta^2$ ns/km dispersion, is a factor of 2 higher than a figure derived by Arnaud[4] in some recent work for the limiting graded-index guide dispersion, in which he quotes the minimum as $150\Delta^2$ ns/km.

Equation 15.4.1 applies to fibers having "perfect profiles." However, examination of the expression for the rms pulse width of a generalized alpha profile fiber (equation 6.3.10) shows that, away from the dispersion minimum, the rms pulse width is almost linearly proportional to the value of Δ. It follows that in many situations where the fiber bandwidth is far removed from the ideal value (which at the time of writing is essentially always the case) the bandwidth is likely to be essentially independent of Δ, since the effects of profile errors will conceal the effects of a change of Δ over the small range that is likely to be experimentally accessible.

15.4.3 Launched Power and Joint Loss

In the event that an LED source is used with efficient optics, the launched power is proportional to the number of guided modes supported by the fiber (see Section 13.4). For the graded fiber the number of modes is given by equation 6.2.14 to be, for an alpha profile with $\alpha = 2$,

$$M' = \tfrac{1}{2} a^2 n_0^2 k_0^2 \Delta \tag{15.4.2}$$

Evidently, more power will be launched from a diffuse large area source into a fiber with a large index difference and core diameter.

Similar considerations apply when power is launched from a GaAs laser source to a fiber. In principle, the asymmetrical beam from the GaAs laser can be matched to perfectly couple to a single mode of the fiber by using a suitable astigmatic lens. However, in practice it is attractive to avoid the use of additional optics wherever possible, and this goal will be greatly facilitated by the use of a fiber having a large core diameter and large acceptance angle. The semiacceptance cone angle, θ_1, is given by (see Figure 7.1)

$$\theta_1 = \sqrt{2n_1^2 \Delta} \qquad (NA = \sin\theta_1) \tag{15.4.3}$$

and the solid acceptance angle by (equation 13.4.2)

$$\psi = 2\pi n_1^2 \Delta \tag{15.4.4}$$

On either count our fiber will be easier to work with if both a and Δ are increased.

Joint loss also follows the same pattern. We saw in Section 13.2.1 that the joint loss in terms of the power immediately lost in the vicinity of the joint is expressible in terms of the joint misalignment in units of core radius. The loss being approximately linear with fractional displacement, we can see immediately that for a given joint displacement (absolute) the loss will be lower for a large core diameter fiber. The theory indicated an approximately linear relation with core radius, although more recently an analysis by Gloge[5] has shown that, when the effect of the stable mode distribution is taken into account, a more accurate description is for the loss to vary as the square of the fractional displacement, and hence inversely as a^2.

382   OPTICAL FIBER SYSTEMS

15.4.4 Microbending Losses

A number of theoretical models have been proposed to describe microbending. To be realistic, they must account for a number of effects. The
stiffness of the fiber itself is a function of its radius, since it can be
considered as a stiff beam bending under the effects of externally applied
forces. The actual forces applied depend critically upon the environment in
which the fiber is contained and might simply be specified in terms of a
rough, hard surface with a given roughness power spectrum, or they might
include the effects of a cushion material between the surface and the fiber,
the effects of the fiber touching only the roughness peaks of the surface
and the fiber itself being an elastic medium indenting as well as deforming.
The sensitivity of a fiber to a given amplitude and frequency of spatial
deformation, then, depends critically upon the fiber parameters such as
core profile, diameter, and index difference, since these factors control the
mode spacings and strong coupling occurs only when the spatial frequency
of the deformation, K, matches the difference in $\Delta^z k$ values between a pair
of guided modes of suitable symmetry (see Appendix 2). The calculation of
the latter group of effects is fairly precise, whereas the former mechanical
effects are very ill defined for lack of valid experimental data on the actual
deformation spectra that are likely to occur in practice.

For a graded-index (parabolic) fiber the critical wavelength for microbending is given by (in wave number). (See Appendix 3.)

$$K = \frac{\sqrt{2\Delta}}{a} = \frac{2\pi}{\Lambda} \tag{15.4.5}$$

The fiber acts as a stiff beam if

$$\Lambda < R \tag{15.4.6}$$

where

$$R = 2\pi \left(\frac{H}{D}\right)^{1/4} \tag{15.4.7}$$

with $H = \pi/4)Ea_c^4$; here E is Young's modulus of fiber, a_c is radius of fiber
cladding, and D is Young's modulus of packaging material surrounding
fiber.

For fairly typical fiber materials we find that $R = 12a_{\text{clad}}$, while for an
index difference of 0.01, $\Lambda = 44a$, so that the fiber will tend to follow an
impressed microbend and the natural stiffness of the fiber will play a
rather small part.

Under these conditions we can derive a simple functional relationship
for the dependence of the microbending upon the fiber parameters. We

start by assuming that the impressed mechanical power spectrum that causes the bending is described, quite generally, by

$$P(K) = P(0)(1 + l^2 K^2)^{-\mu} \tag{15.4.8}$$

in which we assume that the parameter l, effectively a correlation length, is large compared to the spatial frequency K, and the index μ is assumed to be greater than unity.

Using this model, Gloge[1] has derived the following general relationship for the loss (in dB/km) due to microbending:

$$\gamma_m = \mathcal{K}_m \Delta^{1-\mu} a^{2\mu-4} a_c^{1/2-\mu} \tag{15.4.9}$$

where $\mathcal{K}_m$ is a constant. Evidently, we must know the value of μ before we can decide what this result tells us about optimizing our fiber performance. On this subject we have rather limited guidance. We know that μ will be greater than unity, simply on the physical grounds necessary to describe a typical power spectrum over a limited range of frequency (centered on Λ). Gardner[2] has made some experimental measurements which suggest that a value of $\mu = 3$ is appropriate. Using this, we find that

$$\gamma_m = \mathcal{K}_m \Delta^{-2} a^2 a_c^{-5/2} \tag{15.4.10}$$

This would clearly favor large index difference, small core diameter, and large cladding diameter. However, further experimental evidence is needed before firm decisions are made.

15.4.5 Fiber Strength

Reference to Sections 12.2.2 and 12.2.3 reminds us that glass fails when the strain in the sample under test reaches some critical value such that the stress concentration at the tip of the largest surface crack in the sample reaches the failure strength associated with the perfect material. The analysis of such failures assumes that there is a certain probability of a crack of a given size appearing in a given sample which is proportional to the surface area of the sample. Accordingly, the probability of failure of a sample at any particular design strain is proportional to the sample area, the fiber cladding radius, and the strength in terms of failure strain proportional to a_c^{-1}.

The load associated with the same failure is proportional to the strain, the sample cross sectional area, and Young's modulus. Accordingly, the probability of failure at a given load is proportional to $a_c \cdot a_c^{-2} = a_c^{-1}$ and hence the strength in terms of load is proportional to a_c. A choice must clearly be made as to the relevant design parameters. If the fiber is to be designed to carry a maximum load, it will be desirable to increase the fiber diameter. If the fiber is not to act as a load bearing member but is merely to withstand a design strain, it will be advantageous to minimize the cladding diameter.

15.4.6 Summary of Fiber Optimization

Casual inspection of Table 15.4, in which the results of the preceding discussion are summarized, shows that, if we seek the highest performance from our fiber in all attributes, we simultaneously require large and small values of each parameter. We are immediately forced into a position of trading one property against another, with different optima emerging in

TABLE 15.4 TRADE-OFFS THAT CAN BE MADE IN OPTIMIZING FIBER, IN TERMS OF THE FUNCTIONAL DEPENDENCIES OF FILTER PERFORMANCE ON FIBER PARAMETERS

Fiber feature	Increases as		
	a_{core}	a_{clad}	Δ
Mass of fiber material	—	a_{clad}^2	—
Mass of core dopant material	a_{core}^2	—	Δ
Dispersion			
Optimum profile	—	—	Δ^2
Nonoptimum profile	—	—	Δ
Step index	—	—	Δ
Launch power			
LED	a_{core}^2	—	Δ
Laser	$a_{core}(?)$	—	$\Delta^{1/2}(?)$
Joint loss			
All modes excited	a_{core}^{-1}	—	—
Stable mode	a_{core}^{-2}	—	—
Microbending loss			
In general	$a_{core}^{2\mu-4}$	$a_{clad}^{0.5-\mu}$	$\Delta^{1-\mu}$
$\mu = 3$	a_{core}^2	$a_{clad}^{5/2}$	Δ^{-2}
Fiber strength			
Strain			
Pristine	—	—	—
Damaged	—	a_{clad}^{-1}	—
Load			
Pristine	—	a_{clad}^2	—
Damaged	—	a_{clad}	—

TABLE 15.5 GRADED-INDEX FIBER PARAMETER
VALUES ATTRACTING MOST INTEREST

Core diameter	50–70 microns
Cladding diameter	100–150 microns
Numerical aperture	0.15–0.25
Δ	0.004–0.013
Number of guided modes (at 900 nm)	140–900
Pulse dispersion (mode)	0.1–4.0 ns/km
Loss	2.0–5.0 dB/km

different situations. The practical difficulty of actually doing this rapidly
emerges when we examine the difficulty of measuring the numerical scale
factors that are required to actually evaluate the functional relationships
indicated.

Given that the more important of the relationships in Table 15.4 can be
at least crudely evaluated by the use of a few carefully chosen experimental points, it becomes possible to return to the overall system model and
consider the effects of increasing or decreasing a particular parameter. For
example, decreasing the core radius, a, will be expected to increase joint
losses, decrease microbending loss, and thus affect long length bandwidth
and possibly decrease launching efficiency. Similar arguments emerge
immediately for the other parameters.

At the time of writing, the specifications for graded-index fibers appear
to be narrowing to lie within a small range of fiber parameters, as listed in
Table 15.5. No doubt, in time these ranges will shrink further as experience
clarifies the precise nature of the various benefits and penalties.

15.5. CODING

All the discussions of system and receiver design so far have implicitly
assumed that 0 and 1 digits are equally likely, and no consideration has
been given to the necessity of some redundancy in the coding to allow the
signal to be understood after reception, errors to be detected, and so on. A
number of factors need to be considered.

Our analysis of the receiver implicitly assumed that the decision circuit
in the receiver had prior and precise information on the timing intervals
and the start of each. In practice, this information must be extracted from
the incoming optical signal. This requirement sets limitations on the
coding. For example, in a code in which all digits are used for data, and 0's
and 1's are equally likely, the probability of N consecutive identical digits
is 2^{-N}. Assume that N is 10; then, during 10,000 digits, there is a good
chance that such a sequence will occur. But a series of 10 identical digits at

the receiver contains no timing information, and the receiver may mistake them for 9 or 11 digits, unless its timing system is highly stable. However, a highly stable clock implies that, on start-up, it will take some time to pull into synchronization. A compromise thus exists.

In practice, codes are designed in which extra digits are introduced into the raw data stream on a regular and logical basis to achieve several ends. The first is to minimize or reduce the number of possible consecutive identical receive digits, to facilitate timing extraction and to reduce the low frequency response requirement of the optical receiver. A simple code of this type is one in which a 0 can be signaled as 01, and a 1 as 10. This ensures that there are never more than two consecutive identical digits, and it thus produces a high density of decision level crossings favoring timing extraction.

It is also immediately apparent that the same code, by virtue of its freedom from long identical symbol sequences, means that the receiver has a continuous input of the present 0 and 1 levels for the purposes of automatic gain control (AGC) and decision level control. However, this has been done at the expense of a factor of 2 in bandwidth, signaling two bits for one input data bit, and the code offers limited error detection capability. For example, a sequence of 1's could be mistaken for a sequence of 0's by slipping one bit in timing, so that the timing information is not unambiguous.

A similar but slightly more complex code is coded mark inversion (CMI). In this code an input 0 is signaled as 01, and alternate input 1's are signaled as 00 and 11. The coding process remains simple, the possibilities of ambiguity are reduced, and consequently the content of timing information and error detection capability is increased since the receiver can look for violations of the 01 sequence, 10 being illegal, and for violations of the alternate 00-11 code for 1's. Once again, the code is highly redundant, using twice as many signaling bits as input data bits.

More efficient codes, in which for every input M bits extra bits are added to form a sequence of N bits, are widely known and are described as MBNB codes. The codes described above are thus 1B2B codes. In general, the use of more efficient codes, such as a 7B8B code, will involve more complex coding rules to allow for the balancing of numbers of 0's and 1's, the ready extraction of timing data, error detection, and synchronization of the subgroup of N bits at the repeater. We do not propose to discuss these further but refer the interested reader to the many excellent books on the subject,[3] having indicated that here again there is room for a range of possibilities, which should be chosen on a rational basis. In general, optical systems seem to be very insensitive to code choice.

15.6. SUMMARY OF SYSTEM OPTIMIZATION

From our discussion we see that system optimization has rapidly moved from the level of a simple intuitive activity to that of a highly complex study. In the basic study we saw how, with a given source power and detector, we can optimize the bit-rate times length product for a given transmission medium. Having done this, we can now return to the components and start to ask second order questions.

Starting with the fiber itself, we can ask ourselves whether our system would be made better or worse by increasing core diameter, changing the NA or deliberately introducing more microbending in our cable, and so on. Also, for the components at the system termination, we can ask whether we would have a better system by choosing a simpler code, allowing simpler timing extraction but using more bandwidth and power, or whether we should opt for a complex code, to conserve bandwidth and optical power but at the expense of electronics.

Underlying all these thoughts are questions of economics. We have used the term "best system" or "better system" without ever specifying very precisely what it means. In most civil applications it means the cheapest system that will do the required job satisfactorily. Once again, this appears to be a simple criterion, but a moment's thought reveals that complex judgments are immediately involved. Is it better to overrun components, to obtain the highest technical performance at the cost of reliability? Should many spare fibers be installed in a cable initially to provide for future growth and repair situations, or not? Evidently, we must now examine not only the cost and performance of the cable itself, but also the installation costs, the risk of subsequent damage to cables by installing others, the growth rates on the route, which control the time before surplus capacity will be used, and the prevailing cost structures and interest rates, which dictate the effective cost of unused investment.

It would be fruitless to attempt to answer any of these questions here, since they are almost totally dependent on the peculiarities of the particular user's situation. However, airing them in general terms has served to bring them to the reader's notice, and it is hoped that partial or complete answers can be arrived at by the individual for his or her particular problem. Evidently the number of potential variables and of interactions between them is enormous, and some intelligent guesswork is likely to be necessary to arrive quickly at near-optimum solutions. It is hoped that the general discussion presented in this volume of the underlying principles and the ways in which the mechanisms interact, together with the selection of actual results presented, will aid the reader in making such choices.

REFERENCES

1. D. Gloge, *Bell Syst. Tech. J.*, **54**, 243 (1975).

2. W. B. Gardner, *Bell Syst. Tech. J.*, **54**, 457 (1975).

3. For further reading see one of the general textbooks on communications theory or telecommunications, such as these:
    H. Taub and D. L. Schilling, *Principles of Communication Systems*. New York: McGraw-Hill, 1971.
    R. L. Freeman, *Telecommunications Transmission Handbook*. New York: Wiley Interscience, 1975.

4. J. Arnaud, *Opt. Quant. Electron.*, **9**, 111 (1977).

5. D. Gloge, *Bell Syst. Tech. J.*, **55**, 905 (1976).

Appendix 1

Fiber Perturbation Due to Microbending

We assume that the fiber mean lay is perturbed from the straight path by an amount $y = b\sin(\Delta^z kz)$, as in Figure A1.1. This can be expressed in polar coordinates as follows, by reference to Figure A1.2:

$$(r')^2 = y^2 + r^2 + 2ry\cos\phi \qquad (A1.1)$$

If we assume that the perturbation y is very small, this simplifies to give

$$r' \simeq r + ry\cos\phi$$

$$= r + \delta r$$

$$\delta r = b\sin(\Delta^z kz)\cos\phi \qquad (A1.2)$$

If we now examine the effect of this perturbation on an alpha profile fiber, we find that the lateral perturbation, δr, can be expressed as a dielectric constant perturbation, $\delta\epsilon$. For the unperturbed alpha profile within the fiber core, we have

$$\epsilon(r) = n^2(r) = n_1^2\left[1 - 2\Delta\left(\frac{r}{a}\right)^\alpha\right]$$

$$= \epsilon_1\left[1 - 2\Delta\left(\frac{r}{a}\right)^\alpha\right] \qquad (A1.3)$$

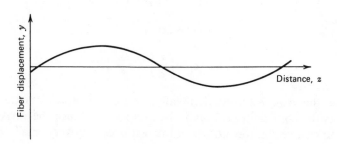

Figure A.1.1. Perturbed fiber lay, shown greatly exaggerated.

389

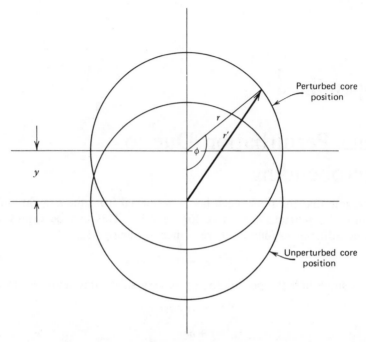

Figure A.1.2. Definition of coordinates for describing the perturbed fiber core position.

Substituting r' for r, we then obtain the following:

$$\epsilon(r') = \epsilon_1\left[1 - 2\Delta\left(\frac{r + \delta r}{a}\right)^{\alpha}\right]$$

$$= \epsilon_1\left[1 - 2\Delta\left(\frac{r}{a}\right)^{\alpha}\left(1 + \frac{\delta r}{r}\right)^{\alpha}\right]$$

$$= \epsilon_1\left[1 - 2\Delta\left(\frac{r}{a}\right)^{\alpha}\left(1 + \frac{\alpha\delta r}{r}\right)\right]$$

$$= \epsilon_1(r) + \delta\epsilon(r)$$

$$\delta\epsilon(r) = -\epsilon_1 2\Delta\left(\frac{r}{a}\right)^{\alpha}\left(\frac{\alpha\delta r}{r}\right) \tag{A1.4}$$

Outside the core $\delta\epsilon(r) = 0$. Evidently, the above approximation breaks down as the core center is approached ($r \rightarrow 0$), but since the variation of ϵ with r in that region for profiles of interest is usually very small, this is not disastrous.

Appendix 2

Mode Coupling Coefficients
Due to Fiber Microbends

We showed in Appendix 1 that microbending results in a fiber perturbation that can be expressed as a function of (r, ϕ, z) in the fiber core dielectric constant. We will now use this result to study the form of the coupling induced between guided modes as a result.

We begin with equations 2.2.1 and 2.2.2, as we did in deriving the wave equation (2.5.6), but we set the dielectric constant ϵ to be of the form $\epsilon + \Delta\epsilon$, where $\Delta\epsilon(r, \phi, z) \ll \epsilon$. We then find that, neglecting products of small quantities, we have for the wave equation

$$\left[\nabla^2 + k_0^2 \epsilon(r) \right] E = -k_0^2 \Delta\epsilon(r, \phi, z) E \qquad (A2.1)$$

We know that the solution of the left-hand side ($\Delta\epsilon = 0$) corresponds to the mode fields, which we will describe in terms of amplitudes a_i and spatial distributions $A_i(r, \phi, z)$, so that the A_i have the property that $\int\int\int (A_i A_i^*) \, dr \, d\phi \, dz = 1$. We now look for the effects of mode coupling by allowing the mode amplitudes to become z dependent.

Thus we write the mode fields in the form

$$a_i(z) A_i = a_i(z) F(r) \exp(\pm il\phi) \exp(-i^z kz) \qquad (A2.2)$$

and substitute into equation A2.1 to obtain the result, neglecting the term in $\partial^2 a_i / \alpha z^2$,

$$\left[\nabla^2 + k_0^2 \epsilon(r) \right] a_i A_i - i^z k_i \left(\frac{\partial a_i}{\partial z} \right) A_i = -k_0^2 \Delta\epsilon(r, \phi, z) a_j A_j \qquad (A2.3)$$

The extreme left-hand term goes to zero, A_i being a solution of the unperturbed wave equation, so that we are left with an expression for the rate of change of mode amplitude, a_i, in terms of the driving term arising

391

from all the other modes:

$$\left(\frac{\partial a_i}{\partial z}\right)A_i = -i\left(\frac{k_0^2}{^z k_i}\right)\Delta\epsilon(r,\phi,z)a_j A_j \qquad \text{(A2.4)}$$

We now multiply by A_i^* and integrate both sides with respect to r and ϕ to obtain

$$\frac{\partial a_i}{\partial z} = -i\left(\frac{k_0^2}{^z k}\right)\int\int \Delta\epsilon A_j A_i^* \, dr \, d\phi \, a_j \qquad \text{(A2.5)}$$

Recalling the functional form of the A_i, we can now examine the various terms in the right-hand side of equation A2.5. It separates into functions of r, ϕ, and z, the former two being integrated out. The r function is

$$\mathcal{R} = \int_0^\infty \Delta\epsilon(r)\cdot F_i^*(r)F_j(r) \, dr \qquad \text{(A2.6)}$$

Noting the value of $\delta\epsilon$ from Appendix 1, and the fact that it is zero for $r>a$, we find that equation A2.6 becomes

$$\mathcal{R} = \int_0^a -2\Delta\epsilon_1\left(\frac{r}{a}\right)^\alpha\left(\frac{\alpha b}{r}\right)F_i(r)F_j(r) \, dr \qquad \text{(A2.7)}$$

Similarly, the ϕ terms take the form

$$\Phi = \int_0^{2\pi} \exp(\pm il_i\phi)\exp(\pm il_j\phi)\cos\phi \, d\phi \qquad \text{(A2.8)}$$

This integral takes the value zero unless

$$|l_i - l_j| = 1$$

so that microbending couples together modes whose azimuthal quantum numbers vary by unity.

Finally, from the z dependence of the perturbation and the fields involved, a term arises of the general form

$$\exp\left[\pm i(^z k_i - {}^z k_j \pm \Delta^z k)z\right] \qquad \text{(A2.9)}$$

so that strong coupling will occur when $|^z k_i - {}^z k_j| = \Delta^z k$.

We should note that the presentation assumes coherent fields and continuous perturbations, rather than the incoherent signals that occur in practice together with the random perturbations. However, our analysis shows the conditions for strong coupling to occur under the situation experienced in microbending. Throughout it has been assumed that very small displacements and power transfers are occurring, per unit length, so that the approximations can be used. Numerous detailed and sophisticated analyses of these effects have been published, and the reader should follow these up.

Appendix 3

Critical Wavelength for Mode Coupling Due to Microbends in a Parabolic Index Fiber

The expression given by Streiffer and Kurtz* for the value of ^zk for a general power law index profile fiber is (equation 7.2.20)

$$^zk^2 = n_1^2 k_0^2 \left\{ 1 - \left[\frac{\sqrt{\pi}\,(2m+l+1)\Gamma\left(\frac{3}{2}+\frac{1}{\alpha}\right)}{n_1 k_0 a \Gamma\left(1+\frac{1}{\alpha}\right)} \right]^{2\alpha/(\alpha+2)} 2\Delta^{2/(2+\alpha)} \right\} \quad (A3.1)$$

Substituting the value $\alpha=2$ and using the simplification that $\Delta \ll 1$, we obtain

$$^zk = n_1 k_0 \left[1 - \frac{(2m+l+1)\sqrt{2\Delta}}{n_1 k_0 a} \right] \quad (A3.2)$$

Microbending is characterized by selection rules of the form

$$\delta(2m+l+1) = \pm 1 \quad (A3.3)$$

(see Appendix 2).

Applying this condition, we find the result required:

$$\delta\,^zk = \frac{\sqrt{2\Delta}}{a} = \frac{2\pi}{\Lambda} \quad (A3.4)$$

where Λ is now the critical spatial wavelength to induce microbending in the parabolic graded fiber.

*W. Streiffer and C. N. Kurtz, *J. Opt. Soc. Am.*, **57**, 779 (1967).

Appendix 4

Power Series Approximation to the Power Law Alpha Profile

We saw in our discussion of the power law alpha profile in Section 6.3 that great interest centers on guides with nearly parabolic core profiles. One useful method of describing such profiles theoretically is the alpha profile, in which the refractive index as a function of radius is given by

$$n(r) = n_0 \left[1 - \Delta \left(\frac{r}{a} \right)^\alpha \right], \qquad 0 < r < a \qquad (A4.1)$$

$$n(r) = n_0(1 - \Delta) \qquad\qquad r > a \qquad (A4.2)$$

However in the examination of experimental profiles and for analytical purposes, it is frequently convenient to have a theoretical model that allows an additional degree of freedom in describing the index variation with position. Such a description is obtained by setting the refractive index as a function of radius equal to

$$n(r) = n_0 \left(1 - \left[A_1 \left(\frac{r}{a} \right)^2 + A_2 \left(\frac{r}{a} \right)^4 \right] \right) \qquad 0 < r < a \qquad (A4.3)$$

$$n(r) = n(a) \qquad\qquad r > a \qquad (A4.4)$$

where A_1 and A_2 are constants.

To relate this power series profile to the alpha profile, we carry out a least squares fitting of equation A4.3 to equation A4.1 and solve for the values of A_1 and A_2 as a function of α to give the minimum rms error. Thus we define the function

$$F = \left(\frac{r}{a} \right)^\alpha - A_1 \left(\frac{r}{a} \right)^2 - A_2 \left(\frac{r}{a} \right)^4 \qquad (A4.5)$$

and calculate, using F, the result

$$\sigma^2 = \int_0^1 F^2 \, dy - \left(\int_0^1 F \, dy \right)^2 \qquad (A4.6)$$

where $y = r/a$.

Then, seeking the conditions for minimum σ^2 by setting $\partial\sigma^2/\partial A_1 = \partial\sigma^2/\partial A_2 = 0$, we obtain two simultaneous equations in A_1 and A_2, which yield the results

$$A_1 = \frac{[1/5(\alpha+1)-1/(\alpha+5)](\frac{1}{9}-\frac{1}{25})^{-1}-[1/3(\alpha+1)-1/(\alpha+3)](\frac{1}{7}-\frac{1}{15})^{-1}}{(\frac{1}{5}-\frac{1}{9})(\frac{1}{7}-\frac{1}{15})^{-1}-(\frac{1}{7}-\frac{1}{15})(\frac{1}{9}-\frac{1}{25})^{-1}}$$

(A4.7)

$$A_2 = \frac{[1/5(\alpha+1)-1/(\alpha+5)](\frac{1}{7}-\frac{1}{15})^{-1}-[1/3(\alpha+1)-1/(\alpha+3)](\frac{1}{5}-\frac{1}{9})^{-1}}{(\frac{1}{7}-\frac{1}{15})(\frac{1}{5}-\frac{1}{9})^{-1}-(\frac{1}{9}-\frac{1}{25})(\frac{1}{7}-\frac{1}{15})^{-1}}$$

(A4.8)

These values for A_1 and A_2 are shown in Figure A4.1 for a range of alpha values in the vicinity of 2. Fiber temporal response for the power series profile in this form has been calculated by a number of workers.

Finally, we should note that the additional freedom given by this power series profile over the alpha profile is somewhat illusory, since by definition, when $r/a = 1$, $n(0) - n(a) = \Delta$, so that we find that the constraint should be applied that $A_1 + A_2 = \Delta$ and we have really only obtained a different group of profiles that largely follow the alpha profile set.

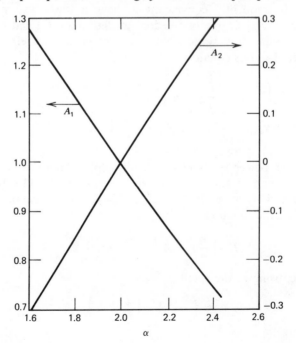

Figure A.4.1. Plot of the parameters A_1 and A_2 versus α for the power series profile.

Appendix 5

Optical and Electrical Power and Bandwidth in Fiber Systems

We showed in Section 14.4 that the electrical current from the detector is given by

$$i = \frac{gm_1 e}{\tau} = gNQe \tag{A5.1}$$

The electrical power generated is $i^2 R_l$ and is related to the optical incident power, P_{opt}, through the relation

$$P_{opt} = Nh\nu \tag{A5.2}$$

so that

$$P_{elec} = (gQNe)^2 R_L$$
$$= \left(\frac{g^2 Q^2 R_L}{h^2 \nu^2} \right) P_{opt}^2 \tag{A5.3}$$

In equations A5.2 and A5.3, h is Planck's constant and ν is the optical frequency in hertz. The important fact to note is that the electrical power generated at the detector varies as the square of the received optical power, so that a change of 3 dB in the optical power leads to a 6 dB fluctuation in the electrical power. This has led to considerable confusion in the specification of fiber transmission links, since circuit designers are usually interested in the electrical power response of the system, while optical measurements usually lead to direct measurement of optical power, in terms of the current from a photodetector. The confusion is particularly acute in specifying the bandwidth of a fiber; one may measure the optical power pulse shape with a photodetector, Fourier-transform the voltage to yield an optical 3 dB and 10 dB bandwidth, or take the electrical power response proportional to i^2 and calculate the electrical response. In an attempt to clarify the situation, we consider below a special case, that of the Gaussian pulse response, which, although particular, gives a good estimate of the performance figures in a wide range of practical situations.

Gaussian Pulse Response

Many fibers, and particularly jointed fiber links, exhibit pulse outputs in time whose temporal variation is closely approximated by a Gaussian response. This is attractive, being simple to analyze, and useful because we can quickly derive some special results which can be quite widely applied.

We assume that the optical power emerging from the fiber can be described in time as

$$f(t) = \frac{1}{\sqrt{2\pi}\,\sigma} \exp\left(-\frac{t^2}{2\sigma^2}\right) \tag{A5.4}$$

This is therefore the shape that would be observed in a voltage display on an oscilloscope when viewing the detector output.

The Fourier transform of this function is given by

$$g(\omega) = \frac{1}{\sqrt{2\pi}} \exp\left(-\frac{\omega^2\sigma^2}{2}\right) \tag{A5.5}$$

It follows directly from equation A5.4 that the time t_e, such that $f(t_e)/f(0) = 1/e$, is given by

$$t_e = \sqrt{2}\,\sigma \tag{A5.6}$$

If we define the full width of the pulse at the $1/e$ points as τ_e, then

$$\tau_e = 2t_e = 2\sqrt{2}\,\sigma \text{ sec} \tag{A5.7}$$

The 3 dB optical bandwidth is the frequency (of modulation) at which the optical received power has fallen to 0.5 of the value at frequency 0. Thus we find from equation A5.5 that

$$\frac{\left[\omega(3 \text{ dB opt})\right]^2\sigma^2}{2} = 0.693$$

$$\omega(3 \text{ dB opt}) = \frac{\sqrt{2} \times 0.8326}{\sigma}$$

$$f(3 \text{ dB opt}) = \frac{\sqrt{2} \times 0.8326}{2\pi\sigma} = \frac{0.53}{\tau_e} \text{ Hz} \tag{A5.8}$$

The 3 dB electrical bandwidth is the bandwidth such that the response of the optical power has fallen to $1/\sqrt{2}$ and is thus given by

$$f(3 \text{ dB elec}) = \frac{0.375}{\tau_e} \text{ Hz} \qquad (A5.9)$$

Finally, we note that many analyses of the pulse responses of optical fibers are phrased in terms of the rms pulse width. This is defined in terms of the fiber temporal response, $h(t)$, as follows [$h(t)$ normalized so that $\int_{-\infty}^{\infty} h(t)\,dt = 1$]:

$$\sigma_p^2 = \int_{-\infty}^{\infty} t^2 h(t)\,dt - \left[\int_{-\infty}^{\infty} th(t)\,dt \right]^2 \qquad (A5.10)$$

In the particular case of the Gaussian response defined by equation A5.4, the rms pulse width defined by equation A5.10 evaluates as

$$\sigma_p = \sigma \qquad (A5.11)$$

It follows that for pulses that are reasonably symmetrical and Gaussian in form, a good first approximation will be to set

$$\tau_e = 2\sqrt{2}\,\sigma_p \qquad (A5.12)$$

and to use this result to calculate the respective 3 dB bandwidths via equation A5.9 or A5.10. The more asymmetrical or non-Gaussian the pulse, the less accurate will be the result of this approach; in particular, fibers yielding multiply peaked responses or having long tails or precursors will give rise to erroneous results, unless an accurate numerical transformation is carried out.

These simple relations between rms pulse width, full width at $1/e$ points, and the 3 dB bandwidths are particularly useful in linking quickly from theoretical pulse response analysis to fiber pulse measurement to receiver design.

Appendix 6

Standard Transmission Rates

The discussion in this book has been largely based upon the transmission of information in binary digital format. This arises from the fact that the telecommunications networks of both Europe and the United States are increasingly using this format for their main transmission cable and carrier systems. In so doing, various standard rates have been evolved, and it seems useful to summarize them here for the benefit of the reader.

Both systems begin by digitizing the single telephone channel. The number of sampling levels and the sampling rates are shown in the Table A6.1. Blocks of these digitized telephone channels are then multiplexed

TABLE A6.1 DATA ON THE DIGITAL RATES USED IN TELECOMMUNICATIONS IN EUROPE AND THE UNITED STATES

Digital multiplexing of telephony

Typical analogue bandwidth	4 kHz
Sampling rate	8 kHz
Number of levels	2^7
7 bit level + 1 synchronizing	8 bits per sample
1 telephone channel	64 kbit/sec

CEPT (European) groups

32 telephone channels	2.048 Mbit/sec
4×30 = 120 channels	8.448 Mbit/sec
4×120 = 480 channels	34.368 Mbit/sec
4×480 = 1920 channels	139.364 Mbit/sec
4×1920 = 7680 channels	565 Mbit/sec

Bell Telephone groups

$T1$		24 channels	1.544 Mbit/sec
$T2$	$4 \times T1$ =	96 channels	6.312 Mbit/sec
$T3$	$7 \times T2$ =	672 channels	46.304 Mbit/sec
$T4$	$6 \times T3$ =	4032 channels	281 Mbit/sec

together, along with some additional bits to allow unscrambling on reception, forming higher bit-rate blocks. These blocks in turn can be multiplexed together to form the next level in the digital hierarchy.

In the United States, four standard rates have been agreed upon, labeled $T1$, $T2$, $T3$, and $T4$. These correspond to the information rates shown in Table A6.1 and are formed by various combinations of telephone channels and other user services. The raw ingredients are also listed in Table A6.1.

The European networks have settled on different signaling rates, which fall into the same general blocks and are shown in Table A6.1, together with the associated block units of which they may be composed. In both cases the attraction of digital transmission stems chiefly from the fact that the transmission medium is largely unaware of the nature or source of the signal being transmitted, be it telephone conversations, data or television (2 T.V. channels$\simeq$1920 telephone channels).

Finally, we should note that the actual bit-rates transmitted over a given cable system—fiber, coaxial, or whatever—will be slightly greater than those listed above since these rates include only sufficient redundancy to enable the terminal equipment to separate the multiplexed channels again. The transmission itself requires some additional bit-capacity to enable it to reliably sense the bit-rate and timing at each regenerator, and to allow signaling to be imposed for monitoring the system operation. The additional capacity required for these operations depends in detail upon the coding design but may be of the order of 20%. (See Section 15.5.)

Appendix 7

Some General Relations on Helices

The helix is a frequently encountered geometrical shape in cables, and a few comments on some of its properties seem appropriate. Strength members in cables are frequently wound around a central core in helical fashion. One result is that, if the strength member has a limited ability to slide over the central member, some additional flexibility is obtained in the completed cable, since, when bent, the strength member slides from the inside to the outside curve region to make up for the excess length on the periphery of the cable. This is done without the member itself being significantly extended, and thus without great stiffness of spring.

The helical lay of strength member must also be considered when the cable is to be tensioned, since stretching the cable will tend to compress the strength member into the central core around which it was wound. Choices must be made regarding helix pitch, to balance flexibility, pressure on the inner member when stretched, and acceptable extension for a given load.

Helices also appear in the analysis of fibers that are held loosely inside a containing structure, such as a tube. If the fiber is longer than the tube (the tube has contracted), the fiber is forced into a helical lay in the tube and is forced out against the tube walls.

Alternatively, the fibers may be deliberately laid in a helical path over a central deformable cushion member so that, when the cable is subjected to longitudinal strain, the fibers themselves experience a very much lower strain level by moving to a smaller diameter helix and compressing the cushion. An awareness of the relative magnitudes of such effects is thus useful in considering the possible trade-offs that may be made in a given situation.

For a helix of diameter R and pitch P the additional length of the helical path over the direct axial path is given by

$$\frac{\Delta L}{L} = \left\{ \cos\left[\tan^{-1}\left(\frac{2\pi R}{P} \right) \right] \right\}^{-1} - 1$$

$$\approx 2\left(\frac{\pi R}{P} \right)^2 \tag{A7.1}$$

The local radius of curvature of the helix is given by

$$\frac{\Delta L}{L} = \left\{ \cos\left[\tan^{-1}\left(\frac{2\pi R}{P} \right) \right] \right\}^{-1} - 1$$

$$\approx 2\left(\frac{\pi R}{P} \right)^2 \tag{A7.1}$$

The local radius of curvature of the helix is given by

$$R_b = R \left\{ \sin^2\left[\tan^{-1}\left(\frac{2\pi R}{P} \right) \right] \right\}^{-1}$$

$$\approx R\left(\frac{P}{2\pi R} \right)^2 \tag{A7.2}$$

If we wish to design a cable the structure of which will be able to extend by a certain fraction $(\delta L/L)$ and in which the helically wound fiber will not be bent to greater than a certain radius of curvature, R_0, and will not be subjected to undesired strain, we can do so by choosing values of the helix parameters, R and P, and the central deformable member for the design of Figure 12.27.

Let us assume that the value of $\delta L/L < \Delta L/L$, so that the helix will not be fully compressed when the cable is stretched. Under compression the helix radius, R', is now related to the unloaded helix radius, R, by the result

$$R' \simeq R\sqrt{1 - \tfrac{1}{2}(\delta L/L)(P/\pi R)^2}$$

$$\simeq R\sqrt{1 - \delta L/\Delta L} \tag{A7.3}$$

As a cable of the type shown in Figure 12.27 is extended, the fiber and its buffer layer will contract into a smaller diameter helix, thus compressing the inner foam layer. Assuming that there are N fibers roughly covering the surface of the foam, we find that the tensile force in the packaged fibers is given approximately by

$$F = \frac{E_f}{2\pi N} \frac{P^2 R}{R'} \left(\frac{R}{R'} - 1 \right) \tag{A7.4}$$

where E_f = Young's modulus of foam.

In a more complicated situation, in which the various members of the cable are comparably compressible, straining the cable results in straining the fiber compressing its buffer coating and compressing the inner sheath of the design, so that evaluation of the strain relief factor for the fiber relative to the cable structure as a whole is a complex matter.

Index